PROCEEDINGS

OF THE

1995 INTERNATIONAL CONFERENCE

ON

PARALLEL PROCESSING

August 14 - 18, 1995

Vol. II Software
C. Polychronopoulos, Editor
University of Illinois at Urbana-Champaign

Sponsored by
THE PENNSYLVANIA STATE UNIVERSITY

CRC Press
Boca Raton New York Tokyo London

Catalog record is available from the Library of Congress
ISSN 0190-3918
ISBN 0-8493-2619-2 (set)
ISBN 0-8493-2615-X (vol. I)
ISBN 0-8493-2616-8 (vol. II)
ISBN 0-8493-2617-6 (vol. III)
ISBN 0-8493-2618-4 (ICPP Workshop)
IEEE Computer Society Order Number RS00027

Additional copies may be obtained from:

CRC Press, Inc.
2000 Corporate Blvd., N.W.
Boca Raton, Florida 33431

PREFACE

We are pleased to introduce you to the proceedings of the 24th International Conference on Parallel Processing to be held from August 14-18, 1995. The technical program consists of 27 sessions, organized as three technical tracks, one in Architecture, one in Software, and one in Algorithms and Architectures. We have also put together three panels, and three keynote speeches. A workshop on "Challenges for Parallel Processing" has been put together by Prof. Dharma Agrawal to be held before the conference. Two tutorials will be held after the conference and have been organized by Prof. Mike Liu.

There will be three keynote speeches at the conference to start each day of the meeting. The first keynote speech will be entitled, "Future Directions of Parallel Processing", by Dr. David Kuck of Kuck and Associates. Prof. Hidehiko Tanaka from the University of Tokyo will present the second keynote speech on "High Performance Computing in Japan". The third keynote speech will be given by Prof. John Rice from Purdue University on "Problem Solving Environments for Scientific Computing".

There are three panels with the conference to end each day. The first panel is entitled, "Heterogeneous Computing" and has been organized by Mary Eshaghian. The second panel "SPMD: On a Collision Course with Portability", has been organized by Tom Casavant and Balkrishna Ramkumar. The third panel on "Industrial Perspective of Parallel Processing", has been organized by A. L. Narasimha Reddy and Alok Choudhary.

The technical program was put together by a distinguished program committee. Each paper was assigned to two program committee members and two external reviewers. Reviews of each paper were handled using an electronic review process. The program committee met on March 17, 1995 in Urbana, IL, and decided on the final program. The decision on acceptance and rejection of each paper was made on the basis of originality, technical content, quality of presentation and the relevance to the theme of ICPP. A summary of the disposition of papers by area is presented in the following table:

Area	Submitted	Accepted	
		Regular	Concise
Architecture	160	14(8.75%)	27(16.8%)
Software	90	18(20.0%)	15(16.7%)
Algorithms	115	16(13.9%)	16(13.9%)
TOTAL	365	48(13.2%)	58(15.9%)

All papers submitted from the University of Illinois were processed by Prof. Chitta Ranjan Das of Penn State University.

We would like to thank all the people responsible for the success of the conference. First, we would like to thank the University of Illinois for providing the infrastructure necessary for preparing the program for the conference including the support staff and the mailing facilities. We would like to express our gratitude to Carolin Tschopp and Donna Guzy for managing the processing of all the papers, for preparing the proceedings, and for arranging the program committee meeting. We would like to thank the entire program committee for doing such a diligent job in reviewing so many papers in such a fine manner. Finally, we are grateful to Prof. Tse Feng for providing the guidance and wisdom for running this conference.

Prith Banerjee, Program Chair
Constantine Polychronopoulos, Program Co-Chair
Kyle Gallivan, Program Co-Chair
University of Illinois
Urbana, IL-61801

Program Committee Members

Keynote Speakers

Speaker: **David Kuck, Kuck and Associates**
Topic: Future Directions of Parallel Processing

Speaker: **Hidehiko Tanaka, Univ. of Tokyo**
Topic: High Performance Computing in Japan

Speaker: **John Rice, Purdue University**
Topic: Problem Solving Environments for Scientific Computing

Panel Sessions

Panel I: Heterogeneous Computing
Moderator: Mary M. Eshaghian, New Jersey Institute of Technology

<u>Panelists:</u>

Gul Agha	Univ. of Illinois at Urbana-Champaign
Ishfaq Ahmad	The Hong Kong University of Science and Technology
Song Chen	New Jersey Institute of Technology
Arif Ghafoor	Purdue University
Emile Haddad	Virginia Polytechnic Institute and State University
Salim Hariri	Syracuse University
Alice C. Parker	University of Southern California
Jerry L. Potter	Kent State University
Arnold L. Rosenberg	University of Massachusetts
Assaf Schuster	Technion-Israel Institute of Technology
Muhammad E. Shaaban	University of Southern California
Howard J. Siegel	Purdue University
Charles C. Weems	University of Massachusetts
Sudhakar Yalamanchili	Georgia Institute of Technology

Panel II: SPMD: on a Collision Course with Portability?
Moderator: Tom Casavant, University of Iowa

<u>Panelists:</u>

Andrew Chien	University of Illinois at Urbana-Champaign
Alex Nicolau	University of California at Irvine
Balkrishna Ramkumar	University of Iowa
Sanjay Ranka	Syracuse University
David Walker	Oakridge National Laboratories

Panel III: Industrial Perspective of Parallel Processing
Moderator: A. L. N. Reddy, IBM Almaden

<u>Panelists:</u>

Tilak Agerwala	IBM
Ken Jacobsen	SGI
Bruce Knobe	Siemens-Nixdorf
Stan Vestal	Thinking Machines Corp.

Conference Awards

Daniel L. Slotnick Best Paper Award

A. Nowatzyk, G. Aybay, M. Browne, E. Kelly, M. Parkin, B. Radke, and S. Vishin, "The S3.mp Scalable Memory Multiprocessor"

Outstanding Paper Awards

S. P. Midkiff, "Local Iteration Set Computation for Block-Cyclic Distributions"
A. Heirich and S. Taylor, "A Parabolic Load Balancing Method"

List of Referees- Full Proceedings

Abdelrahman, T.
Abraham, S.
Adve, S.
Adve, V.
Ahamad, M.
Alexander, W. E.
Almquist, K.
Al-yami, A. M.
Armstrong, J. B.
Arrouye, Y.
Babbar, D.
Bagherzadeh, N.
Baker, J. W.
Baruah, S.
Bayoumi, M.
Beckman, C.
Beguelin, A.
Bekerle, M.
Bhuyan, L.
Bianchini, R.
Blough, D.
Boppana, R.
Bose, B.
Bruck, J.
Buddihikot, M. M.
Burr, J.
Carver, D. L.
Casavant, T.
Celenk, M.
Chae, S-H.
Chalasani, S.
Chalmers, A.
Chamberlain, R. D.
Chandy, J.
Chao, L-F.
Chatterjee, A.
Chatterjee, S.
Chaudhary, A.
Chaudhary, V.
Chen, P.

Chen, Y-L.
Cheng, A.M.K.
Cheng, K-H.
Chern, M-Y.
Chien, A.
Choudhary, A.
Clarke, E.
Conley, W.
Conte, T.
Cormen, T.H.
Craig, D.
Culler, D.
Cuny, J.E.
Das, C.R.
Das, S. K.
Das Sharma, D.
Davis, J.A.
Davis, T.
DeRose, L.
Deshmuth, R. G.
Dewan, G.
Dhagat, M.
Dhall, S.K.
Dietz, H. G.
Dincer, K.
Dowd, P.
Duato, J.
Dubois, M.
Dutt, S.
Dykes, S. G.
Efe, K.
El-Amawy, A.
Enbody, R.
Felten, E.
Feng, M. D.
Foster, I.
Fraigniaud, P.
Franklin, M.
Fu, J.
Fujimura, K.

Gallagher, D.
Ganapathy, K.
Gao, G.
Ge, Y.
Ghose, K.
Ghosh, J.
Ghosh, K.
Ghozati, S.
Gibson, G.
Gorda, B.
Grunwald, D.
Gupta, A.
Gupta, R.
Gupta, S.
Gyllenhal, J.
Haddad, E.
Haghighat, M.
Hanawa, T.
Hank, R.
Harper, M.
Harper, III, D.
Hassen, S. B.
Heath, M.
Herbordt, M.
Hermenegildo, M.
Hill, M.
Hirano, S.
Ho, C-T.
Holm, J.
Hou, R.
Hsu, D. F.
Huang, S.
Huang, Y-M.I.
Hwang, K.
Ibel, M.
Iwashita, H.
Jacob, J. C.
Jayasimha, D. N.
Jha, N. K.
Joe, K.

Johnson, D.
Jordan, H.
Joshi, B. S.
Kale, L.
Kavianpour, A.
Killeen, T.
Kim, H.
Kim, J-Y.
Kothari, S. C.
Krantz, D.
Krishnaswamy, D.
Krishnamoorthy, M. S.
Lai, A. I-C.
Latifi, S.
Lau, F.
Leathrum, J.
LeBlanc, T. J.
Lee, C-H.
Lee, C.
Lee, G.
Lee, J. D.
Lee, S.
Levy, H.
Lew, A.
Li, Q.
Lilja, D.
Lin, R.
Lin, W.
Lin, W-M.
Livingston, M.
Lo, V.
Lombardi, F.
Long, J.
Lough, I.
Loui, M.
Lu, M.
Lu, Y-W.
Maeng, S.
Mahgoub, I.
Mahmood, A.
Makki, K.
Marsolf, B.
Martel, C. U.
Mazumder, P.

McKinley, P.
McMillan, K.
Menon, J.
Michallon, P
Midkiff, S. P.
Mohapatra, P.
Moore, L.
Moreira, J.
Mouftah, H. T.
Mowry, T.
Mudge, T.
Mukherjee, A.
Mukherjee, B.
Munson, D.
Mutka, M.
Netto, M. L
Ngo, V. N
Nowatzyk, A.
Olariu, S.
Oruc, Y.
Padmanabhan, K.
Pai, M. A.
Palermo, D.
Palis, M. A.
Pan, Y.
Panda, D.
Park, J. S.
Parkes, S.
Passos, N. L.
Patil, S.
Patt, Y.
Pic, M. M.
Pinkston, T. M.
Pradhan, D.
Prasad, S. K.
Prasanna, V. K.
Quinn, M.
Radiyam, A.
Raghavendra, C. S.
Ramaswamy, S.
Ramkumar, B.
Rau, B.
Reeves, A.
Robertazzi, T.

Rodriguez, B.
Rogers, A.
RoyChowdhury, V.
Saha, A.
Saletore, V.
Samet, H.
Sarkar, V.
Scherson, I. D.
Schimmel, D.
Schouten, D.
Schwabe, E. J.
Schwan, K.
Sen, A.
Sengupta, A.
Seo, S-W.
Seznec, A.
Sha, E. H-M.
Shang, W.
Shi, H.
Shi, W.
Shih, C. J.
Shin, K.
Shoari, S.
Shu, W. W.
Siegel, H. J.
Singh, A.
Singh, J.
Singhal, M.
Sinha, B.
Siu, K.-Y.
Sohi, G.
Somani, A.
Srimani, P.
Stasko, J. T.
Stavrakos, N.
Stunkel, C.
Sun, T.
Sunderam, V.
Sundaresan, N.
Surma, D. R.
Suzaki, K.
Szymanski, T.
Taylor, S.
Teng, S-H.

Thakur, R.
Thapar, M.
Theel, O.
Theobald, K.B.
Thirumalai, S.
Tripathi, A.
Trivedi, K.
Tseng, Y-C.
Tzeng, N-F.
Ulm, D. R.
Vaidya, N.
Varma, A.
Varvarigos, E.
Vetter, J.
Wah, B.
Wang, F.
Wang, P. Y.
Wang, Y-F.
Warren, D. H. D.
Watson, D. W.
Weems, C.
Wen, C-P.
Wilsey, P.
Wittie, L.
Wojciechowski, I.
Wu, J.
Wu, K-L.
Wu, M-Y.
Wyllie, J.
Yalamanchili, S.
Yan, J.
Yang, Q.
Yang, T.
Yang, Y.
Yap, T. K.
Yew, P.
Yoo, S-M.
Youn, H-Y.
Yousif, M.
Zhang, X.
Zheng, S. Q.
Zwaenepoel, W.

Author Index - Full Proceedings

I = Architecture
II = Software
III = Algorithms and Applications

TABLE OF CONTENTS
VOLUME II - SOFTWARE

(R): Regular Papers
(C): Concise Papers

Session 1B: Parallelizing Compilers
Chair: Dr. Milind Girkar, Sun Microsystems, Inc.

Session 2B: Languages & Parallelism Support
Chair: Prof. Alex Nicolau, University of California at Irvine

Session 3B: Synchronization and Debugging
Chair: Dr. Nawaf Bitar, Silicon Graphics, Inc.

LINEAR TIME EXACT METHODS FOR
DATA DEPENDENCE ANALYSIS IN PRACTICE

Kleanthis Psarris

Division of Computer Science
The University of Texas at San Antonio
San Antonio, TX 78249
psarris@ringer.cs.utsa.edu

Abstract -- Data Dependence Analysis is of fundamental importance to parallelizing and vectorizing compilers. The proposed dependence analysis techniques in the literature fall into two different categories, either efficient and approximate tests or exact but exponential. In this paper we show that exact data dependence information can be computed efficiently in practice. The Banerjee inequality and the GCD test are the two tests traditionally used to determine statement data dependence in automatic vectorization / parallelization of loops. These tests are approximate in the sense that they are necessary but not sufficient conditions for data dependence. In an earlier work we formally studied the accuracy of the Banerjee and the GCD tests and derived a set of conditions which can be tested along with the Banerjee inequality and the GCD test to obtain exact data dependence information. In this work we perform an empirical study to explain and demonstrate the accuracy of the Banerjee and GCD tests in actual practice. Our experiments indicate that exact data dependence information can be computed in linear time in practice.

1. INTRODUCTION

In automatic parallelization of sequential programs, parallelizing compilers [2, 4, 12, 18, 24, 27] perform subscript analysis [3, 5, 6, 26, 29] to detect data dependences between pairs of array references inside loop nests. A number of subscript analysis tests have been proposed in the literature [6, 9, 10, 15, 16, 21, 22, 28]. In each test there is a different tradeoff between accuracy and efficiency. The most widely used approximate subscript analysis tests are the Banerjee inequality and the GCD test [6, 26, 29]. The major advantage of these tests is their low computational cost, which is linear in the number of variables in the dependence equation. However, both the Banerjee and GCD tests are necessary but not sufficient conditions for data dependence. When independence can not be proved, both tests approximate on the conservative side by assuming dependence, so that their use never results in unsafe parallelization.

In an earlier work [19, 20] we formally studied the accuracy of the Banerjee inequality and GCD tests. We derived a set of conditions, which can be tested along with the Banerjee inequality and the GCD test, to prove data dependence. The cost of testing these conditions is linear in the number of variables in the dependence equation. In this paper we perform an empirical study on the Perfect Club benchmarks [7] to evaluate the formal results and explain the accuracy of the Banerjee inequality and GCD tests in actual practice. We show that our formal results indeed prove to be always accurate in practice. Our empirical study indicates that the Banerjee inequality extended with the accuracy conditions, becomes an exact test, i.e., a necessary as well as sufficient condition for data dependence.

In Section 2 we briefly discuss the data dependence problem and review the Banerjee inequality and GCD tests. In Section 3 we present the formal results about the accuracy of the Banerjee inequality and GCD tests. In Section 4 we present the results of our empirical study. Finally, in Section 5 we discuss related work, compare our results with other methods and present our conclusions.

2. THE DEPENDENCE PROBLEM

Let $A(b_0 + b_1 J_1 + b_2 J_2 + ... + b_r J_r)$ and $A(c_0 + c_1 J_1 + c_2 J_2 + ... + c_r J_r)$ be two potentially conflicting references to an array A, embedded in a nest of r DO loops as in Figure 1, where b_i and c_i are integers, $1 \le i \le r$.

The potential vectorization/parallelization of the loops in the nest depends upon the *data dependence problem*, i.e., the question of whether the equation

$$b_0 + b_1 j_1 + b_2 j_2 + ... + b_r j_r$$
$$= c_0 + c_1 j'_1 + c_2 j'_2 + ... + c_r j'_r \qquad (1)$$

has an integer solution which satisfies the constraints

$$L_k \le j_k, j'_k \le U_k, 1 \le k \le r \qquad (2)$$

```
DO J_1 = L_1, U_1
   DO J_2 = L_2, U_2
   .      .
   .      .
      DO J_r = L_r, U_r
         .
         .
         A(b_0 + b_1 J_1 + b_2 J_2 + ... + b_r J_r) = ...
            .
            .
            ... = A(c_0 + c_1 J_1 + c_2 J_2 + ... + c_r J_r)
               .
               .
      END
   .
   .
   END
END
```

Figure 1 A Nest of Loops

If the answer to this question is *no*, then this particular pair of references presents no encumbrance to the transformation of any of the DO loops in the nest into a parallel DOALL loop. Dropping terms with coefficients of zero from equation (1), and simplifying the notation for both coefficients and variables, we see that data dependence testing amounts to determining whether a linear equation of the form

$$a_1 I_1 + a_2 I_2 + ... + a_n I_n = a_0 \qquad (3)$$

has an integer solution satisfying constraints of the form

$$M_k \leq I_k \leq N_k, \text{ for } 1 \leq k \leq n \qquad (4)$$

The Banerjee inequality and the GCD test [6, 26, 29] are the two tests traditionally used in parallelizing compilers to determine whether an equation of the form (3) has an integer solution satisfying constraints of the form (4). Neither the GCD test nor the Banerjee test actually checks the equation for the existence of a *constrained integer* solution. The GCD test ignores the loop limits constraints entirely and it simply determines whether the equation has an *unconstrained integer* solution. The Banerjee test, on the other hand, takes constraints into account, but determines whether the equation has a *constrained real* solution.

The GCD test is based upon a theorem of elementary number theory which says that equation (3) has an integer solution iff gcd $(a_1, a_2, ... , a_n)$ is a divisor of a_0. The test consists of checking for the indicated divisibility. If the gcd does not divide a_0, then there is definitely *no* dependence; otherwise there *maybe* a dependence.

The Banerjee test computes the extreme values *min* and *max*, assumed by the expression on the left hand side of equation (3) when the variables are subject to the constraints in (4). By the Intermediate Value Theorem, equation (3) has a real solution within the indicated limits in (4), iff min $\leq a_0 \leq$ max. If the inequality min $\leq a_0 \leq$ max is not true, then there is definitely *no* dependence; otherwise there *maybe* a dependence.

Since neither test is, in theory, more accurate than the other, it is common practice to perform both. If both the GCD test and the Banerjee test return a *maybe* answer, then we assume a data dependence. However, in that case we do not know whether an approximation was made or not.

There are also more complicated instances of the data dependence problem that we encounter in practice. When testing multidimensional arrays for data dependence the problem is to determine whether a system of linear equations, as opposed to a single equation of the form (3), has an integer solution subject to constraints of the form (4). One suggested approach, known as *subscript by subscript testing*, is to test one equation at a time. The Banerjee inequality and the GCD test can be applied to test one single equation at a time. In that case we assume that there is a data dependence unless at least one of the equations can be shown not to have a constrained integer solution. This method introduces a conservative approximation if we are testing multidimensional arrays with coupled subscripts; we say that two different subscript pairs are coupled [10, 14, 15] if they contain at least one loop iteration variable in common. In case of coupled subscripts, polynomial time techniques such as constraint propagation [10] and the lambda test [15] can be applied to reduce the problem to testing single equations for constrained integer solutions.

Based on the fact that the Banerjee inequality and the GCD test are approximate tests and they can not provide exact data dependence information a number of tests with exponential cost such as the Omega Test, the Power Test, etc., [9, 22, 28] have been proposed in the literature. They are all based on integer programming techniques, such as Fourier Motzkin variable elimination, and they deal with the most general form of the dependence problem. Namely, they determine whether there exists an integer solution to a system of linear equalities and inequalities. We will show that exact data dependence information can be computed in polynomial time for most cases we encounter in actual practice. Techniques with exponential costs need only be used in the few cases in which the polynomial time techniques can not provide an exact answer.

3. THE ACCURACY OF THE BANERJEE AND GCD TESTS.

In a formal study [19, 20] of the accuracy of the Banerjee and GCD tests we derived the following results stated here as Corollaries 1 and 2. Corollaries 1 and 2 provide conditions under which the Banerjee test and a combination of the GCD test and the Banerjee test become necessary and sufficient conditions for data dependence. We show in this empirical study, that these conditions are almost invariably satisfied in practice and thus making the Banerjee and GCD tests exact tests in practice.

Corollary 1
Consider equation (3) together with the constraints in (4). Without loss of generality assume $|a_1| \leq |a_2| \leq ... \leq |a_n|$.

If

• $|a_1| = 1$

• for each j, $2 \leq j \leq n$,

$$|a_j| \leq 1 + \sum_{k=1}^{j-1} |a_k|(N_k - M_k)$$

then

$min \leq a_0 \leq max$

iff

equation (3) has an integer solution satisfying the constraints in (4). ∎

Corollary 1 states that if the coefficients of the dependence equation are small enough to satisfy its hypothesis, then the Banerjee inequality is an exact test, i.e., a necessary and sufficient condition for data dependence. Testing the conditions in the hypothesis of Corollary 1 has a cost which is linear in the number of variables in the equation. Since the coefficients are integers, it takes linear time to sort them by applying the bin sort algorithm [1]. Once the coefficients have been sorted, it also takes linear time to evaluate the inequalities in the hypothesis. Hence, in case the conditions of Corollary 1 are satisfied, we are able to derive exact data dependence information in linear time.

Corollary 2
Consider equation (3) together with the constraints in (4). Without loss of generality assume $|a_1| \leq |a_2| \leq ... \leq |a_n|$.
Let $d = gcd(a_1, a_2, ..., a_n)$.

If

• $|a_1| = d$

• for each j, $2 \leq j \leq n$,

$$|a_j| \leq d + \sum_{k=1}^{j-1} |a_k|(N_k - M_k)$$

then

d divides a_0 and $min \leq a_0 \leq max$

iff

equation (3) has an integer solution satisfying the constraints in (4). ∎

Corollary 2 states that if the coefficients of the dependence equation satisfy its hypothesis, then a combination of the GCD test and the Banerjee inequality is an exact test, i.e., a necessary and sufficient condition for data dependence. The application cost of the conditions in the hypothesis of Corollary 2 is, as in Corollary 1, linear in the number of variables in the equation.

The Corollaries 1 and 2, presented in this section, apply to the general data dependence problem. In [19] these Corollaries have also been extended to the case of data dependence testing subject to directions vectors [26].

4. EMPIRICAL RESULTS

In this section we present empirical results on how often the conditions of Corollaries 1 and 2 guarantee exact answers in practice. For the experimental evaluation we used the Perfect benchmarks [7], a representative collection of programs executed on parallel computers. Table 1 provides information about the number of lines, the number of subroutines and the number of loops in each program of the Perfect benchmarks.

We have implemented the conditions in Corollaries 1 and 2 in the dependence analyzer of the Tiny program restructuring research tool [27]. The original version of Tiny was developed by Michael Wolfe and has been extended with additional features at University of Maryland. Both the Banerjee inequality and the GCD test are implemented in Tiny. The Fortran 77 versions of the Perfect benchmarks were preprocessed and converted into Tiny syntax using *f2t*, the Fortran 77 to Tiny converter.

Intraprocedural constant propagation and dead code elimination were carried out before the application of the data dependence tests. Induction variable recognition was also performed, before applying the tests, to remove induction variables and convert array subscripts, containing induction variables, into functions of the loop index variables. No interprocedural or symbolic analysis [11, 13] was performed. We tested only for potential dependences caused by array references. Furthermore, only dependences within the same loop nest were considered. Subscript pairs were not tested if they could not be expressed as linear functions of the loop indices.

Table 1: Benchmark Characteristics

Program	Lines	Routines	Loops
adm (AP)	6105	97	108
arc2d (SR)	3965	39	121
bdna (NA)	3980	43	85
dyfesm (SD)	7608	89	37
flo52q (TF)	1986	28	53
mdg (LW)	1238	16	17
mg3d (SM)	2812	28	53
ocean (OC)	4343	36	14
qcd2 (LG)	2327	34	62
spec77 (WS)	3885	60	63
track (MT)	3735	34	8
trfd (TI)	485	5	11
Total	42469	509	632

If both subscripts in a pair of array references, tested for data dependence, are loop invariant, then existence of dependence can be determined by simply comparing their values. This is known as the constant or ZIV test [10]. We do not report any results for these trivial cases. Results about the application frequency and independence rate of the constant test are reported in [10, 16, 17, 23].

We considered the case of statically unknown loop limits. In some programs a number of loop limits can be statically unknown, even after applying intraprocedural constant propagation and induction variable recognition and elimination. An experiment in [17] shows that overall 12% of the lower loop limits and 71% of the upper loop limits are unknown on the Perfect benchmarks. In an interactive system such as ParaScope [12] or PAT [24], information about statically unknown loop limits can be provided by the user. If such information can not be made available at compile time, the Banerjee inequality can still be applied making the following conservative assumption. Whenever a lower loop limit is unknown we assume $-\infty$ as its value, and

whenever an upper loop limit is unknown we assume $+\infty$ as its value.

Three experiments were performed on the Perfect benchmarks to study the effectiveness of the conditions in Corollaries 1 and 2 on both unknown and known loop limits. In the first experiment (Table 2), unknown lower and upper loop limits, that were not linear functions of the loop indices in the enclosing loops, were assumed to be $-\infty$ and $+\infty$ respectively. In the second experiment (Table 3), any unknown lower loop limit, that was not a linear function of the loop indices in the enclosing loops, was replaced by 1. Similarly, any unknown upper loop limit, that was not a linear function of the loop indices in the enclosing loops, was replaced by 40. The choice of those numbers was to maintain consistency with earlier experiments [17, 23]. In the third experiment (Table 4) such unknown lower and upper loop limits were replaced by 1 and 5 respectively. One can observe that the hypothesis of Corollaries 1 and 2 is less likely to be satisfied when the loop iteration domains are small. The choice of 1 and 5 as the loop limits in this case was made to demonstrate the effectiveness of the conditions, in Corollaries 1 and 2, even in very small iteration domains.

The dependence tests were applied dimension by dimension to all proper [29] direction vectors for a pair of array references. Direction vector pruning [16] was applied to prune away all unused variables. If an independence was proved in a given dimension, the tests were not applied in further dimensions. In each dimension the dependence tests were carried out in the following order. First the Banerjee inequality was applied. If the Banerjee inequality was not satisfied, then the counter for Banerjee-No was incremented and we stopped; otherwise the counter for Banerjee-Maybe was incremented. In the latter case Corollary 1 was applied. If the conditions in Corollary 1 were true, then the counter for Corollary 1-Yes was incremented and we stopped; otherwise we continued by applying the GCD Test. If the gcd did not divide the right hand side of the dependence equation, then the counter for GCD-No was incremented and we stopped; otherwise the counter for the GCD-Maybe was incremented. In the latter case Corollary 2 was applied. If the conditions in Corollary 2 were true, then the counter for Corollary-2-Yes was incremented; otherwise we can not conclusively resolve a dependence and an answer of *maybe* should be reported.

To measure the success of the conditions in Corollaries 1 and 2 we introduce the following definitions:

Success Rate 1 indicates the success rate of Corollary 1, which is the percentage of the cases a Banerjee inequality *maybe* is converted into *yes* by applying Corollary 1. In our experiment it is computed as the ratio:

Success Rate 1 = Corollary 1-Yes / Banerjee-Maybe.

Success Rate 2 indicates the combined success rate of Corollaries 1 and 2, which is the percentage of the cases a Banerjee inequality and a GCD test *maybe*, is converted into *yes* by applying either Corollary 1 or Corollary 2. In our experiment it is computed as the ratio:

Success Rate 2 = (Corollary 1-Yes + Corollary 2-Yes) /

(Banerjee-Maybe - GCD-No).

Several important observations can be derived from our experiments. First as we can see from Tables 2-4 the success rate of Corollary 1 is 100% in all the benchmarks but one (AP) and in that one the combined success rate of Corollary 1 and 2 is 100%. This demonstrates that the Banerjee inequality extended with the conditions in Corollary 1 or at least a combination of the Banerjee inequality and the GCD test extended with the conditions in Corollary 2 are always exact in practice.

Because of the perception that the Banerjee inequality and the GCD test were approximate methods, possibly inaccurate in practice, a number of exponential dependence tests [9, 16, 22, 28] have been proposed in the literature. Our empirical study has shown that exact answers can be derived in linear time in practice. This obviates the need for more expensive, potentially exponential tests or special case approaches [10, 16]. The only cases were more expensive exponential tests might be helpful is in the presence of multidimensional coupled subscripts. The number of coupled subscripts in the Perfect benchmarks has shown though to be only about 4% of the total number of subscript pairs [10]. In case of coupled subscripts, as we discussed before, a system of equations has to be tested for a simultaneous constrained integer solution and the subscript by subscript Banerjee inequality and GCD tests may introduce approximations. In those cases polynomial time techniques such as constraint propagation [10], wherever applicable, or the lambda test [15] can also be applied to reduce the problem to testing single equations for constrained integer solutions. Future work will address the issue of simultaneous constrained integer solutions.

Our experiments also indicate that the GCD test does not have to be applied at all in most cases. Only in the AP benchmark a few of the cases were resolved by the application of the GCD test and the conditions in Corollary 2. We can also see that the results of Tables 2 and 3 are almost identical. Only in the AP and SR benchmarks a few additional independences were discovered when the unknown lower and upper loop limits were replaced by 1 and 40 respectively. Comparing the results of Tables 3 and 4 though, we see that when we replaced the upper loop limit by 5 more independences in four benchmarks (AP, SR, LW, SM) were discovered. In LW the number of independences increased from 31.8 % to 41.0%. This indicates that even though the Banerjee inequality can be successfully performed without the need

for complete information about the loop limits, in certain extreme cases this information may help discover additional independences. In all three experiments though the success rates of the Corollaries remain the same.

5. RELATED WORK & CONCLUSIONS

A number of empirical studies have been performed evaluating the performance of different data dependence tests. They are all based though on the notion that the Banerjee inequality and the GCD test are approximate tests and they can not conclusively determine a data dependence. Our empirical study in this paper focuses on how often the conditions for exactness developed for the Banerjee test and a combination of the GCD and Banerjee tests occur in practice, i.e., how often they guarantee exact answers in practice.

Shen et al. [23] in a preliminary empirical study present information about the usage frequency and independence detection rates of various dependence tests, including the constant test, the GCD test and different implementations of the Banerjee inequality. Their empirical study was performed on a number of Fortran numerical packages.

Petersen and Padua [17] perform an experimental evaluation on the Perfect benchmarks of a proposed sequence of dependence tests. This sequence consists of the constant test, the generalized GCD test, Banerjee's inequalities, integer programming and the Omega test [22]. In case of a Banerjee inequality *maybe* answer, based on the perception that the Banerjee inequality can not prove dependence, exponential tests such as integer programming and Omega test were applied to elicit a *yes* or *no* answer. Their results also show that Banerjee inequality is sufficiently accurate in practice and most of the applicability of the integer programming and Omega tests is in proving a dependence that would otherwise be assumed dependent. Here we have shown that this can be done in linear time by applying the conditions in Corollaries 1 and 2 rather than exponential dependence tests.

Maydan et al [16] propose a different sequence of exact dependence tests for special case inputs and they show that they derive an exact answer in all cases in the Perfect benchmarks. Their sequence includes Fourier-Motzkin variable elimination, an exponential method, as a back up test. It has been shown in their experiments that Fourier-Motzkin has to be applied in a number of cases. Triolet [25] found that using Fourier-Motzkin variable elimination takes from 22 to 28 times longer than conventional dependence testing.

Another approach taken, is based on the fact that most array references in scientific programs are fairly simple. Goff et al [10] propose a dependence testing scheme based on classifying pairs of subscripts. Efficient and exact tests are presented for certain classes of commonly occurring array references involving single index variables (SIV). Our results in Corollaries 1 and 2 are more general than their special SIV cases. In case of multiple index variable (MIV) subscripts their techniques, in fact, rely on the Banerjee and GCD tests.

As one can see from the above discussion these approaches fall into two categories. First, developing exponential tests for solving more general instances of the problem and second, applying sequences of different dependence tests each one exact for special instances of the problem. In this work we demonstrate that conventional dependence testing techniques are sufficiently accurate as well as efficient in practice and, therefore, there is no need for exponential or special case approaches.

REFERENCES

[1] A. Aho, J. Hopcroft, and J. Ullman. *Data Structures and Algorithms*. Addison-Wesley, 1983.

[2] F. Allen, M. Burke, P. Charles, R. Cytron, and J. Ferrante. An Overview of the PTRAN Analysis System for Multiprocessing. In *Proceedings of the 1987 International Conference on Supercomputing*, Athens, Greece, June 1987.

[3] J. R. Allen. Dependence Analysis for Subscripted Variables and its Application to Program Transformations. Ph.D. Thesis, Dept. of Computer Science, Rice University, April 1983.

[4] J. R. Allen and K. Kennedy. PFC: A program to convert Fortran to parallel form. *Supercomputers: Design and Applications*, IEEE Computer Society Press, Silver Spring, MD, 1984.

[5] J. R. Allen and K. Kennedy. Automatic Translation of Fortran Programs to Vector Form. *ACM Transactions on Programming Languages and Systems*, Vol. 9, No. 4, October 1987.

[6] U. Banerjee. *Dependence Analysis for Supercomputing*. Kluwer Academic Publishers, Norwell, MA, 1988.

[7] M. Berry, et al. The Perfect Club Benchmarks: Effective Performance Evaluation of Supercomputers. *The International Journal of Supercomputer Applications*, Vol. 3, 1989.

[8] M. Burke and R. Cytron. Interprocedural Dependence Analysis and Parallelization. In *Proceedings of SIGPLAN '86 Symposium on Compiler Construction*, Palo Alto, CA, June 1986.

[9] C. Eisenbeis, J.-C. Sogno. A General Algorithm for Data Dependence Analysis. In *Proceedings of the Sixth ACM International Conference on Supercomputing*, Washington, D.C., July 1992.

[10] G. Golf, K. Kennedy, and C. W. Tseng. Practical Dependence Testing. In *Proceedings of the SIGPLAN '91 Conference on Programming Language Design and Implementation*, Toronto, Canada, June 1991.

[11] M. Haghighat and C. Polychronopoulos. Symbolic Dependence Analysis for High-Performance Parallelizing Compilers. In *Proceedings of the Third Annual Workshop on Languages and Compilers for Parallel Computing*, Irvine, CA, August 1990.

[12] K. Kennedy, K. McKinley, C.-W. Tseng. Interactive Parallel Programming Using the ParaScope Editor. *IEEE Transactions on Parallel and Distributed Systems*, Vol. 2, No. 3, July 1991.

[13] A. Lichnewsky and F. Thomasset. Introducing Symbolic Problem Solving Techniques in the Dependence Testing Phases. In *Proceedings of the Second ACM International Conference on Supercomputing*, Saint-Malo, France, July 1988.

[14] Z. Li, P. Yew. Some Results on Exact Data Dependence Analysis. In *Proceedings of the 2nd Workshop on Languages and Compilers for Parallel Computing*, Urbana, Illinois, August 1989.

[15] Z. Li, P. Yew, and C. Zhu. An Efficient Data Dependence Analysis for Parallelizing Compilers. *IEEE Transactions on Parallel and Distributed Systems*, Vol. 1, No. 1, January 1990.

[16] D. Maydan, J. Hennesy, and M. Lam. Efficient and Exact Data Dependence Analysis for Parallelizing Compilers. In *Proceedings of the SIGPLAN '91 Conference on Programming Language Design and Implementation*, Toronto, Canada, June 1991.

[17] P. Petersen, D. Padua. Static and Dynamic Evaluation of Data Dependence Analysis. In *Proceedings of the Seventh ACM International Conference on Supercomputing*, Tokyo, Japan, July 1993.

[18] C. Polychronopoulos, et al. Parafrase-2 : An Environment for Parallelizing, Partitioning, Synchronizing and Scheduling Programs on Multiprocessors. *International Journal of High Speed Computing*, Vol. 1, No. 1, May 1989.

[19] K. Psarris. On Exact Data Dependence Analysis. In *Proceedings of the Sixth ACM International Conference on Supercomputing*. Washington, D.C., July 1992.

[20] K. Psarris, D. Klappholz, and X. Kong. On the Accuracy of the Banerjee Test. *Journal of Parallel and Distributed Computing, Special Issue on*

Shared Memory Multiprocessors, Vol. 12, No. 2, June 1991.

[21] K. Psarris, X. Kong, and D. Klappholz. The Direction Vector I Test. *IEEE Transactions on Parallel and Distributed Systems*, Vol. 4, No. 11, November 1993.

[22] W. Pugh. A Practical Algorithm for Exact Array Dependence Analysis. *Communications of the ACM*, Vol. 35, No. 8, August 1992.

[23] Z. Shen, Z. Li, and P. Yew. An Empirical Study of Fortran Programs for Parallelizing Compilers. *IEEE Transactions on Parallel and Distributed Systems*, Vol. 1, No. 3, July 1990.

[24] K. Smith and W. Appelbe. PAT--An Interactive Fortran Parallelizing Assistant Tool. *Proceedings of the 1988 International Conference on Parallel Processing*, Saint-Charles, IL, August 1988.

[25] R. Triolet. Interprocedural analysis for program restructuring with Parafrase. CSRD Report No. 538. Department of Computer Science, University of Illinois at Urbana-Champaign, December 1985.

[26] M. Wolfe. *Optimizing Supercompilers for Supercomputers*. Pitman, London and The MIT Press, Cambridge, MA, 1989.

[27] M. Wolfe. The Tiny Loop Restructuring Research Tool. In *Proceedings of the 1991 International Conference on Parallel Processing*, St Charles, IL, August 1991.

[28] M. Wolfe and C.-W. Tseng. The Power Test for Data Dependence. *IEEE Transactions on Parallel and Distributed Systems*, Vol. 3, No. 5, September 1992.

[29] H. Zima and B. Chapman. *Supercompilers for Parallel and Vector Computers*. ACM Press, New York, NY, 1991.

Table 2: Dependence Results on Perfect Benchmarks.

Program	Instances	Banerjee NO	Banerjee MAYBE	Corollary 1 YES	Success Rate 1	GCD NO	GCD MAYBE	Corollary 2 YES	Success Rate 2
adm (AP)	5715	3073 (53.8%)	2642 (46.2%)	2561 (44.8%)	96.9%	22 (0.4%)	59 (1.0%)	59 (1.0%)	100%
arc2d (SR)	16524	4765 (28.8%)	11759 (71.2%)	11759 (71.2%)	100%	0	0	0	100%
bdna (NA)	5124	2146 (41.9%)	2978 (58.1%)	2978 (58.1%)	100%	0	0	0	100%
dyfesm (SD)	1947	978 (50.2%)	969 (49.8%)	969 (49.8%)	100%	0	0	0	100%
flo52q (TF)	8928	4378 (49.0%)	4550 (51.0%)	4550 (51.0%)	100%	0	0	0	100%
mdg (LW)	2778	883 (31.8%)	1895 (68.2%)	1895 (68.2%)	100%	0	0	0	100%
mg3d (SM)	3009	560 (18.6%)	2449 (81.4%)	2449 (81.4%)	100%	0	0	0	100%
ocean (OC)	243	98 (40.3%)	145 (59.7%)	145 (59.7%)	100%	0	0	0	100%
qcd2 (LG)	2115	1055 (50.0%)	1060 (50.0%)	1060 (50.0%)	100%	0	0	0	100%
spec77 (WS)	6015	4578 (76.1%)	1437 (23.9%)	1437 (23.9%)	100%	0	0	0	100%
track (MT)	108	84 (77.8%)	24 (22.2%)	24 (22.2%)	100%	0	0	0	100%
trfd (TI)	909	132 (14.5%)	777 (85.5%)	777 (85.5%)	100%	0	0	0	100%

Table 3: Dependence Results on Perfect Benchmarks (Unknown Upper Loop Limits assumed 40).

Program	Instances	Banerjee NO	Banerjee MAYBE	Corollary 1 YES	Success Rate 1	GCD NO	GCD MAYBE	Corollary 2 YES	Success Rate 2
adm (AP)	5715	3084 (54.0%)	2631 (46.0%)	2553 (44.7%)	97.0%	20 (0.3%)	58 (1.0%)	58 (1.0%)	100%
arc2d (SR)	16524	4958 (30.0%)	11566 (70.0%)	11566 (70.0%)	100%	0	0	0	100%
bdna (NA)	5124	2146 (41.9%)	2978 (58.1%)	2978 (58.1%)	100%	0	0	0	100%
dyfesm (SD)	1947	978 (50.2%)	969 (49.8%)	969 (49.8%)	100%	0	0	0	100%
flo52q (TF)	8928	4378 (49.0%)	4550 (51.0%)	4550 (51.0%)	100%	0	0	0	100%
mdg (LW)	2778	883 (31.8%)	1895 (68.2%)	1895 (68.2%)	100%	0	0	0	100%
mg3d (SM)	3009	560 (18.6%)	2449 (81.4%)	2449 (81.4%)	100%	0	0	0	100%
ocean (OC)	243	98 (40.3%)	145 (59.7%)	145 (59.7%)	100%	0	0	0	100%
qcd2 (LG)	2115	1055 (50.0%)	1060 (50.0%)	1060 (50.0%)	100%	0	0	0	100%
spec77 (WS)	6015	4578 (76.1%)	1437 (23.9%)	1437 (23.9%)	100%	0	0	0	100%
track (MT)	108	84 (77.8%)	24 (22.2%)	24 (22.2%)	100%	0	0	0	100%
trfd (TI)	909	132 (14.5%)	777 (85.5%)	777 (85.5%)	100%	0	0	0	100%

Table 4: Dependence Results on Perfect Benchmarks (Unknown Upper Loop Limits assumed 5).

Program	Instances	Banerjee NO	Banerjee MAYBE	Corollary 1 YES	Success Rate 1	GCD NO	GCD MAYBE	Corollary 2 YES	Success Rate 2
adm (AP)	5715	3229 (56.5%)	2486 (43.5%)	2438 (42.7%)	98.1%	37 (0.6%)	11 (0.2%)	11 (0.2%)	100%
arc2d (SR)	16524	5151 (31.2%)	11373 (68.8%)	11373 (68.8%)	100%	0	0	0	100%
bdna (NA)	5124	2146 (41.9%)	2978 (58.1%)	2978 (58.1%)	100%	0	0	0	100%
dyfesm (SD)	1947	978 (50.2%)	969 (49.8%)	969 (49.8%)	100%	0	0	0	100%
flo52q (TF)	8928	4378 (49.0%)	4550 (51.0%)	4550 (51.0%)	100%	0	0	0	100%
mdg (LW)	2778	1138 (41.0%)	1640 (59.0%)	1640 (59.0%)	100%	0	0	0	100%
mg3d (SM)	3009	578 (19.2%)	2431 (81.8%)	2431 (81.8%)	100%	0	0	0	100%
ocean (OC)	243	98 (40.3%)	145 (59.7%)	145 (59.7%)	100%	0	0	0	100%
qcd2 (LG)	2115	1055 (50.0%)	1060 (50.0%)	1060 (50.0%)	100%	0	0	0	100%
spec77 (WS)	6015	4578 (76.1%)	1437 (23.9%)	1437 (23.9%)	100%	0	0	0	100%
track (MT)	108	84 (77.8%)	24 (22.2%)	24 (22.2%)	100%	0	0	0	100%
trfd (TI)	909	132 (14.5%)	777 (85.5%)	777 (85.5%)	100%	0	0	0	100%

Construction of Representative Simple Sections*

Aart J.C. Bik and Harry A.G. Wijshoff

High Performance Computing Division

Department of Computer Science, Leiden University

P.O. Box 9512, 2300 RA Leiden, the Netherlands

ajcbik@cs.leidenuniv.nl

Abstract

In this paper, we present a method to find a number of regions in a matrix such that all accesses induced by an arbitrary occurrence of this matrix in a program are confined to one of these regions. The index set of regions will be described in terms of two-dimensional simple sections. Furthermore, simple loop transformations are used to increase the number of resulting regions.

1 Introduction

Restructuring compilers are used to convert programs into a form exploiting certain characteristics of the target machine. For example, a so-called vectorizing compiler transforms serial software into vector code to enable the effective use of pipelined functional units or parallel processors in a target machine. Other compilers generate code for shared or distributed memory multiprocessors after the automatic detection of concurrency. Frequently, such objectives can be met by means of program transformations only, which has been well-studied in the past and still is an important research topic (see e.g. [3, 12, 14, 16, 17]). However, in some cases the application of data structure transformations can also be useful. Simple examples are formed by the transposition or linearization of arrays to enhance vector performance. However, more complex data structure transformations are possible.

In earlier work [5, 7, 8], for example, we studied the automatic generation of sparse codes [10, 11, 13, 18], where the programmer defines the source program for the dense case, i.e. all matrix operations are coded using two-dimensional arrays. Subsequently, program and data structure transformations are applied to this program by the compiler to exploit the sparsity of some of these matrices. Nonzero structure information can be used by the compiler to exploit specific characteristics of the data operated on [6].

*Support was provided by the Foundation for Computer Science (SION) of the Dutch Organization for Scientific Research (NWO) and the EC Esprit Agency DG XIII (Grant No. APPARC 6634 BRA III).

One important step for a restructuring compiler performing such data structure transformations, is the construction of a number of regions in an array such that all accesses induced by an arbitrary occurrence of this array in the program are confined to one of these regions. Given these regions, the compiler can select different storage organizations for the regions and convert the code accordingly. Because each occurrence only accesses one region, we prevent the requirement of run-time tests to determine which of the selected storage organizations must be accessed. We will consider regions of which the index set can be described in terms of a simple section [1, 2]. In this manner, most access shapes found in numerical programs are supported.

In this paper we discuss the construction of a number of representative simple sections, referred to as **representatives** for short, such that accesses of each occurrence are confined to a region described by one of the representatives. Furthermore, because usually the number of resulting representatives is small, since partially overlapping simple sections have to be combined, we also discuss how simple loop transformations can be used to increase the resulting degree of fragmentation. The method has been developed in the context of automatic generation of sparse codes. Therefore, we focus on the construction of representative two-dimensional simple sections for matrices stored in two-dimensional arrays. However, in principle, the techniques are also applicable to arrays of arbitrary dimension using the general definition of simple sections given in [1, 2]. This paper is organized as follows. In section 2, we give some preliminaries. In section 3, we discuss how representatives can be selected by a compiler and discuss the use of loop transformations to increase the number of resulting representatives. Subsequently, in section 4 an application of the method is discussed. Finally, we state some conclusions in section 5.

2 Two-Dimensional Simple Sections

A convenient representation of parts of the index set of a matrix consists of the two-dimensional **simple section** [1, 2].

A two-dimensional simple section S consists of all discrete points $(x, y) \in \mathbf{Z}^2$ satisfying the inequalities imposed by four boundary pairs as shown below, where only the values $l_i, u_i \in \mathbf{Z}$ have to be stored:

$$S = \{ (x,y) \in \mathbf{Z}^2 \mid \begin{bmatrix} l_1 & \leq & x & \leq & u_1 \\ l_2 & \leq & y & \leq & u_2 \\ l_3 & \leq & x+y & \leq & u_3 \\ l_4 & \leq & x-y & \leq & u_4 \end{bmatrix} \}$$

Simple sections are closed under intersection, and we can construct the intersection $S \cap S'$ of two simple sections S and S' by taking the most interior values for all boundary pairs. Unfortunately, simple sections are not closed under union. The *smallest* simple section that envelops the union of these simple sections is obtained by taking the most exterior values for the boundary pairs [1], as illustrated in figure 1. We use the notation $S \uplus S'$ to denote this kind of union.

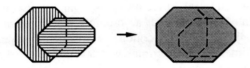

Figure 1: Smallest Enveloping Simple Section

These observations give rise to the routines `intersect` and `combine`. Procedure `intersect` constructs the intersection of two simple sections stored in `s1` and `s2`. Constructs 's.l[i]' and 's.u[i]' are used to access the `i`th component of the vectors $\vec{l}$ and $\vec{u}$ of simple section `s` respectively:

```
proc intersect(s1, s2, var s)
begin
  for i := 1, 4 do
    s.l[i] := max(s1.l[i],s2.l[i]);
    s.u[i] := min(s1.u[i],s2.u[i]);
  enddo
end
```

A similar definition with the roles of `min` and `max` interchanged can be given for procedure `combine` to compute the smallest enveloping simple section of two non-empty simple sections stored in `s1` and `s2`.

Function `empty` determines whether a simple section stored in `s` is empty.

```
boolean function empty(s)
begin
  empty := ((s.l[1]>s.u[1]) or (s.l[2]>s.u[2])
         or (s.l[3]>s.u[3]) or (s.l[4]>s.u[4]));
end
```

Likewise, function `overlap` can be used to detect a non-empty intersection of two simple sections:

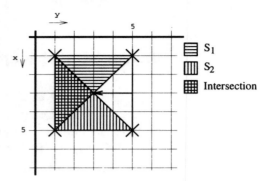

Figure 2: Possible Refinement

```
boolean function overlap(s1, s2)
begin
  intersect(s1, s2, tmp);
  overlap := not empty(tmp);
end
```

Each boundary pair in a two-dimensional simple section is defined by two simple boundaries, which are lines of the form $x = c$, $y = c$, $x + y = c$ or $x - y = c$. A boundary is called **tight** if it contains at least one point of the simple section, and **non-tight** otherwise.

A disadvantage of the previous given definition of procedure `intersect` is that, even if all boundaries defining the original simple sections are tight, some *non-tight* boundaries may arise in the resulting intersection. For example, using procedure `intersect` to compute $S_1 \cap S_2$ for the simple sections shown in figure 2 results in the following simple section, in which boundary $y = 5$ becomes non-tight:

$$\{ (x,y) \in \mathbf{Z}^2 \mid \begin{bmatrix} 1 & \leq & x & \leq & 5 \\ 1 & \leq & y & \leq & 5 \\ 2 & \leq & x+y & \leq & 6 \\ 0 & \leq & x-y & \leq & 4 \end{bmatrix} \}$$

Although the intersection is properly defined, non-tight boundaries are undesirable because they may affect the outcome of the construction of the smallest simple section enveloping a number of simple sections in which this intersection is involved [1]. Therefore, as suggested in [1] we have constructed a procedure `refine` to refine such non-tight boundary pairs, based on the fact that a boundary of a boundary pair can be refined using pair-wise combinations of other boundary pairs [8]. If, for example, `refine` is called after $S_1 \cap S_2$ has been computed, $y = 5$ is refined into $y = 3$ because the combination $x + y \leq 6$ and $0 \leq x - y$ implies $y \leq 3$.

The following algorithm, based on the method in [1, 2], is used to approximate the index set of a region in a matrix accessed by an occurrence of this matrix in terms of a simple section.

We assume that the subscripts of this occurrence, appearing at nesting depth d in a loop with indices $I_1, \ldots, I_d$, can be expressed as an affine mapping $F(\vec{I}) = \vec{v} + W \cdot \vec{I}$, where $\vec{I} = (I_1, \ldots, I_d)^T$ and $v_i, w_{ij} \in \mathbf{Z}$:

$$F(\vec{I}) = \begin{pmatrix} v_1 \\ v_2 \end{pmatrix} + \begin{pmatrix} w_{11} & \cdots & w_{1d} \\ w_{12} & \cdots & w_{2d} \end{pmatrix} \cdot \vec{I} \quad (1)$$

Computing the simple section S for this occurrence consists of determining the extremal values of the four expressions x, y, $x+y$ and $x-y$ under the correspondence $(x,y)^T = F(\vec{I})$:

$$
\begin{aligned}
x &= v_1 + \sum_{j=1}^{d} w_{1j} \cdot I_j \\
y &= v_2 + \sum_{j=1}^{d} w_{2j} \cdot I_j \\
x+y &= v_1 + v_2 + \sum_{j=1}^{d} (w_{1j} + w_{2j}) \cdot I_j \\
x-y &= v_1 - v_2 + \sum_{j=1}^{d} (w_{1j} - w_{2j}) \cdot I_j
\end{aligned}
\quad (2)
$$

A lower bound of the ith expression in (2) defines a value of the ith component of $\vec{l}$ belonging to S. Likewise, an upper bound of the ith expression defines a value of the ith component of $\vec{u}$. Such lower and upper bounds can be obtained by successively replacing the loop indices by extremal values *in decreasing order of nesting depth*.

Starting with $k = d$, each expression in (2) can be expressed as follows:

$$a_0 + \sum_{j=1}^{k} a_j \cdot I_j \quad (3)$$

If the kth loop index is bounded as $L_k \le I_k \le U_k$ then a lower bound of this expression is defined by the following inequality, in which the definitions $a^+ = \max(a, 0)$ and $a^- = \max(-a, 0)$ from [3] are used:

$$a_0 + \sum_{j=1}^{k-1} a_j \cdot I_j + (a_k^+ \cdot L_k - a_k^- \cdot U_k) \le a_0 + \sum_{j=1}^{k} a_j \cdot I_j$$

Because either $a_k^+ \ne 0$ or $a_k^- \ne 0$, only the lower or upper bound of a loop index is required. If we assume that the required bound can be expressed as $b_0 + \sum_{j=1}^{k-1} b_j \cdot I_j$ where all $b_i \in \mathbf{Z}$, then the lower bound defined by the previous inequality can be expressed as (3) again for a lower value of k. Repetitively replacing loop indices by extremal values according to this inequality for $k = d, \ldots, 1$, eventually results in a constant forming one of the four components of $\vec{l}$.

Likewise, an upper bound of (3) is defined by the following inequality:

$$a_0 + \sum_{i=1}^{k} a_i \cdot I_i \le a_0 + \sum_{i=1}^{k-1} a_i \cdot I_i + (a_k^+ \cdot U_k - a_k^- \cdot L_k)$$

Successively replacing loop indices in one of the expression in (2) by extremal values for $k = d, \ldots, 1$ eventually results in the corresponding component of $\vec{u}$.

Consider, for example, the following fragment:

```
DO I = 1, 10
   DO J = 1, I
      ... A(I,J) ...
   ENDDO
ENDDO
```

The boundary values are computed by determining the lower and upper bound for each of the expressions x, y, $x+y$ and $x-y$, where $(x,y) = (I, J)$:

1			$\le$	I	$\le$			10
1			$\le$	J	$\le$	I	$\le$	10
2	$\le$	I+1	$\le$	I+J	$\le$	I+I	$\le$	20
0	$=$	I-I	$\le$	I-J	$\le$	I-1	$\le$	9

This example also illustrates the importance of eliminating loop indices in decreasing order of nesting depth. If, for example, at each elimination, *all* indices are replaced at once by extremal values defined by the loop bounds until no indices remain, the lower bound of the last expression would be computed as $I - J \ge 1 - I \ge -9$.

In [7, 8], we discuss how the assumption that subscript bounds are not violated can be used to obtain a reasonable accurate simple section in case some complex subscripts or loop bounds are used.

3 Representatives

Computing the simple sections for all occurrences of a matrix A in a program gives rise to a bag $\Sigma_A = \{S_1, \ldots, S_p\}$, in which multiple occurrences of the same simple section may appear. An important step in the automatic data structure selection is the construction of a set $R_A = \{S'_1, \ldots, S'_q\}$ of disjunct representatives with the property that for each $S \in \Sigma_A$, we have $S \subseteq S'$ for some $S' \in R_A$. This implies that each simple section in Σ_A precisely has one representative.

Because the number of different regions accessed in a program tends to be small, $q \ll p$ will hold in most cases. Typical regions that may arise are, for instance, strict lower- and upper-triangular parts of a matrix together with the diagonal. Using the following simple approach to construct representative, it is even very likely that only *one* representative results, consisting of the whole index set of the matrix.

3.1 A Simple Approach

One way to construct representatives for a matrix is to combine overlapping simple sections until a number of non-overlapping representatives remains. Starting with $R_A = \emptyset$, for each $S \in \Sigma_A$, the following actions are performed:

1. If $S \cap S' = \emptyset$ for all $S' \in R_A$, which can be tested with function `overlap`, then S is added to R_A.

2. Otherwise, we have $S \cap S' \ne \emptyset$ for some $S' \in R_A$.

(a) If $S \subseteq S'$, S' is used as representative of S.

(b) Otherwise, $\overline{S} = S \uplus S'$ is computed and S' is deleted from R_A. If $\overline{S} \cap S' \neq \emptyset$ for some (other) $S' \in R_A$, then this step is repeated until $\overline{S} \cap S' = \emptyset$ for all $S' \in R_A$. The resulting $\overline{S}$ is inserted into R_A.

Continuing in this fashion eventually results in a set $R_A = \{S'_1, \ldots, S'_q\}$, where $S'_i \cap S'_j = \emptyset$ for $i \neq j$ and for each $S \in \Sigma_A$ we have $S \subseteq S'$ for some $S' \in R_A$. Consider, for example, the three occurrences of a matrix A in the following program fragment:

```
DO I = 1, 100
    DO J = 1, I-1
        A₁(I,J) = C(I,J)
    ENDDO
    B(I) = A₂(I,I)
    DO J = I, 100
        A₃(I,J) = D(I,J)
    ENDDO
ENDDO
```

Starting with $R_A = \emptyset$ and $\Sigma_A = \{S_1, S_2, S_3\}$ consisting of the simple sections computed by the compiler, we first consider the simple section S_1:

$$S_1 = \left\{ (x,y) \in \mathbf{Z}^2 \;\middle|\; \begin{bmatrix} 2 & \leq & x & \leq & 100 \\ 1 & \leq & y & \leq & 99 \\ 3 & \leq & x+y & \leq & 199 \\ 1 & \leq & x-y & \leq & 99 \end{bmatrix} \right\}$$

Because R_A is empty, no overlap is detected and S_1 is inserted into this set. Subsequently, the next simple section is considered:

$$S_2 = \left\{ (x,y) \in \mathbf{Z}^2 \;\middle|\; \begin{bmatrix} 1 & \leq & x & \leq & 100 \\ 1 & \leq & y & \leq & 100 \\ 2 & \leq & x+y & \leq & 200 \\ 0 & \leq & x-y & \leq & 0 \end{bmatrix} \right\}$$

Because $S_1 \cap S_2 = \emptyset$, this simple section can also be used as representative and we obtain $R_A = \{S_1, S_2\}$ thereafter. In the final step, the following simple section is considered:

$$S_3 = \left\{ (x,y) \in \mathbf{Z}^2 \;\middle|\; \begin{bmatrix} 1 & \leq & x & \leq & 100 \\ 1 & \leq & y & \leq & 100 \\ 2 & \leq & x+y & \leq & 200 \\ -99 & \leq & x-y & \leq & 0 \end{bmatrix} \right\}$$

Because $S_2 \cap S_3 \neq \emptyset$ and $S_3 \not\subseteq S_2$, we set $\overline{S} = S_2 \uplus S_3 = S_3$ and delete S_2 from R_A. Since $S_1 \cap \overline{S} = \emptyset$ for the only remaining representative S_1, we add $\overline{S}$ to R_A. The final set $R_A = \{S_1, S_3\}$ represents the index set of the lower triangular and strict upper triangular part of A.

The disadvantage of this approach is that it is very likely that we obtain only a few representatives for each matrix. Therefore, iteration space partitioning is used to increase the degree of resulting fragmentation.

3.2 Iteration Space Partitioning

Before we compute $S_2 \uplus S_3$ in step (2)b for the previous example, the following actions can be performed. First, we substitute the subscripts of one of the occurrences to which these simple sections belong for (x,y) in the inequalities defining the intersection $S_2 \cap S_3$. The resulting inequalities define the part of the iteration space in which elements with indices in this intersection are referenced.

For example, substituting the subscripts (I,I) of the second occurrence for (x,y) in the inequalities defining the intersection yields a system of inequalities that can be simplified into $1 \leq \mathrm{I} \leq 100$, defining the whole iteration space (viz. $S_2 \subseteq S_3$). On the other hand, substituting the subscripts (I,J) of the third occurrence for (x,y) results (after simplification) in the following system:

$$1 \leq \mathrm{I} \leq 100 \quad \mathrm{I} \leq \mathrm{J} \leq \mathrm{I}$$

Confining the overlap to $S_2 \cap S_3$ can be done by isolating the loop-body in which the third occurrence appears for all iterations satisfying $1 \leq \mathrm{I} \leq 100$ and $\mathrm{I} \leq \mathrm{J} \leq \mathrm{I}$, using a method which we will refer to as **iteration space partitioning** [8]:

```
DO I = 1, 100              DO I = 1, 100
    DO J = 1, I-1              DO J = 1, I-1
        A₁(I,J) = C(I,J)          A₁(I,J) = C(I,J)
    ENDDO                     ENDDO
    B(I) = A₂(I,I)            B(I) = A₂(I,I)
    DO J = I, 100      →      A₃ₐ(I,I) = D(I,I)
        A₃(I,J) = D(I,J)      DO J = I+1, 100
    ENDDO                         A₃ᵦ(I,J) = D(I,J)
ENDDO                         ENDDO
                          ENDDO
```

Hence, after application of a relatively simple transformation, the main diagonal together with the strict lower and upper triangular part can be selected as representatives, so that more representatives result than in the previous section. A potential disadvantage of this approach is that the code size may increase substantially. Another disadvantage is due to the fact that redundant fragmentation of other simple sections may occur if we have several occurrences of a matrix in the same loop-body. Although loop distribution [12, 16, 17] can be used to limit the amount of resulting redundant fragmentation, this effect still occurs if loop distribution is invalid, or if several occurrences appear in the same statement. Therefore, in an effective implementation of representatives construction, a balance between code duplication and redundant fragmentation on one side, and the degree of fragmentation on the other side must be found. Since, in general, a simple section $S' \in R_A$ may represent the simple sections of *many* occurrences, it is only feasible to partition the iteration space of a loop surrounding an occurrence of which the simple section has not been considered yet. Assume that for a certain simple section $S \in \Sigma_A$ associated with an occurrence at nesting depth d in a loop with iteration space $IS \subseteq \mathbf{Z}^d$, we have $S \cap S' \neq \emptyset$ and $S \not\subseteq S'$ for some representative $S' \in R_A$.

Then, our goal is to isolate the loop-body of this loop for all iterations lying in the following set $I \subseteq \mathbf{Z}^d$, where $F : \mathbf{Z}^d \to \mathbf{Z}^2$ represents the subscripts of the occurrence to which S belongs:

$$I = \{\vec{\mathbf{I}} \in \mathbf{Z}^d | F(\vec{\mathbf{I}}) \in S \cap S'\} \qquad (4)$$

Hence, $I \cap IS$ defines the part of the iteration space in which elements with indices in $S \cap S'$ are referenced. For simple subscripts, the mapping is as an affine mapping $F(\vec{\mathbf{I}}) = \vec{v} + W \cdot \vec{\mathbf{I}}$, where $\vec{\mathbf{I}} = (\mathbf{I}_1, \ldots, \mathbf{I}_d)^T$ and $v_i, w_{ij} \in \mathbf{Z}$. If we denote the boundary values of $S \cap S'$ by $\vec{l}$ and $\vec{u}$, substituting $F(\vec{\mathbf{I}})$ for (x, y) in the pairs of inequalities defining $S \cap S'$ yields the following system, consisting of four pairs of linear inequalities:

$$
\begin{array}{llll}
l_1 \leq & v_1 + & \sum_{j=1}^{d} w_{1j} \cdot \mathbf{I}_j & \leq u_1 \\
l_2 \leq & v_2 + & \sum_{j=1}^{d} w_{2j} \cdot \mathbf{I}_j & \leq u_2 \\
l_3 \leq & v_1 + v_2 + & \sum_{j=1}^{d} (w_{1j} + w_{2j}) \cdot \mathbf{I}_j & \leq u_3 \\
l_4 \leq & v_1 - v_2 + & \sum_{j=1}^{d} (w_{1j} - w_{2j}) \cdot \mathbf{I}_j & \leq u_4
\end{array}
$$

For appropriate $l^k, a_j^k, u^k \in \mathbf{Z}$, the kth pair of inequalities in this system, where $1 \leq k \leq 4$, gives rise to the definition of a slice $C^k \subseteq \mathbf{R}^d$ that can be written in the following form:

$$\{(\mathbf{I}_1, \ldots, \mathbf{I}_d) \in \mathbf{R}^d | l^k \leq a_1^k \cdot \mathbf{I}_1 + \ldots + a_d^k \cdot \mathbf{I}_d \leq u^k\}$$

A step towards the isolation of a loop-body for all iterations in (4) is to slice the iteration space according to one of these slices C^k (details of this form of iteration space partitioning are given in [8]). Using only *one* pair of inequalities at the time enables the efficient incremental construction of the simple sections into which the simple section S is fragmented, thereby preventing re-computation of the simple sections by means of subscripts and loop bounds analysis. Central to this construction is the following procedure alter, in which a simple section stored in new is obtained from a simple section stored in old by refining all boundaries after the boundary values of the kth boundary pair have been replaced by 1 and u:

```
proc alter(old, k, l, u, var new)
begin
  new := old;
  new -> l[k] := l;
  new -> u[k] := u;
  refine(new)
end
```

We can slice an iteration space by converting one DO-loop in the nested loop into (at most) three new consecutive DO-loops. The three simple sections associated with the resulting copies of the original occurrences into which S becomes fragmented by slicing the iteration space according to the kth pair can be computed by calling the following procedure new with S, $S \cap S'$ as first two parameters. The three simple sections are computed in s1, s2 and s3:

```
proc new(s, t, k, swp, var s1, var s2, var s3)
begin
  alter(s, k, s -> l[k],    t -> l[k]-1, s1);
  alter(s, k, t -> l[k],    t -> u[k],   s2);
  alter(s, k, t -> u[k]+1, s -> u[k],    s3);
  if (swp) then swap(s1, s3);
end
```

If t_k denotes the index of the last nonzero coefficient defining C^k, we assign the boolean value $a_{t_k}^k < 0$ to swp. If swp holds, the association between these simple sections and the occurrences in the resulting DO-loops must be reversed. No accuracy of description is lost (viz. $|s| = |s1| + |s2| + |s3|$) due to the characteristics of simple sections.

For example, assume that during construction of the representative simple sections for a 100×100 matrix A, we have $R_A = \{S'\}$:

$$S' = \left\{ (x, y) \in \mathbf{Z}^2 \;\middle|\; \begin{array}{ccccc} 1 & \leq & x & \leq & 100 \\ 1 & \leq & y & \leq & 100 \\ 2 & \leq & x+y & \leq & 200 \\ 0 & \leq & x-y & \leq & 0 \end{array} \right\}$$

Now, assume that the simple section associated with the following occurrence is considered next:

```
DO I = 0, 80
  DO J = 1, 10+I
    A(10+I,J) = ...
  ENDDO
ENDDO
```

The simple section is shown below:

$$S = \left\{ (x, y) \in \mathbf{Z}^2 \;\middle|\; \begin{array}{ccccc} 10 & \leq & x & \leq & 90 \\ 1 & \leq & y & \leq & 90 \\ 11 & \leq & x+y & \leq & 180 \\ 0 & \leq & x-y & \leq & 89 \end{array} \right\}$$

Because $S \not\subseteq S'$ and $S \cap S' \neq \emptyset$, we first determine the part of the iteration space in which elements with indices in this intersection are referenced. Substituting $(\mathbf{I}+10, \mathbf{J})$ for (x, y) in the inequalities defining $S \cap S'$ yields the following pairs of inequalities:

$$
\begin{array}{llccccc}
(1) & & 0 & \leq & \mathbf{I} & \leq & 90 \\
(2) & & 10 & \leq & \mathbf{J} & \leq & 90 \\
(3) & & 10 & \leq & \mathbf{I}+\mathbf{J} & \leq & 170 \\
(4) & & -10 & \leq & \mathbf{I}-\mathbf{J} & \leq & -10
\end{array}
$$

Subsequently, we slice the iteration space of the double loop according to one of these pairs. For instance, after multiplication with -1, the 4th pair can be written as $10 \leq -\mathbf{I} + \mathbf{J} \leq 10$, inducing the following iteration space slicing:

```
DO I = 0, 80                 DO I = 0, 80
  DO J = 1, 10+I               DO J = 1, 9+I
    A(10+I,J) = ...    →          A1(10+I,J) = ...
  ENDDO                         ENDDO
ENDDO                          A2(10+I,10+I) = ...
                             ENDDO
```

The zero trip loop with execution set $[11 + I, 10 + I]$ is not generated. The simple sections associated with the three resulting occurrences in this loop are computed with the following call:

$$\texttt{new}(S, S \cap S', 4, \texttt{true}, S_1, S_2, S_3);$$

Effectively, the three resulting simple sections are obtained by refining the boundaries of S after the 4th pair has been replaced with $0 \leq x - y \leq -1$, $0 \leq x - y \leq 0$ and $1 \leq x - y \leq 89$. Furthermore, the first and third resulting simple section are interchanged to account for the multiplication with -1. This yields the simple section S_1 shown below, $S_2 = S \cap S'$ and $S_3 = \emptyset$ corresponding to the occurrence in the zero-trip loop:

$$S_1 = \{\, (x,y) \in \mathbf{Z}^2 \mid \begin{bmatrix} 10 & \leq & x & \leq & 90 \\ 1 & \leq & y & \leq & 89 \\ 11 & \leq & x+y & \leq & 179 \\ 1 & \leq & x-y & \leq & 89 \end{bmatrix} \,\}$$

If, for instance, the iteration space of the original loop would be sliced according to $0 \leq I \leq 90$, no transformation would be applied. Indeed, after the following call, we have $S_1 = \emptyset$, $S_2 = S$ and $S_3 = \emptyset$:

$$\texttt{new}(S, S \cap S', 1, \texttt{false}, S_1, S_2, S_3);$$

Using only one pair of inequalities to slice an iteration space instead of using all inequalities at once enables the efficient incremental construction of the simple sections into which the simple section under consideration becomes fragmented, preventing re-computation of simple sections by means of subscripts and loop bounds analysis. Furthermore, we will explain that this step-wise approach provides a convenient method to control the amount of resulting fragmentation by determining at each step whether further slicing is useful.

3.3 Implementation

First, we initialize the set of representatives. In general, we start with $R_A = \emptyset$. However, if we want to isolate particular regions of the matrix in a program because these regions have certain exploitable properties, then we add the simple sections describing the index set of these regions to R_A. Subsequently, we scan over all simple sections in the bag Σ_A containing all simple sections computed for the occurrences of the matrix. Because smaller simple sections tend to induce a fragmentation of bigger simple sections, simple sections in Σ_A are considered in increasing order of size. For each $S \in \Sigma_A$, one of the steps (1), 2(a), or 2(b) is performed. However, before we combine S with an overlapping simple section $S' \in R_A$ in step 2(b), we use procedure new to determine the simple sections S_1^k S_2^k and S_3^k for $1 \leq k \leq 4$.

These are the simple sections into which S becomes fragmented if we slice the iteration space according to the kth pair of inequalities obtained by substituting the subscripts of the occurrence to which S belongs in the inequalities defining $S \cap S'$. We determine k' such that $|S_2^{k'}| = \min_k |S_2^k|$.

If we slice the iteration space according to the k'th pair, overlap will be confined to the smallest region. However, this slicing is only performed if $|S \uplus S'| - |S_2^{k'}| \geq t$ holds for some threshold t, testing whether the potential saving justifies further iteration space partitioning. To keep fragmentation proportional to the size of access patterns, using $t \approx \min(m,n)$ seems appropriate for a $m \times n$ matrix. If this saving is less than the threshold, another overlapping representative is considered until either a suitable iteration space partitioning is found or all representatives have been considered. In the latter case, we combine S and S' as in step 2(b). In the former case an iteration space is sliced according to the k'th pair of inequalities, the following steps are taken to deal with the corresponding program transformations.

We associate $S_1^{k'}$, $S_2^{k'}$ and $S_3^{k'}$ with the resulting duplicates of the occurrence to which S belong and replace $S \in \Sigma_A$ by all non-empty simple sections. Because $|S_i^{k'}| \leq |S|$, one of these simple sections will be considered next. Furthermore, we have not discussed yet how to deal with the fact that slicing the iteration space of a nested loop to fragment a simple section belonging to a particular occurrence, also affects the simple sections associated with other occurrences in the loop-body. We simply avoid re-computation of such simple sections by associating the simple section S^o belonging to an occurrence with all resulting duplicates of this occurrence. Thereafter, we replace S^o in the bag Σ_A with these new simple sections. In case we already have a representative of S^o, this representative can be used for all new simple sections. If S^o has not been considered yet, all the new simple sections participate in the method explained above. In this manner, any required iteration space slicing will be applied to each nested loop in which we have a duplicate of the occurrence to which S^o belongs. Furthermore, if slicing an iteration space to fragment a particular simple section S actually results in the appropriate fragmentation of another simple section S^o, then any subsequent iteration space slicing to fragment S^0 has no effect on the code, but induces the appropriate incremental construction of the simple sections into which S^0 was actually fragmented.

Using a threshold provides some control on the degree of fragmentation, since for increasing value of the threshold less iteration spaces will be sliced. Consider, for instance, the following code fragment with two occurrences of a matrix A:

```
DO I = 1, 50
  B(I) = A_1(I+25,I+10)
  DO J = 1, 50
    C(I,J) = A_2(I+J,J)
  ENDDO
ENDDO
```

The simple section computed for the first occurrence consists of the index set of 50 elements along the 15th diagonal below the main diagonal:

$$S_1 = \{ (x,y) \in \mathbf{Z}^2 \mid \begin{bmatrix} 26 & \leq & x & \leq & 75 \\ 11 & \leq & y & \leq & 60 \\ 37 & \leq & x+y & \leq & 135 \\ 15 & \leq & x-y & \leq & 15 \end{bmatrix} \}$$

The simple section computed for the second occurrence reflects the fact that 2500 elements in a trapezoidal part of the matrix are referenced:

$$S_2 = \{ (x,y) \in \mathbf{Z}^2 \mid \begin{bmatrix} 2 & \leq & x & \leq & 100 \\ 1 & \leq & y & \leq & 50 \\ 3 & \leq & x+y & \leq & 150 \\ 1 & \leq & x-y & \leq & 50 \end{bmatrix} \}$$

Hence, we have $\Sigma_A = \{S_1, S_2\}$ for this program. The regions described by both simple sections are shown in figure 3.

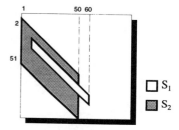

Figure 3: Regions in the Matrix

If initially $R_A = \emptyset$, the smallest simple section S_1 can be inserted directly into this set. Consideration of the second simple section reveals the following partial overlap $S_1 \cap S_2$ (note the refinement of some boundaries after the intersection has been computed):

$$\{(x,y) \in \mathbf{Z}^2 \mid \begin{bmatrix} 26 & \leq & x & \leq & 65 \\ 11 & \leq & y & \leq & 50 \\ 37 & \leq & x+y & \leq & 115 \\ 15 & \leq & x-y & \leq & 15 \end{bmatrix} \}$$

At this stage we have a choice between combining these simple sections into a new representative $S_1 \uplus S_2$ of size 2890, or slicing the iteration space of the loop in which the second occurrence appears to confine overlap. Below, we show the four pairs of inequalities on the loop indices arising from substituting the subscripts (I+J , J) for (x,y) in the inequalities defining the intersection. Furthermore, we present the sizes of the three simple sections S_1^k, S_2^k and S_3^k into which S_2 is fragmented. Note that for all k, $|S_1^k| + |S_2^k| + |S_3^k| = |S_2|$:

k	Inequalities			$|S_1^k|$	$|S_2^k|$	$|S_3^k|$
1	$26 \leq$	I+J	≤ 65	300	1570	630
2	$11 \leq$	J	≤ 50	500	2000	0
3	$37 \leq$	I+2J	≤ 115	306	1870	324
4	$15 \leq$	I	≤ 15	700	50	1750

Obviously, $|S_2^4|$ is minimal. Therefore, we use the 4th inequality, yielding a potential saving of $|S_1 \uplus S_{A_2}| - 50 = 2840$. Hence, for a threshold $2840 < t$, the simple sections are combined into one representative simple section, as illustrated in the first matrix of figure 4. Otherwise, iteration space slicing yields the following code:

```
DO I = 1, 14
  B(I) = A_1a(I+25,I+10)
  DO J = 1, 50
    C(I,J) = A_2a(I+J,J)
  ENDDO
ENDDO
B(15) = A_1b(40,25)
DO J = 1, 50
  C(15,J) = A_2b(J+15,J)
ENDDO
DO I = 16, 50
  B(I) = A_1c(I+25,I+10)
  DO J = 1, 50
    C(I,J) = A_2c(I+J,J)
  ENDDO
ENDDO
```

Redundant fragmentation of S_1 is simply ignored by associating this simple section with the three duplicates of the first occurrence. The simple sections into which S_2 becomes fragmented are obtained by refining the boundaries after replacement of the last pair with $1 \leq x-y \leq 14$, $15 \leq x-y \leq 15$ and $16 \leq x-y \leq 50$ respectively.

Subsequently, because S_{2a} and S_{2c} do not overlap with any representative simple section, these simple sections are inserted into R_A. However, the simple section S_{2b}, shown below, and S_1 still partially overlap:

$$S_{2b} = \{(x,y) \in \mathbf{Z}^2 \mid \begin{bmatrix} 16 & \leq & x & \leq & 65 \\ 1 & \leq & y & \leq & 50 \\ 17 & \leq & x+y & \leq & 115 \\ 15 & \leq & x-y & \leq & 15 \end{bmatrix} \}$$

Again, we have the choice between combining these simple sections, yielding a simple section of size 60 or slicing the iteration space. The sizes of the simple sections that result if we slice the iteration space according to one of the pairs obtained by substituting (15+J , J) for (x,y) in the inequalities defining $S_1 \cap S_{2b}$ are shown below:

k	Inequalities			$|S_1^k|$	$|S_2^k|$	$|S_3^k|$
1	$11 \leq$	J	≤ 50	10	40	0
2	$11 \leq$	J	≤ 50	10	40	0
3	$22 \leq$	2J	≤ 100	10	40	0
4	$0 \leq$	0	≤ 0	0	50	0

Overlap is confined to the smallest region if we slice the iteration space according to one of the first three pairs, giving a potential saving of $60 - 40 = 20$. Hence, for $20 < t$ we combine the simple sections which gives rise to the representative simple sections shown in the second matrix of figure 4. For a smaller threshold, the second J-loop is replaced by the following fragment:

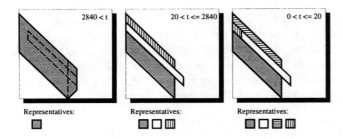

Figure 4: Regions Described by Representatives

```
DO J = 1, 10
  C(15,J) = A_{2b'}(J+15,J)
ENDDO
DO J = 11, 50
  C(15,J) = A_{2b''}(J+15,J)
ENDDO
```

The simple sections into which S_{2b} is fragmented are constructed incrementally by refining boundaries after the first pair has been replaced by $16 \leq x \leq 25$, $26 \leq x \leq 65$ and $66 \leq x \leq 65$, yielding two non-empty simple sections $S_{2b'}$ and $S_{2b''}$. Because $S_{2b'}$ is added directly to R_A and $S_{2b''} \subseteq S_1$, the representatives shown in the last matrix of figure 4 result. This figure clearly illustrates how the use of a threshold can be used to control the resulting degree of fragmentation. Accumulated effects of redundant fragmentation of S_1 are avoided by simply associating this original simple sections with all duplicates of the occurrence to which this simple section belongs after slicing an iteration space.

The following example illustrates that if several occurrences in a loop require the same iteration space partitioning to confine overlap, this is correctly dealt with by the presented method. Below, the code for LU-factorization of a general unsymmetric 150×150 matrix A is shown, in which pivoting for stability is not performed:

```
DO J = 2, 150
  DO I = 1, J-1
    A_1(J,I) = A_2(J,I) / A_3(I,I)
    DO K = I+1, 150
      A_4(J,K) = A_5(J,K) - A_6(J,I) * A_7(I,K)
    ENDDO
  ENDDO
ENDDO
```

The simple sections computed for all occurrences of the matrix A are shown in increasing size at the left side of figure 5. Obviously, some of the simple sections are identical. Starting with $R_A = \emptyset$, we obtain $R_A = \{S_3, S_1, S_7\}$ after the simple sections S_3, S_1, S_2, S_6 and S_7 have been processed. However, if subsequently S_4 is considered, the following overlap with representative S_3 is detected:

$$S_3 \cap S_4 = \{(x,y) \in \mathbf{Z}^2 \mid \begin{bmatrix} 2 & \leq & x & \leq & 149 \\ 2 & \leq & y & \leq & 149 \\ 4 & \leq & x+y & \leq & 298 \\ 0 & \leq & x-y & \leq & 0 \end{bmatrix} \}$$

Substituting subscripts (J,K) for (x,y) in the inequalities defining this intersection gives rise to the following data:

k	Inequalities			$|S_1^k|$	$|S_2^k|$	$|S_3^k|$
1	$2 \leq$	J	≤ 149	0	22052	149
2	$2 \leq$	K	≤ 149	0	22052	149
3	$4 \leq$	$J+K$	≤ 298	0	22198	3
4	$0 \leq$	$J-K$	≤ 0	11026	149	11026

Because overlap is confined to the smallest region if the 4th pair is used, we slice the iteration space of the loop in which A_4 appears according to $0 \leq -J + K \leq 0$. The following code results after unrolling the second resulting K-loop:

```
DO J = 2, 150
  DO I = 1, J-1
    A_1(J,I) = A_2(J,I) / A_3(I,I)
    DO K = I+1, J-1
      A_{4a}(J,K) = A_{5a}(J,K) - A_{6a}(J,I) * A_{7a}(I,K)
    ENDDO
    A_{4b}(J,J) = A_{5b}(J,J) - A_{6b}(J,I) * A_{7b}(I,J)
    DO K = J+1, 150
      A_{4c}(J,K) = A_{5c}(J,K) - A_{6c}(J,I) * A_{7c}(I,K)
    ENDDO
  ENDDO
ENDDO
```

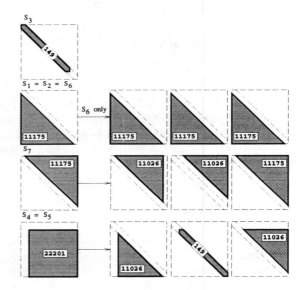

Figure 5: Fragmented Regions

As illustrated in figure 5, S_4 becomes fragmented along the main diagonal into the simple sections S_{4a} and S_{4b} and S_{4c}, which are constructed incrementally. Furthermore, this transformation does not affect S_6 and induces fragmentation of S_7 and S_5. However, these effects are simply ignored by assigning S_5 to S_{5a}, S_{5b} and S_{5c}, S_6 to S_{6a}, S_{6b} and S_{6c}, and S_7 to S_{7a}, S_{7b} and S_{7c}. Because S_6 and S_7 already have been considered, no further actions are required for the simple sections associated with the duplicates of occurrences A_6 and A_7.

Moreover, since $S_{4a} \subseteq S_1$ and $S_{4c} \subseteq S_7$, representatives of these simple sections also have been found. For any reasonable value of the threshold we combine the partially overlapping simple section S_{4b} and S_3 into $S_{4b} \uplus S_3$, because potential saving is limited to 2.

However, consideration of S_{5a} reveals a overlap with this main diagonal, despite the fact this simple section actually has been fragmented appropriately. Fortunately, substituting the subscripts $(\mathtt{J}, \mathtt{K})$ for (x, y) in the inequalities defining $S_{5a} \cap (S_{4b} \uplus S_3)$ gives rise to slicing the iteration space of the loop in which the corresponding occurrence appears according to $0 \leq -\mathtt{J} + \mathtt{K} \leq 0$. Obviously, this slicing has *no effect on the code*, but results in the incremental construction of three simple sections. Only the first is associated with occurrence A_{5a} (the other two are associated with the occurrences in the zero-trip loops that are not generated). Likewise, the iteration spaces of the loops in which A_{5b} and A_{5c} appear are sliced according to the trivial pair $0 \leq 0 \leq 0$ (because the K-loop has been unrolled, the correspondence $(x, y) = (\mathtt{J}, \mathtt{J})$ holds in this case) and $0 \leq -\mathtt{J} + \mathtt{K} \leq 0$ respectively. Again, these inequalities do not effect the code but result in the appropriate adaptation of S_{5b} and S_{5c}.

Consequently, the compiler is able to detect the fact that separate storage of the strict triangular parts and the main diagonal matches the kind of operations performed in the code. Indeed, many storage schemes for general sparse matrices are based on this fragmentation [15, 18, 19].

4 Application of the Method

We present an application of our method for a compiler performing data structure transformations. Consider, for example, the following fragment computing the matrix vector product $\vec{b} = A\vec{x}$:

```
DO I = 1, 100
  DO J = 1, 100
    B(I) = B(I) + A(I,J) * X(J)
  ENDDO
ENDDO
```

Furthermore, assume that a 100×100 matrix A having the nonzero structure shown in figure 6 is operated on. Because the loop-body of the previous fragment only has to be executed for nonzero elements [5, 7], it may be useful to isolate regions in A containing nonzeros. The simple sections describing such regions, which can be determined using nonzero structure analysis [6], are added to R_A before any simple section in Σ_A is considered. For the nonzero structure shown in figure 6, for instance, we may start with $R_A = \{S', S''\}$ describing the index set of the main diagonal and a border in the lower triangular part of A respectively. Application of the method presented in the paper results in the partitioning of the execution set of the J-loop into $[1, \mathtt{I} - 1]\,[\mathtt{I}, \mathtt{I}]\,[\mathtt{I} + 1, 100]$.

Figure 6: Nonzero Structure

Subsequently, the execution set of the I-loop is partitioned into $[1, 90]$ and $[91, 100]$ in case the border starts at the 91th row. Thereafter, we eliminate all statements operating on regions that only contain zero elements. Consequently, if a data structure is selected by the compiler in which all elements along the diagonal and in the border are stored in the arrays ADIAG and ABRD respectively, where rows of the matrix are stored along the columns of ABRD to enhance spatial locality, the following code can be generated automatically [7]:

```
DO I = 1, 90
  B(I) = B(I) + ADIAG(I) * X(I)
ENDDO
DO I = 91, 100
  DO J = 1, I-1
    B(I) = B(I) + ABRD(J,I-90) * X(J)
  ENDDO
  B(I) = B(I) + ADIAG(I) * X(I)
ENDDO
```

In table 1, we present the CPU-time required to execute the original dense fragment, equivalent code using sparse row-wise storage of the matrix in which only nonzero elements are stored (this version can also be derived automatically [7]) and the previous fragment on an HP-UX 9000/720 and one processor of a Cray C98/4256 for the previous presented 100×100 matrix and a 1000×1000 matrix having a similar nonzero structure. The programs for the HP and the Cray are compiled with optimization and vectorization enabled respectively. The loops in the dense version have been interchanged to enhance spatial locality.

Table 1: CPU-Time in Seconds

	Dense	General Sparse	Specific Sparse
HP			
n=100	$7.9 \cdot 10^{-4}$	$8.8 \cdot 10^{-5}$	$8.8 \cdot 10^{-5}$
n=1000	$1.6 \cdot 10^{-1}$	$9.6 \cdot 10^{-4}$	$8.9 \cdot 10^{-4}$
Cray			
n=100	$5.4 \cdot 10^{-5}$	$1.1 \cdot 10^{-4}$	$1.8 \cdot 10^{-5}$
n=1000	$3.8 \cdot 10^{-3}$	$1.1 \cdot 10^{-3}$	$5.1 \cdot 10^{-5}$

On the HP, we see that even for a small matrix, the use of sparse code is more efficient. Furthermore, although the sparse version accounting for the nonzero structure avoids the overhead caused by indirect addressing, only a slight reduction in execution time is achieved with respect to the general sparse version, because additional zero elements are operated on. However, because indirect addressing induces substantial overhead on the Cray, the nonzero structure of A must be exploited to obtain an acceptable reduction in execution time with respect to the dense version. This clearly illustrates the importance to account for the characteristics of *both* the target machine and the data operated on.

5 Conclusions

In this paper we have presented a method to identify and isolate a number of regions in a matrix. Regions of which the index set can be represented with two-dimensional simple sections have been considered, because this representation can be manipulated efficiently. After all simple sections belonging to the occurrences of a matrix in the program have been computed, representative simple sections can be found in a single pass over these simple sections in increasing order of size. Iteration space partitioning can be used to increase the degree of resulting fragmentation. The method has been incorporated in a prototype restructuring compiler MT1 [4, 9] to support the automatic conversion of a dense program into sparse code.

Acknowledgments The authors would like to thank Peter Knijnenburg and Arnold Niessen for proofreading this article.

References

[1] Vasanth Balasundaram. *Interactive Parallelization of Numerical Scientific Programs*. PhD thesis, Department of Computer Science, Rice University, 1989.

[2] Vasanth Balasundaram. A mechanism for keeping useful internal information in parallel programming tools: The data access descriptor. *Journal of Parallel and Distributed Computing*, Volume 9:154–170, 1990.

[3] U. Banerjee. *Loop Transformations for Restructuring Compilers: The Foundations*. Kluwer Academic Publishers, Boston, 1993.

[4] Aart J.C. Bik. A prototype restructuring compiler. Master's thesis, Utrecht University, 1992. INF/SCR-92-11.

[5] Aart J.C. Bik and Harry A.G. Wijshoff. Compilation techniques for sparse matrix computations. In *Proceedings of the International Conference on Supercomputing*, pages 416–424, 1993.

[6] Aart J.C. Bik and Harry A.G. Wijshoff. Nonzero structure analysis. In *Proceedings of the International Conference on Supercomputing*, pages 226–235, 1994.

[7] Aart J.C. Bik and Harry A.G. Wijshoff. On automatic data structure selection and code generation for sparse computations. In Utpal Banerjee, David Gelernter, Alex Nicolau, and David Padua, editors, *Lecture Notes in Computer Science, No. 768*, pages 57–75. Springer-Verlag, 1994.

[8] Aart J.C. Bik and Harry A.G. Wijshoff. On strategies for generating sparse codes. Technical Report no 95-01, Department. of Computer Science, Leiden University, 1995.

[9] Peter Brinkhaus. Compiler analysis of procedure calls. Master's thesis, Utrecht University, 1993. INF/SCR-93-13.

[10] Iain S. Duff, A.M. Erisman, and J.K. Reid. *Direct Methods for Sparse Matrices*. Oxford Science Publications, 1990.

[11] Alan George and Joseph W.H. Liu. *Computer Solution of Large Sparse Positive Definite Systems*. Prentice-Hall Inc., 1981.

[12] David A. Padua and Michael J. Wolfe. Advanced compiler optimizations for supercomputers. *Communications of the ACM*, pages 1184–1201, 1986.

[13] Sergio Pissanetsky. *Sparse Matrix Technology*. Academic Press, London, 1984.

[14] C.D. Polychronoupolos. *Parallel Programming and Compilers*. Kluwer Academic Publishers, Boston, 1988.

[15] Youcef Saad and Harry A.G. Wijshoff. Spark: A benchmark package for sparse computations. In *Proceedings of the 1990 International Conference on Supercomputing*, pages 239–253, 1990.

[16] Michael J. Wolfe. *Optimizing Supercompilers for Supercomputers*. Pitman, London, 1989.

[17] H. Zima. *Supercompilers for Parallel and Vector Computers*. ACM Press, New York, 1990.

[18] Zahari Zlatev. *Computational Methods for General Sparse Matrices*. Kluwer Academic Publishers, 1991.

[19] Zahari Zlatev, Jerzy Wasniewski, and Kjeld Schaumburg. Y12M - solution of large and sparse systems of linear algebraic equations. In *Lecture Notes in Computer Science, No. 121*. Springer-Verlag, Berlin, 1981.

Fusion of Loops for Parallelism and Locality*

Naraig Manjikian and Tarek S. Abdelrahman
Department of Electrical and Computer Engineering
The University of Toronto
Toronto, Ontario, Canada M5S 1A4
email: {nmanjiki,tsa}@eecg.toronto.edu

Abstract—*Loop fusion improves data locality and reduces synchronization in data-parallel applications. However, loop fusion is not always legal. Even when legal, fusion may introduce loop-carried dependences which reduce parallelism. In addition, performance losses result from cache conflicts in fused loops. We present new, systematic techniques which: (1) allow fusion of loop nests in the presence of fusion-preventing dependences, (2) allow parallel execution of fused loops with minimal synchronization, and (3) eliminate cache conflicts in fused loops. We evaluate our techniques on a 56-processor KSR2 multiprocessor, and show improvements of up to 20% for representative loop nest sequences. The results also indicate a performance tradeoff as more processors are used, suggesting careful evaluation of the profitability of fusion.*

1 Introduction

The performance of data-parallel applications on cache-coherent shared-memory multiprocessors is significantly affected by data locality and by the cost of synchronization. Loop fusion is a code transformation which is used to combine multiple parallel loops into a single loop, enhancing data locality and reducing synchronization. Fusion is not always legal in the presence of dependences between the loops being fused. Even when it is legal, fusion may introduce loop-carried dependences which reduce existing parallelism and introduce additional synchronization. Fusing a large number of loops also increases the number of arrays referenced in the fused loop, which gives rise to cache conflicts that degrade performance.

In this paper, we propose a new set of related loop and data transformations to address the above difficulties. The loop transformation technique fuses loops, even in the presence of fusion-preventing dependences, by *shifting* iteration spaces prior to fusion. Parallel execution of a fused loop with minimal synchronization is then enabled by *peeling* iterations to remove serializing dependences. We refer to this loop transformation as *shift-and-peel*. The data

*This research is supported by grants from NSERC (Canada) and ITRC (Ontario). The use of the KSR2 was provided by the University of Michigan Center for Parallel Computing.

$$L_a : \quad \text{doall } i_1 = \ldots$$
$$\ddots$$
$$\text{doall } i_k = \ldots$$
$$\ddots$$
$$\text{do } i_n = \ldots$$
$$A_1[F_1^a(\vec{\imath})], A_2[F_2^a(\vec{\imath})], \ldots$$
$$L_b : \quad \text{doall } i_1 = \ldots$$
$$\ddots$$
$$\text{doall } i_k = \ldots$$
$$\ddots$$
$$\text{do } i_n = \ldots$$
$$A_1[F_1^b(\vec{\imath})], A_2[F_2^b(\vec{\imath})], \ldots$$
$$\vdots$$

Figure 1: Program model

transformation technique adjusts the array layout in memory after fusion to eliminate mapping conflicts in the cache. We refer to this data transformation as *cache partitioning*.

The domain of the shift-and-peel transformation is program segments consisting of a sequence of loop nests that reuse a number of arrays, as shown in Figure 1. Each loop nest is assumed to be in a canonical form in which at least $k \geq 1$ outermost loops are fully parallel loops. A dependence *between* any pair of loop nests is assumed to be uniform, i.e., with a constant distance. In addition, there should be no intervening code between the loop nests. We are interested in fusing the sequence of loop nests such that the resulting loop nest has k outer loops that are fully parallel. We do not address the loss of parallelism in any of the original loop nests with more than k parallel loops. This issue has been addressed in previous work [6]. Our data transformation technique requires in addition that array accesses be *compatible*, i.e, accesses to a given array must have the same stride and direction across all loop nests. Although compatible accesses imply uniform dependences, the reverse is not necessarily true, hence compatibility is a stricter requirement. Nonetheless, compatible array accesses are typical in many scientific applications.

Program transformations may be used to obtain a program segment that lies in the domain of our techniques.

L1: do i=2,n−1
    a[i] = ...
    b[i] = ...
  end do
L2: do i=2,n−1
    ... = a[i+1]
    ... = b[i−1]
  end do

(a) Adjacent loops
    to be fused

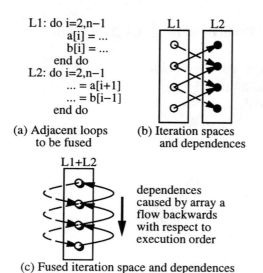

(b) Iteration spaces
    and dependences

L1+L2

dependences
caused by array a
flow backwards
with respect to
execution order

(c) Fused iteration space and dependences

Figure 2: Illustration of illegal fusion of adjacent loops

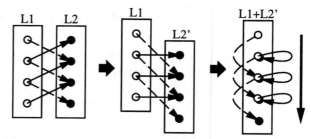

Figure 3: Shifting iteration spaces to permit legal fusion

Interprocedural analysis can identify candidate loops for fusion-enabling transformations such as loop extraction and loop embedding[9]. Loop distribution [3] may be used to produce parallel loop nests that adhere to the model. Since the loops will be eventually fused together, there is no loss of locality. Code motion [1] may be employed to obtain a sequence of parallel loops with no intervening code. Loop and/or array dimension permutation [3, 7] may be used to ensure compatible access patterns.

2 Loop Fusion

Fusion of loops from adjacent loop nests combines their respective loop bodies into a single body and collapses their respective iteration spaces into one combined space. In so doing, the number of iterations separating references to the same array is reduced, and array reuse can then be exploited to enhance register and cache locality. Register locality is enhanced through reuse of register-allocated array values in a single iteration of the fused iteration space [6]. Cache locality is enhanced through reuse of array elements across multiple iterations of the fused iteration space [3]. In either case, exploiting the reuse avoids costly references to main memory, thereby improving performance. In addition, fusion permits a reduction in the number of barrier synchronizations needed between parallel loops.

Fusion is not always legal. Array reuse between adjacent loops implies the existence of data dependences between different loops. These dependences are initially loop-independent since their source and sink iterations are in different iteration spaces. Fusion places the source and sink iterations of each dependence in the same iteration space. Fusion is legal only if it does not result in a loop-carried dependence that flows backwards with respect to

the iteration execution order [10, 14]. For example, Figure 2 illustrates how the fusion of two loops may result in backward dependences.

We propose a simple technique to enable legal fusion of multiple loops in the presence of backward loop-carried dependences, based on the alignment techniques described in [4, 11]. The only necessary condition for this technique is uniform dependences. The key idea is to make backward dependences loop-independent in the fused loop by *shifting* the iteration space containing the sink of the dependences with respect to the iteration space containing the source of the dependence. The amount by which to shift is determined by the dependence distance. Other dependences between the loops are affected, but do not prevent fusion. For example, the iteration space of loop $L2$ in Figure 3 is shifted by one iteration because of the backward dependence with a distance of one. Note that the shift increases the distance of the forward dependence by one, but this does not prevent fusion.

In general, more than two loops may be considered for fusion, and fusion-preventing dependences may result from *any* pair of candidate loops. Complex dependence relationships may exist between candidate loops in the form of *dependence chains* passing through iterations in different loops. These dependence chains are dictated by the array reuse and constitute iteration ordering requirements which must be preserved for correctness. If one loop is shifted, subsequent loops along all dependence chains passing through this loop must also be shifted. Hence, shifts must be propagated along dependence chains. Consequently, it is advantageous to treat candidate loops collectively rather than incrementally one pair at a time.

We present a systematic method to determine the amount of shift needed for each iteration space to permit legal fusion of multiple loops. The technique is presented assuming one-dimensional loop nests for simplicity. However, it should be emphasized that it is applicable to multidimensional loop nests. An acyclic *dependence chain multigraph* is constructed to represent dependence chains. Each loop nest is represented by a vertex, and each dependence between a pair of loops is represented by a directed edge weighted by the dependence distance. A forward depen-

TraverseDependenceChainGraph(G)::
    foreach $v \in V[G]$ **do** $weight(v) = 0$ **endfor**
    foreach $v \in V[G]$ in topological order **do**
      foreach $e = (v, v_c) \in E[G]$ **do**
        if $weight(e) < 0$ **then**
          $weight(v_c) = \min(weight(v_c),$
                $weight(v) + weight(e))$
        endif
      endfor
    endfor

Figure 4: Algorithm for propagating shifts

dence has a positive distance, and results in an edge with a positive weight. Conversely, a backward dependence has a negative distance, and results in an edge with a negative weight. A multigraph is required since there may be multiple dependences between the same two loops. The multigraph is reduced to a simpler *dependence chain graph* by replacing multiple edges between two vertices by a single edge whose weight is the *minimum* of the original set of edges between these two vertices. When this minimum is negative, it determines the shift required to remove all backward dependences between the two corresponding loops. This reduction preserves the structure of the original dependence chains. A traversal algorithm is then used to propagate shifts along the dependence chains in this graph. Each vertex is assigned a weight, which is initialized to zero, and the vertices are visited in topological order to accumulate shifts along the chains. Note that this topological order is given directly by the original loop order, hence there is no need to perform a topological sort. Only edges with a negative weight contribute shifts; all other edges are treated as having a weight of zero. The algorithm is given in Figure 4. The complexity of the algorithm is linear in the size of the graph, and upon termination, the final vertex weights are interpreted as the amount by which to shift each loop *relative to the first loop* to enable legal fusion. Figure 5 illustrates the above procedure for representing dependence chains and deriving shifts.

Once the required shifts have been derived, the loops must be transformed to complete the legal fusion. There are two methods to implement the shifts for fusion. In a direct approach, the loops are fused, and subscript expressions in statements from shifted loops are adjusted according to the shift. An advantage of this direct approach is the potential for register locality. However, shifting increases the reuse distance and hence limits reuse which may be exploited in registers. Furthermore, shifting makes the resulting iteration space nonuniform, requiring guards in the fused loop. Iterations at both the start and end of the fused loop can be peeled out of the iteration space to avoid these guards, as shown in Figure 6(a). Nonetheless, guards may still be required if the original iteration spaces differ.

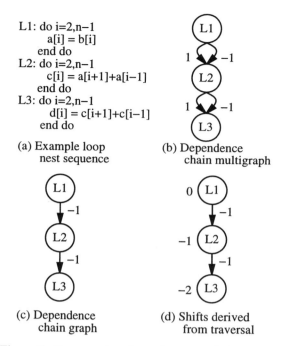

L1: do i=2,n−1
    a[i] = b[i]
  end do
L2: do i=2,n−1
    c[i] = a[i+1]+a[i−1]
  end do
L3: do i=2,n−1
    d[i] = c[i+1]+c[i−1]
  end do

(a) Example loop nest sequence

(b) Dependence chain multigraph

(c) Dependence chain graph

(d) Shifts derived from traversal

Figure 5: Representing dependences to derive shifts

A simple alternative to the direct approach is to strip-mine [3] the original loops by a factor of s, then fuse the resulting outer controlling loops to interleave iterations in groups of s, as shown in Figure 6(b). Implementing shifts in this method only requires adjusting the inner loop bound expressions with the amount of the shift, leaving the subscript expressions unchanged. Peeling is required only for the end iterations in the original iteration spaces of shifted loops. Differing iteration spaces are easily accommodated by suitable modifications to the inner loop bounds. Note that with a strip size $s = 1$, each inner loop performs at most one iteration and effectively serves as a guard. Strip-mining may incur some overhead in comparison to the direct approach, but the strip size may be used to determine the amount of data loaded into the cache for each array referenced in the inner loops. This flexibility is desirable in controlling the extent of cache conflicts, as will be described later.

In this section, shifting has been described for one-dimensional loop nests. It should be emphasized that shifting is equally applicable when fusing more than one loop across multidimensional loop nests. The procedure described above is simply applied at each loop level, beginning with the outermost loop and working inward. The dependences carried at each level determine the dependence chains which are then used to derive the required shifts. Although shifting an outer loop may change dependences with respect to inner loops, shifts are derived independently at each level using the original dependences

```
do i=2,3
    a[i] = b[i]
end do
c[2]=a[3]+a[1]

do i=4,n−1
    a[i] = b[i]
    c[i−1] = a[i]+a[i−2]
    d[i−2] = c[i−1]+c[i−3]
end do

c[n−1] = a[n]+a[n−2]
do i=n−2,n−1
    d[i] = c[i+1]+c[i−1]
end do
```

(a) Direct method

```
do ii=2,n−1,s
    do i=ii,min(ii+s−1,n−1)
        a[i] = b[i]
    end do
    do i=max(ii−1,1),min(ii+s−2,n−2)
        c[i] = a[i+1]+a[i−1]
    end do
    do i=max(ii−2,1),min(ii+s−3,n−3)
        d[i] = c[i+1]+c[i−1]
    end do
end do

c[n−1] = a[n]+a[n−2]
do i=n−2,n−1
    d[i] = c[i+1]+c[i−1]
end do
```

(b) Strip−mined method

Figure 6: Implementing shifts for fusion

to remove backward dependences at all levels and permit subsequent parallelization in a uniform manner. Implementation of the shifts at each level is straightforward using the strip-mined approach described above.

3   Parallelizing Fused Loops

Loop fusion enhances locality but may result in loop-carried dependences which prevent synchronization-free parallel execution of the fused loop. This is illustrated in Figure 7. The loops *L1* and *L2* in Figure 7(a) have no loop-carried dependences; the iterations of each loop may be executed in parallel without synchronization. Only a barrier synchronization is required to ensure that all iterations of *L1* have executed before any iterations of *L2*. However, when the iteration spaces of the two loops are fused as shown in Figure 7(b), loop-carried dependences result, requiring synchronization that serializes execution of blocks of iterations assigned to different processors.

We propose a simple method to remove serializing dependences which result from fusion, assuming static, blocked loop scheduling. The central idea of the method

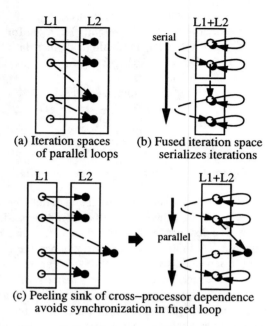

(a) Iteration spaces of parallel loops

(b) Fused iteration space serializes iterations

(c) Peeling sink of cross−processor dependence avoids synchronization in fused loop

Figure 7: Parallel execution of fused loops

is to identify iterations which become the sinks of cross-processor dependences in the fused loop, and then *peel* these iterations from the iteration space. The only necessary condition for this technique is uniform dependences, which force the peeled iterations to be located at block boundaries. The number of iterations which must be peeled is determined by the dependence distance. Peeling removes synchronization between blocks after fusion, as shown in Figure 7(c). Loop-carried dependences still exist, but are contained entirely within a block executed by the same processor. The peeled iterations are executed only after all other iterations within each block have been executed.

In programs where more than two loops are fused, serializing dependences arise from *any* pair of the original loops. Complex dependence relationships may exist between candidate loops in the form of dependence chains similar to those considered when shifting iteration spaces. The peeling of an iteration from one loop requires peeling of all subsequent iterations along all dependence chains passing through that iteration, i.e., peeling must also propagate along dependence chains.

To determine the number of iterations to peel for each loop, we use the same graph-based framework used for shifting iteration spaces. However, only the forward dependences resulting from fusion need to be considered, since shifting removes backward dependences. In the dependence chain multigraph, such dependences are identified by edges with positive weights. The multigraph is reduced to a simpler graph by replacing multiple edges between two vertices with a single edge whose weight is the *maximum* from the original set of edges between the vertices

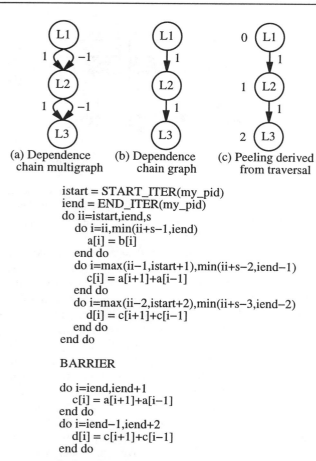

(a) Dependence chain multigraph (b) Dependence chain graph (c) Peeling derived from traversal

```
istart = START_ITER(my_pid)
iend = END_ITER(my_pid)
do ii=istart,iend,s
    do i=ii,min(ii+s−1,iend)
        a[i] = b[i]
    end do
    do i=max(ii−1,istart+1),min(ii+s−2,iend−1)
        c[i] = a[i+1]+a[i−1]
    end do
    do i=max(ii−2,istart+2),min(ii+s−3,iend−2)
        d[i] = c[i+1]+c[i−1]
    end do
end do

BARRIER

do i=iend,iend+1
    c[i] = a[i+1]+a[i−1]
end do
do i=iend−1,iend+2
    d[i] = c[i+1]+c[i−1]
end do
```

(d) Transformed parallel code

Figure 8: Derivation and implementation of peeling

(as opposed to the minimum for shifting). When the maximum weight is positive, it determines the required number of iterations which must be peeled to remove serializing dependences. The graph traversal algorithm used to propagate shifts is used again to propagate the required amounts of peeling along the dependence chains. The only modification is to consider edges with a positive weight. All other edges are treated as having a weight of zero. Upon termination, vertex weights correspond to the number of iterations to peel relative to the first loop. Figure 8(a-c) illustrates this procedure using the dependence chain multigraph shown in Figure 5.

Implementation of peeling to parallelize the fused loop is straightforward using the previously-described strip-mined method. The loop bounds of the outer strip controlling loop determine the subset of the full iteration space assigned to each processor for parallel execution. Peeling is accomplished simply by adjusting the lower bounds in the inner strip loops where necessary. For the transformed code in Figure 8(d), which executes iterations *istart...iend*, one iteration from the second loop must be peeled, based on

Figure 8(c), as well as two iterations from the third loop. The upper bounds, which resulted from shifting for legal fusion, are unaffected.

The iterations peeled from the start of each block of iterations may only be executed after all preceding iterations along dependence chains have been executed; a barrier synchronization is inserted to enforce this synchronization. Iterations peeled from the same block may have dependences between them. However, there are no dependences between sets of iterations peeled from different blocks because the dependences are uniform. As a result, these sets may be executed in parallel without synchronization following the barrier.

However, shifting for legal fusion also results in peeled iterations, and these iterations are always peeled from the end of a block. There may be dependences between the iterations peeled from the end of one block as a result of shifting and the iterations peeled from the start of an adjacent block after parallelization. We opt to place all adjacent peeled iterations in the same set. In this manner, dependences are contained entirely within each set, hence synchronization-free parallel execution of these sets is still possible. For example, in the transformed code shown in Figure 8(d), the peeled iterations include those peeled from the end of block *istart...iend* as a result of shifting, and also those peeled after parallelization from the start of the next block, beginning at *iend+1*. These iterations are executed in parallel with peeled iterations from other blocks following the barrier. The transformed code for blocks at the boundaries of the full iteration space is slightly different; for brevity, it is not shown here.

The combined shift-and-peel transformation is applied to a loop nest sequence assuming that the number of iterations per loop nest is greater than the number of processors. The number of peeled iterations is determined by the dependence distances. If this number exceeds the number of iterations per processor after loop scheduling, the transformation cannot be applied directly. In such cases, the loop nest sequence is divided to isolate the endpoints of the dependences with the greatest distance in separately-fused subsequences and reduce the number of peeled iterations.

A final observation is that peeling to remove serializing dependences can also be applied to any single loop with forward dependences. By identifying the source and sink statements for each dependence, then applying loop distribution to place statements in separate loops, a set of loops is obtained which adheres to our model. The shift-and-peel transformation can then be used to fuse the loops back together again such that serializing dependences are removed. This approach compares favorably with that of Callahan [4] and Appelbe and Smith [2] in that it does not require expensive replication of code and in that it uses

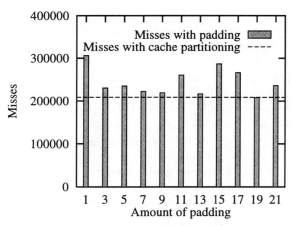

Figure 9: Misses from padding and cache partitioning

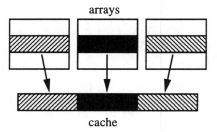

Figure 10: Avoiding conflicts in the cache

algorithms with lower complexity.

In this section, peeling has been described for one-dimensional loop nests. It should be emphasized that peeling is applicable for multidimensional loop nests. The procedure described above is simply applied at each loop level, beginning with the outermost loop and working inward. The number of iterations to peel at each level is derived from the chains of forward dependences at each level. It is important to emphasize that both shifting and peeling are performed independently at each loop level using the original dependence information. In this manner, peeled iterations are grouped together in a systematic manner without requiring a more involved, case-specific dependence analysis, thus simplifying the transformation.

4 Cache Partitioning

Conflicts among elements of various arrays in the cache cause misses that can negate the locality benefits obtained by fusion, and can also reduce parallel performance by increasing memory contention. Conflicts occur when portions of data from different arrays map into overlapping regions of the cache. Reuse of array data may extend across multiple iterations, and shifting to enable legal fusion increases the distance between reuses. The net effect is an increase in the amount of data from each array which must remain cached in order to enhance locality, and hence an increase in the potential for conflicts.

A common solution to this problem is to *pad* array dimensions to perturb the mapping of data into the cache and reduce the occurrence of conflicts [3]. However, it is difficult to predict the amount of padding which minimizes the number of conflicts, particularly when the number of arrays is large. Figure 9 depicts the impact of various amounts of padding on the number of cache misses for the execution of a fused loop referencing nine arrays whose original dimensions are 512×512 (the details of this experiment will be described in a later section). The number of misses varies erratically with the amount of padding, making it difficult to select the amount of padding to use, and the minimum occurs for the relatively large padding of 19.

We propose *cache partitioning* as a means of avoiding conflicts without the "guesswork" of padding. The basic idea behind cache partitioning is to logically divide the cache into nonoverlapping partitions, one for each array, and to adjust the starting address of each array in memory such that each array maps into a different partition, as shown in Figure 10. The starting addresses of the arrays are adjusted by inserting appropriately-sized gaps between the arrays. Cache partitioning relies on compatible data access patterns such that a conflict-free mapping of array starting addresses into the cache ensures that the mapping for the remaining array data will also be conflict-free. Note that the logical cache partitions move through the cache during loop execution, but never overlap. Evidence of the effectiveness of cache partitioning can be seen in Figure 9, which compares the number of misses from applying cache partitioning to the original arrays with the number of misses for various amounts of padding. Cache partitioning directly minimizes the number of misses and prevents erratic cache behavior. In this paper, we limit our presentation to *one-dimensional* partitioning; higher-dimensional partitioning is described in [7]. In one-dimensional partitioning, the data in each partition is made contiguous by limiting the number of indices from only the outermost array dimension to make the data from an array fit in a partition.

The introduction of gaps between arrays is required to force each array to map into a separate partition of the cache. These gaps represent memory overhead which should be minimized. We employ the the greedy layout algorithm shown in Figure 11 to reduce the size of these gaps for a set of n_a arrays, assuming a direct-mapped cache with a typical address mapping function CACHEMAP(). The arrays are selected in an arbitrary order. A set of available partitions P is maintained, and each array to be placed in memory is assigned to a cache partition of size s_p which minimizes the distance between the starting address required for that partition and the end of the array most recently placed in memory. Although multiple memory addresses map into the selected partition, the address in

GREEDYMEMORYLAYOUT(A, c)::
  $n_a = |A|$ // A = set of arrays
  $s_p = c/n_a$ // partition size (c = cache size)
  $P = \{0, 1, \ldots, n_a - 1\}$ // available partitions
  $q = q_0$ // start of storage in memory
  do
    select $a \in A$ // selection is arbitrary
    $mapped_address = $ CACHEMAP(q)
    foreach $p \in P$ **do**   // determine gaps
      $target\_address(p) = p \cdot s_p$
      $gap(p) = target\_address(p) - mapped\_address$
      if $target\_address(p) < mapped\_address$ **then**
        $gap(p) = gap(p) + c$ // "wraparound"
      endif
    endfor
    select $p_{opt} \in P$ **where** $gap(p_{opt}) = \min_{p \in P} gap(p)$
    $P = P \setminus \{p_{opt}\}$
    $START(a) = q + gap(p_{opt})$ // insert gap
    $q = START(a) + SIZE(a)$ // adjust start
    $A = A \setminus \{a\}$
  while $A \neq \emptyset$

Figure 11: Layout algorithm for cache partitioning

free memory closest to the end of the most recently placed array is always used. Each partition selected in this manner is removed from the set of available partitions to ensure that two arrays are not assigned to the same partition. The complexity of the algorithm is $O(n_a^2)$.

The partition size directly determines the maximum strip size for fusion in the previously-discussed strip-mined approach, where the largest possible strip size reduces the overhead of strip-mining. The strip size must be selected such that the total data referenced for each array in an inner loop fits within a cache partition. Selection of a larger strip size is legal, but causes data to overflow into neighboring partitions in the cache, leading to unnecessary conflicts and reducing performance.

The advantage of cache partitioning is that it results in predictable cache behavior by avoiding conflicts. The overhead which cache partitioning introduces as memory gaps between arrays is comparable to the unused memory introduced within arrays with padding. However, these gaps enable a predictable reduction in the number of misses, unlike the unpredictable outcome of padding.

5 Experimental Results

Results of experiments conducted on a Kendall Square Research KSR2 multiprocessor system [5] using two representative loop nest sequences are presented to illustrate the performance advantages that can be obtained from our techniques. From the Livermore Loops, Kernel 18 (LL18) is considered, which is an excerpt of three loop nests from a

hydrodynamics code. These loop nests cannot be fused directly, as backward loop-carried dependences result. From the qgbox ocean modelling code [8], a sequence of five loop nests is considered from the calc subroutine. These loop nests also cannot be fused directly due to backward loop-carried dependences. In addition, differences between the iteration spaces prevent direct fusion; one of the loops has a larger iteration space than the other four. A problem size of 512×512 double-precision (8-byte) floating point values is used for both examples. LL18 has 9 arrays, resulting in a total data size of 18 Mbytes. There are 6 arrays in calc, for a total data size of 12 Mbytes.

The outermost loops in each loop nest sequence are manually fused and parallelized using the techniques described in this paper. Fusion for LL18 requires a shift of 1 for the second loop relative to the first loop, and a shift of 2 for the third loop. Parallelization after fusion requires peeling one iteration from the third loop. In calc, two of the loops require shifting to enable fusion of all five loops; the largest shift is 2. Differences in iteration spaces are accommodated by adjusting the min and max expressions within the inner loops after fusion. Parallelization requires peeling iterations from two of the original loops; the largest amount of peeling is two iterations.

Our experiments focus on the impact of misses in the data cache on performance. Execution time and the number of cache misses are measured with the pmon performance monitoring hardware on each KSR processor. Loops are executed repeatedly to permit reliable measurement of execution time, particularly for parallel execution. In our experiments, cache partitioning is applied assuming a direct-mapped cache. The KSR data cache is 2-way set-associative, but maintains only one address tag for each set of 32 contiguous cache lines and employs a random replacement policy for these sets [12]. The associativity permits two distinct addresses to map to the same cache location without incurring conflict misses. However, random replacement ejects 32 cache lines at a time. Hence, the benefit of the associativity is limited, and we opt to consider the cache as direct-mapped.

The first set of results demonstrates the predictable cache behavior obtained with cache partitioning. Fused and unfused versions of the loops in LL18 were executed on 16 processors. Figure 12 compares the number of misses obtained for various amounts of padding with the number of misses obtained from applying cache partitioning to the original arrays. These numbers reflect the misses on one processor; the results are similar for the other processors in the same parallel run. The cache behavior with padding is erratic and unpredictable. On the other hand, cache partitioning directly results in the smallest number of misses. Results for 4 and 8 processors also yield the same conclu-

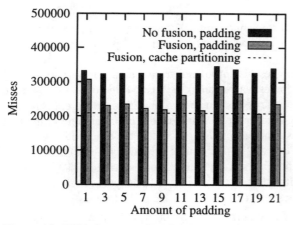

Figure 12: Effectiveness of cache partitioning for LL18

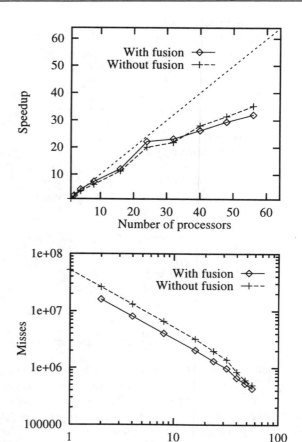

Figure 13: Benefit of fusion for LL18

sions. It is interesting to observe that the potential benefit of fusion is easily lost due to conflict misses; it is only when such misses are eliminated that the full benefit of fusion can be realized. The results indicate that padding cannot guarantee the avoidance of conflicts after fusion. This can be seen for certain values of padding for which the number of misses after fusion is comparable to the number of misses without fusion. Consequently, all our subsequent results and comparisons are based on cache-partitioned memory layout. The gaps introduced using this technique constitute an overhead of less than 2% for both LL18 and calc, which is comparable to the overhead from padding.

The next set of results show the parallel performance of fused and unfused versions of LL18 and calc using cache-partitioned memory layout for up to 56 processors on the KSR2. For each parallel run (with or without fusion), the speedup is computed with respect to the unfused version of the code executing on a single processor (using cache-partitioned layout as mentioned above). The uniprocessor execution time is 78 seconds for LL18, and 64 seconds for calc. The number of cache misses is also measured for each run. Figure 13 shows the speedup curves and cache misses for LL18, and Figure 14 shows the same results for calc. For LL18, fusion improves performance by 7% to 15% for up to 32 processors, beyond which the unfused version performs better. Similarly, fusion improves the performance for calc by 11% to 20% up to 24 processors, beyond which the unfused version performs better. It is important to note that the unfused versions are already benefitting from cache partitioning, hence these results reflect the benefit from fusion alone. As such, they provide a lower bound on the performance improvement.

The results indicate a tradeoff when applying fusion and parallelization. Enhancing locality and reducing the number of barrier synchronizations with fusion must be weighed against the overhead of strip-mining used to implement shifting and peeling. For a given data size, using

more processors increases the likelihood of data fitting in the cache of each processor, which reduces the relative gains from improving locality by fusion. The larger data size for LL18 causes the benefit of fusion to overcome the overhead up to a larger number of processors than for calc. Furthermore, grouping peeled iterations from adjacent blocks to permit them to be executed in parallel may increase the number of misses for nonlocal data. The results indicate that this effect becomes significant for a large number of processors, as can be seen for calc; the number of misses with fusion exceeds the number of misses without fusion beyond 40 processors. These observations suggest using knowledge of data sizes and cache sizes at compile-time to determine the profitability of fusion.

The final set of results compares our peeling transformation with the alignment and replication transformation proposed by Callahan [4] and Appelbe and Smith [2] (to be discussed in section on Related Work). Figure 15 compares the performance of the fused LL18 loop nest parallelized using peeling with the performance of the same loop nest parallelized using direct application of alignment and replication. In this case, it was necessary to replicate two arrays

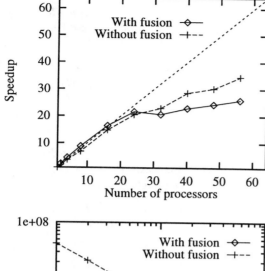

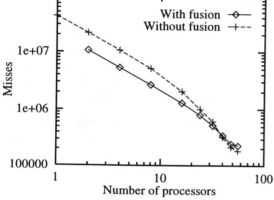

Figure 14: Benefit of fusion for `calc`

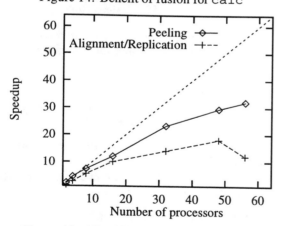

Figure 15: Alignment/replication for `LL18`

and two statements to enable synchronization-free parallel execution. The figure clearly indicates that superior performance is achieved using peeling, which is attributed to the overhead associated with replication of code and data.

The results of this section have demonstrated the ability of our techniques to overcome the limitations of loop-carried dependences for fusion and parallelism. Although

the improvement reported here for fusion is modest, it is expected that a larger improvement will result for faster processors. In addition, the relative cost of synchronization increases with faster processors. As a result, the benefit of reducing the number of barrier synchronizations, in conjunction with locality enhancement, will increase.

6 Related Work

Warren [13] presents an algorithm for incrementally adding candidate loops to a fusible set to enhance vector register reuse, and to permit contraction of temporary arrays into scalars. However, fusion is not permitted with loop-carried dependences or incompatible loop bounds.

Callahan [4] proposes loop alignment to allow synchronization-free parallel execution of a loop nest, and uses code replication to resolve conflicts in alignment requirements. His approach potentially results in exponential growth in the number of statements in the loop, and can result in significant overhead because of the redundant execution of replicated code.

Porterfield [11] suggests a "peel-and-jam" transformation in which iterations are peeled from the beginning or end of one loop nest to allow fusion with another loop nest. However, no systematic method is described for fusion of multiple loop nests, nor is the parallelization of the fused loop nest considered.

Appelbe and Smith [2] present a graph-based algorithm for deriving the required alignment, replication, and re-ordering to permit parallelization of an individual loop nest with forward dependences. Their approach is similar to Callahan's approach in that it incurs overhead due to redundant execution of replicated code.

Kennedy and McKinley [6] use loop fusion and distribution to enhance locality and maximize parallelism. They focus on register reuse, and describe a fusion algorithm for avoiding the fusion of parallel loops with serial loops. However, they disallow fusion when loop-carried dependences result or when the iteration spaces of candidate loops are not identical.

In contrast, the shifting and peeling techniques described in this paper allow the fusion and parallelization of multiple loops in the presence of loop-carried dependences. The adjustments needed for shifting and peeling are derived for multiple loops using a simple graph-based framework. The analysis and transformation algorithms for both shifting and peeling are efficient and can be easily automated in a compiler. The techniques are simpler and result in better performance than the alignment and replication techniques of both Callahan [4] and Appelbe and Smith [2]. A particularly unique aspect of this work is that it addresses cache conflicts that result from bringing references to many different arrays together in a single fused loop, an aspect which has not been adequately addressed in the past.

7 Concluding Remarks

In this paper, we presented new techniques to improve data locality in data-parallel programs using fusion. We presented a shifting transformation to fuse loop nests, even in the presence of fusion-preventing dependences. The fusion of loop nests may result in loop-carried dependences, which prevent synchronization-free parallel execution. We presented a peeling transformation to overcome such dependences and allow parallel execution with minimal synchronization and without the overhead of code replication. Conflicts in the cache among array elements can negate the benefits of fusion. We presented cache partitioning as a method to eliminate such conflicts in a predictable manner.

The above techniques have been evaluated using two real applications, and experimental results on a 56-processor KSR2 multiprocessor show up to 20% improvement in performance. Larger improvements can be expected as processor speeds continue to increase with respect to memory speeds. The results also indicate that performance tradeoffs exist. Performance gains resulting from fusion become smaller as the number of processors becomes larger, and the data used by each processor is more likely to remain cached. When the number of processors is sufficiently large, the overhead of fusion outweighs its benefits, and performance losses result. We conclude that the profitability of fusion should be evaluated with knowledge of the data size with respect to the cache size.

Future work includes the implementation of the techniques in an experimental compiler system being developed by the authors at the University of Toronto, and conducting a more comprehensive study of the profitability of fusion using a larger number of applications.

Acknowledgements

We wish to thank the anonymous reviewers for their useful comments.

References

[1] A. V. Aho, R. Sethi, and J. D. Ullman. *Compilers: Principles, Techniques, and Tools.* Addison-Wesley, Reading, MA., 1986.

[2] W. Appelbe and K. Smith. Determining transformation sequences for loop parallelization. In U. Banerjee et al., editors, *Languages and Compilers for Parallel Computing—Fifth International Workshop*, pages 208–222. Springer-Verlag, 1993.

[3] D. F. Bacon, S. L. Graham, and O. J. Sharp. Compiler transformations for high-performance computing. Tech. Rep. UCB/CSD-93-781, Computer Science Division, University of California, Berkeley, 1993.

[4] C. D. Callahan. *A Global Approach to the Detection of Parallelism.* PhD thesis, Dept. of Computer Science, Rice University, March 1987.

[5] Kendall Square Research. *KSR1 Principles of Operation.* Waltham, Mass., 1991.

[6] K. Kennedy and K. S. McKinley. Maximizing loop parallelism and improving data locality via loop fusion and distribution. In U. Banerjee et al., editors, *Languages and Compilers for Parallel Computing—Sixth International Workshop*, pages 301–320. Springer-Verlag, Berlin, 1994.

[7] N. Manjikian and T. Abdelrahman. Reduction of Cache Conflicts in Loop Nests. Tech. Rep. CSRI-318, Computer Systems Research Institute, University of Toronto, Ontario, Canada, March 1995.

[8] J. D. McCalpin. Quasigeostrophic box model–revision 2.3. Technical report, College of Marine Studies, University of Delaware, 1992.

[9] K. S. McKinley. *Automatic and Interactive Parallelization.* PhD thesis, Dept. of Computer Science, Rice University, April 1992.

[10] D. A. Padua and M. J. Wolfe. Advanced compiler optimizations for supercomputers. *Comm. ACM*, 29(12):1184–1201, December 1986.

[11] A. K. Porterfield. *Software Methods for Improvement of Cache Performance on Supercomputer Applications.* PhD thesis, Dept. of Computer Science, Rice University, April 1989.

[12] R. H. Saavedra, R. S. Gaines, and M. J. Carlton. Micro benchmark analysis of the KSR1. In *Supercomputing '93*, pages 202–213, November 1993.

[13] J. Warren. A hierarchical basis for reordering transformations. In *Proc. 11th ACM Symposium on the Principles of Programming Languages*, pages 272–282, June 1984.

[14] M. J. Wolfe. *Optimizing Supercompilers for Supercomputers.* The MIT Press, Cambridge, MA., 1989.

INTEGRATING TASK AND DATA PARALLELISM IN UC*

Maneesh Dhagat   Rajive Bagrodia
Computer Science Department
University of California at Los Angeles
Los Angeles, CA 90024
{manu,rajive}@cs.ucla.edu

Mani Chandy
Computer Science Department
California Institute of Technology
Pasadena, CA 91125
mani@vlsi.cs.caltech.edu

Abstract

This paper presents an approach for integrating task and data parallelism in the context of a set-based parallel language called UC. We describe the addition of a channel data-type to the language to support asynchronous composition of data-parallel modules. The ideas presented have been implemented in a prototype UC compiler for the IBM SP2. The compiler has been used to implement several integrated task/data-parallel programs for numerical applications and biological simulations. For these applications, we have found a decrease in execution time of as much as 75% when compared to the purely data-parallel implementation.

1   INTRODUCTION

A large number of parallel languages and notations have been designed using the imperative programming paradigm[18]. Most existing languages adopt either the task or data-parallel paradigm. Task-parallel languages provide constructs to create and destroy asynchronous threads and to allow the asynchronous threads to communicate and synchronize as needed. As each thread typically operates on its private data, the data space of the program must be explicitly subdivided among the multiple threads. Data-parallel languages use globally addressable memory together with a synchronous programming model where the computation is implicitly synchronized at some typically pre-determined level of granularity. The global data may be optionally mapped on the memory hierarchy of the parallel architecture using language directives.

Although data parallelism has been applied successfully to solve a large number of scientific problems, a number of applications (e.g. biological simulations and global climate models) may be naturally and sometimes more efficiently designed using both forms of

parallelism. For some applications that can be programmed naturally using data parallelism, the computation granularity of an application or the communication latency of an architecture may favor a decomposition of the program into a task-parallel collection of data-parallel components.

This paper describes the integration of task and data parallelism in an extension of C called UC. We show that a variety of data-parallel and combined task/data-parallel applications can be designed in UC with performance benefits. The next section is a brief description of the language. We present the data-parallel and task-parallel portions of UC and discuss their integration using channels. In section 3 we present several examples and their performance on the IBM SP2. Section 4 discusses related work and section 5 is the conclusion.

2   THE UC LANGUAGE

The central concepts of the UC notation are the well-known mathematical concepts of sets and quantification which have been used in other programming languages [15, 14]. This section gives a brief description of the UC notation relevant to this paper; a complete description can be found in [2].

2.1   Set and Range

Although UC notation supports specification of general relations and sets[2], for brevity we restrict our attention in this paper to dense sets, also referred to as ranges. A *range* is a set of integers with constant lower and upper bounds on the values of its members. For instance, the following declaration

range I:i = {0..N-1};

declares I to be a *range*, and i to be a dummy variable that runs over I. Range I is initialized with all integers in the interval [0, N-1]. Set operations may be applied to a range with the restriction that elements of a range must lie within the constant bounds specified in its declaration.

*This research was partially supported under NSF PYI Award No. ASC-9157610, ONR Grant No. N00014-91-J-1605, Rockwell International Award No. L911014, NSF Center for Research in Parallel Computing Contract CCR-8809615, and ARPA/CSTO Contract No. F30602-94-6-0273.

2.2 Reduction

In reductions, we apply an associative, binary operation on a set of expressions. Our syntax for reductions is a compromise between the syntax for quantification in mathematics and C-syntax. In UC, a reduction is specified as:

$op(I st (b(i)) e(i))

where *op* is an associative, commutative operator, I is a range, i is a dummy variable associated with I, b(i) is a boolean expression, and e(i) is an expression. The reduction returns the result of combining with operator *op* values represented by e(i) for each i for which b(i) is true. The notation supports arithmetic operators such as addition ($+), multiplication ($*), etc.; logical operators such as *and* ($&&), *or* ($||), etc.; and selection operators such as those to return the maximum ($>), the minimum ($<), a subset ($select), an arbitrary element ($,).

Example: The following are some simple examples of reductions. The first reduction computes the average of the first ten elements of array a and the second computes the position of one of the maximum elements in the first ten elements of a.

```
index_set I:i = {0..9}, J:j = I;
int a[N], pos_max_in_a;
float avg_of_a;

avg_of_a = $+(I; a[i]) / $+(I; 1.0);
pos_max_in_a
    = $,(I st (a[i] == $>(J; a[j])) i);
```

2.3 Asynchronous Composition

The idea of quantification can be extended to control-flow operators that are associative and commutative. The keyword **arb** (for arbitrary) is the operator for composing programs asynchronously. The operation *s* **arb** *t*, where *s* and *t* are statements, specifies concurrent execution of *s* and *t* using *arbitrary interleaving*.

Example: A program to transpose a matrix a is:

```
range I:i = {0..N-1}, J:j = I;
arb (I,J) tmp[i][j] = a[i][j];
arb (I,J) a[i][j] = tmp[j][i];
```

This program fragment consists of sequential composition of two statements, each of which is an asynchronous composition of N^2 threads consisting of an assignment (We assume that a and tmp are N^2-size matrices.). The first statement copies matrix a to matrix tmp, and the second assigns the transpose of tmp to a. On the other hand, the result of the fragment "arb (I,J) a[i][j] = a[j][i];" is not specified because the atomicity and interleaving of each of the N^2 statements is not specified.

2.4 Parallel Assignment Composition

Assignment statements can be composed synchronously using the parallel composition operator **par**. The execution of { x = e; } **par** { y = f; } is as follows: the expressions *e* and *f* are first evaluated, and then their values are assigned to x and y in parallel. The parallel composition operator is associative and commutative.

A **par** statement may contain multiple clauses, each guarded by a different boolean expression. These clauses execute asynchronously. For instance, the following fragment sets the even elements of a to 0 and the odd elements to 1.

```
par (I) st (i%2 == 0) a[i] = 0;
       st (i%2 != 0) a[i] = 1;
```

The **par** quantification may include iterative constructs like the **while** statement:

```
par (I) st (b[i])
    while (c[i]) s;
```

is equivalent to

```
while ($||(I; b[i]&&c[i]))
    par (I) st (b[i]&&c[i]) s;
```

For every iteration, the loop condition is evaluated for all elements in I. The loop is terminated if the condition evaluates to false for every element; otherwise, statement *s* is executed synchronously for every enabled element. The meaning of other iterative constructs is similar. If a **par** statement includes a function call, the arguments are first evaluated synchronously followed by the asynchronous execution of the multiple instances of the function.

Note that, asynchronous execution due to **arb** and parallel function calls will cause non-determinism if two threads *interfere*. Two threads are said to interfere if a variable modified in one thread is referenced by another thread.

Example: The UC program in figure 1 illustrates the use of set operations on ranges, reductions, and the **par** statement. It implements the LU decomposition phase of the Gaussian elimination scheme for solving a set of linear equations of the form $Ax = b$, where A is a square matrix of coefficients, and x and b are column vectors of unknowns and constants, respectively. The two reductions in lines 9-10 find the position of the maximum element in the zth column of A to be used an the pivot position. If an appropriate pivot is found, the **par** statement in lines 13-17 performs a row swap. The set delete operations (indicated with the minus operator in UC) in line 20 are then used to limit the parallelism is the subsequent **par** statements (lines 22 and 24) and the reductions (lines 9-10).

2.5 Data Distribution

UC provides a set of data mapping constructs, similar to HPF[6], to describe various types of data

```
1   int main(void) {
2       double A[N][N], tmp[N];
3       range L:l = {0..N-1}, I:i=L, J:j=L;
4       int z, pivot;
5       double max;
6
7       for (z = 0; z < N; z++) {
8           /* good pivot chosen */
9           max = $>(I; fabs(A[i][z]));
10          pivot = $<(I st (max == fabs(A[i][z])) i);
11
12          if (pivot < N)
13              par (L) st (pivot != z) {
14                  tmp[l] = A[z][l];
15                  A[z][l] = A[pivot][l];
16                  A[pivot][l] = tmp[l];
17              }
18          else break;
19
20          I = I - z;  J = J - z;
21          /* Calculation of Multiplier Column */
22          par (I) A[i][z] /= A[z][z];
23          /* Elimination */
24          par (I,J)  A[i][j] -= A[i][z] * A[z][j];
25      }
26  }
```

Figure 1: UC program to implement the LU decomposition phase of Gaussian elimination.

distributions[1]. Three types of data mappings are supported: **perm** to align a data structure with respect to another; **fold** to decompose aggregate data structures (e.g. to specify block partitioning for arrays), and **copy** to replicate parts of a data structure. Any declaration block of a UC program may contain a data distribution statement, allowing UC mappings to be dynamic; data structures can be redistributed in deeper scopes.

In the interest of brevity, we demonstrate some simple uses of mappings in figure 2. In the permute mapping the ith element of array a is aligned with the $(i+2)^{th}$ element of array b. The copy mapping replicates array c along an additional axis. The fold mapping puts size-B blocks of array d on each processor and cyclically places elements of array e on C processors.

2.6 Task Parallelism

Although the asynchronous constructs provided in UC (such as **par** st-clauses, **arb**, and parallel function calls) provide coarse granularity synchronization, their semantics are well-defined only if a variable modified in one thread is not accessed in concurrent threads. This is overly restrictive and we need a construct to specify a weaker form of synchronization within these

```
int a[N], b[N], c[N], d[N], e[N];
perm (I) a[i] :- b[(i+2)%N];  /* alignment */
copy (I,J) c[i] :- c[i][j];  /* replication */
fold (I) {
    d[i] :- d[i%B];  /* block partioning */
    e[i] :- e[i/C];  /* cyclic partioning */
}
```

Figure 2: Data mapping statements in UC.

asynchronous constructs. We use message-passing channels for this purpose.

Channel Variables A *channel* type is a queue of messages. The messages in the queue can be of arbitrary type. The primary operations supported on a channel include depositing a message and retrieving a message. The function **send**($c, expr$) is an asynchronous send that deposits a message containing *expr* on channel c. The function **receive**(c) provides a blocking receive which retrieves items from channel c. In addition to the send and receive operations, two other boolean functions, **empty**(c) and **full**(c), are supported with their standard semantics.

A channel is a type and can be used in most places where types are used. For instance, it is possible to declare an array of channels. A channel can be mapped and aligned using data mappings in exactly the same way as arrays. Additionally, each channel is associated with a buffer size. If unspecified, the buffer size of a channel is assumed to be infinite. A finite size may however be specified in its declaration as in:

channel x[N][N] size *expr*;

The preceding fragment indicates that all channels in variable **x** have a maximum capacity of *expr* bytes. A send operation on a channel of infinite capacity is always non-blocking (limited by the total local memory). If a channel is declared with a finite capacity, the implementation will ensure that a send operation is blocked until the buffer has sufficient capacity to store the data.

Semantics As the send and receive operations on a channel variable are provided as function calls, they can occur anywhere in a UC program where they do not violate other syntactic restrictions. Table I presents the semantics of channel operations within **par** and **arb** statements. Both scalar channel variables (**channel sx;**) and array channel variables (**channel ax[N];**) are considered. The **par** statements are illegal because they allow multiple values (a[i], **receive**(ax[i])) to be written to a single variable (**sx, val**).

Syntactic Convenience: The **arb_stmts** construct may be used to asynchronously compose statements listed explicitly; these statements may be com-

Table I: Semantics of send and receive operations on channel variables.

assignment	semantics
`par(I) send(sx, a[i]);`	illegal
`par(I) a[i]=receive(sx);`	value dequeued from `sx` is broadcast to all `a[i]`
`arb(I) send(sx, a[i]);`	in arbitrary order, each `a[i]` is enqueued on `sx`
`arb(I) a[i]=receive(sx);`	in arbitrary order, each `a[i]` is assigned a dequeued value from `sx`
`par(I) send(ax[i], val);`	copy of `val` is enqueued on each `ax[i]`
`par(I) val=receive(ax[i]);`	illegal
`arb(I) send(ax[i], val);`	copy of `val` is enqueued on each `ax[i]`
`arb(I) val=receive(ax[i]);`	each `ax[i]` is dequeued once and values are assigned to `val` in arbitrary order

pound statements (or blocks). Statements are separated by commas (","). For instance, the following fragment

```
arb_stmts {
    f(...); ,
    (x = 3, y = x); ,
    { ... }
}
```

is the asynchronous composition of a function call, an assignment-expression statement and a compound statement. Note that, as in argument lists of functions calls, comma has been given a special meaning here. If a statement contains the comma operator, it must appear in a parenthetical grouping.

Example: The UC program in figure 3 illustrates the use of channels in connecting data-parallel modules. The program implements iterated polynomial multiplication which consists of three data-parallel modules. Two modules perform forward FFT's on two polynomials. A third module takes their output, performs pointwise multiplication on them and calculates the inverse FFT of the product.

Polynomial coefficients are complex numbers and are represented in the program by two arrays for each polynomial - for the real and imaginary parts. The program uses the **arb_stmt** construct to asynchronously compose three data-parallel subprograms. The first two subprograms use the routine **fft**, a data-parallel program, to read in a polynomial and perform a FFT on it. Similarly, the third subprogram uses the routine **mult_and_ifft** to perform pointwise multiplication and compute the inverse FFT of the product. Channel variables **axr**, **axi** and **bxr**, **bxi** are used by the first two subprograms to communicate with the third subprogram. Note that, although channel variables are being modified in more than one asynchronously executing statements, due to channel variable semantics determinism is preserved. This application is described in more detail in section 3.2.

3 PERFORMANCE STUDIES

Several data-parallel and combined task/data-parallel examples were coded in UC. These are described below and performance results are presented. UC programs were compiled into C with calls to MPL[9], IBM SP2's message passing and collective communications library, and were run on a 24-node IBM SP2 using the high-performance switch with the lightspeed communications protocol. (Compilation is described in detail in [5].) All parallel arrays in the following programs were block mapped along the first axis; all other axes were stored serially.

3.1 Small Data-Parallel Examples

A number of small examples were coded using only the data-parallel portion of the compiler. For brevity, we present only two of them here; a complete treatment can be found in [5]. The speedup obtained with these examples was computed as a function of the number of processors and the size of the dataset. Among the algorithms coded in UC is Gaussian elimination which was discussed earlier (figure 1) and systolic matrix multiplication (figure 6). Their performance graphs are shown in figure 4 and figure 5. All matrices in the examples are of size N^2. For Gaussian Elimination the coefficient matrix was chosen such that pivoting would be required at every iteration.

3.2 Combined Task/Data-Parallel Examples

Several applications that could benefit from an integrated task/data-parallel approach were programmed in UC. Two of them are presented briefly below; a more complete treatment can be found it [5]. Each consists of two or more subproblems that form stages of a pipeline. The stages execute concurrently as data-parallel subprograms in a task-parallel program.

Polynomial Multiplication (FFT) The program is a pipelined computation that performs a sequence

```
float ar[N],ai[N], br[N],bi[N], cr[N],ci[N], 1;
float tmp_ar[N],tmp_ai[N], tmp_br[N],tmp_bi[N];
channel axr[N],axi[N], bxr[N],bxi[N];
range I:i = {0..N-1};

int main(void) {
    arb_stmts {
        for (1 = 0; 1 <= L; 1++) {
            fft(ar, ai);
            arb (I) {
                send(axr[i], ar[i]);
                send(axi[i], ai[i]); }
        },

        for (1 = 0; 1 <= L; 1++) {
            fft(br, bi);
            arb (I) {
                send(bxr[i], br[i]);
                send(bxi[i], bi[i]); }
        },

        for (1 = 0; 1 <= L; 1++) {
            arb (I) {
                tmp_ar[i] = receive (axr[i]);
                tmp_ai[i] = receive (axi[i]);
                tmp_br[i] = receive (bxr[i]);
                tmp_bi[i] = receive (bxi[i]);  }
            mult_and_ifft(cr, ci,
                    tmp_ar, tmp_ai, tmp_br, tmp_bi);
        }
    } }
```

Figure 3: Combined task/data-parallel UC program to implement iterated polynomial multiplication using three FFT's.

of polynomial multiplications using three fast Fourier transforms. Finding the product of two n degree polynomials can be broken down into subproblems as follows:

1. Extend the first polynomial to degree $2n - 1$ by padding with zeros. Convert it to point-value representation by evaluating this extended polynomial at the $2n$ $2n^{th}$ roots of unity. These can be computed efficiently using FFT.

2. Do the same as above for the second polynomial.

3. Multiply the two resulting $2n$-tuples of complex numbers element-wise. This represents the product polynomial evaluated at the $2n$ roots of unity. Convert the tuples to coefficient form by applying an inverse FFT.

For brevity, the details of computing FFT in parallel are not presented here; a recent reference is [4].

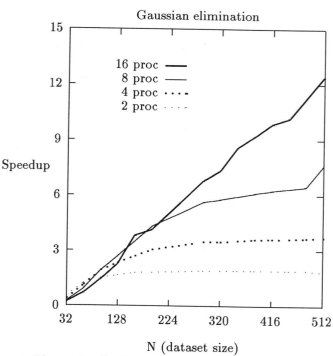

Figure 4: Performance of Gaussian elimination on IBM SP2.

The UC program for the combined task/data-parallel implementation was presented earlier in figure 3. The purely data-parallel program is shown in figure 8. In the data-parallel implementation each subproblem executes on all processors, whereas in the combined implementation each subproblem executes on a third of the processors. The program for each implementation multiplies 20 polynomials. The execution time for the two implementations was plotted as function of the number of processors. These curves are shown in figure 7 for three different dataset sizes (The combined code executes on numbers of processors that are multiples of three, whereas for the purely data-parallel code the number of processors is a multiple of two.).

The graph shows that the combined implementation performs better for these dataset sizes. This can be explained as follows. Each FFT module is communication intensive, performing a series of all-to-all communications. The communication overhead for each module dominates the local computation. FFT modules benefit from being executed at a smaller number of processors as they send fewer (although longer) messages. The purely data-parallel implementation performs each of three FFT's on all processors, whereas the combined implementation performs each FFT on a third of the processors. The combined implementation shows better performance as it benefits from pipelining of FFT modules.

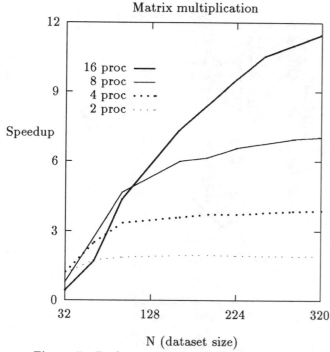

Figure 5: Performance of systolic matrix multiplication on IBM SP2.

```
int main (void) {
    int a[N][N], b[N][N], c[N][N], k;
    range I:i = {0..N-1}, J:j=I;

    par (I,J) {
        /* Alignment Phase */
        a[i][j] = a[i][(i+j)%N];
        b[i][j] = b[(i+j)%N][j];

        /* Multiplication Phase */
        c[i][j] += a[i][j] * b[i][j];
        for (k = 1; k < N; k++) {
            a[i][j] = a[i][(j+1)%N];
            b[i][j] = b[(i+1)%N][j];
            c[i][j] += a[i][j] * b[i][j];
        }

        /* Re-alignment Phase -- omitted */
    } }
```

Figure 6: UC program to implement systolic matrix multiplication.

Pattern Recognition Moments constitute an important set of parameters for image analysis; for example, the lower-order moments indicate the location, size, and orientation of the object in the image[3]. The $(K,L)^{th}$ moment of an image is defined by

$$m_{KL} = \sum_{i=1}^{N} i^K \sum_{j=1}^{N} j^L x_{ij}$$

where x_{ij} represents the gray-scale level at pixel (i,j) in an image of size N^2. In a straight-forward implementation this computation can be divided into two stages. The first stage performs the inner summation over the jth axis for a particular value L. If this axis is stored serially within each processor, this summation loop can be executed locally. The matrix x is then transposed and given for processing to the second stage. The second stage performs the outer sum-reduction over the ith axis which is now stored serially. One pass over the two stages calculates m_{kL} for $k = 1..N$. The two stages are looped over $L = 1..N$ to calculate all moments.

In the purely data-parallel implementation each stage executes on all processors; in the combined task/data-parallel implementation each stage uses half the processors. The execution time of the two implementations was plotted as a function of the problem size, N, where the input image is a matrix of size N^2. In figures 9, the performance of the two implementations is compared for two different machine sizes: 16

processors and 8 processors.

The graphs show that the combined implementation performs better for smaller problem sizes ($N \leq 256$), whereas the purely data-parallel implementation performs better for larger problems sizes. It is informative to analyze the results. In the combined implementation the transpose operation is merged with the transfer of data between the two stages. The purely data-parallel implementation performs the matrix transpose over twice as many processors as the combined implementation. The latter sends fewer messages, although they are longer. For smaller problem sizes communication overhead dominates computation cost and hence the combined implementation performs better. For larger problem sizes, local computation dominates communication overhead and the purely data-parallel implementation performs better.

4   RELATED WORK

UC is related to collection-oriented languages like SETL and Paralation Lisp as well as data-parallel languages like C*, DINO, FortranD, and Kali. This work is also related to recent work in paradigm integration.

Languages like C*[12] specify strict synchronization at the expression level, while other languages such as FortranD[8] and Kali[11] weaken the synchronization granularity and are synchronized at the block level. The code executed between synchronization points is not allowed to access non-local data. DINO[13] provides a more flexible SPMD model similar to UC, but unlike UC it requires local and remote data references to be distinguished statically. The primary difference

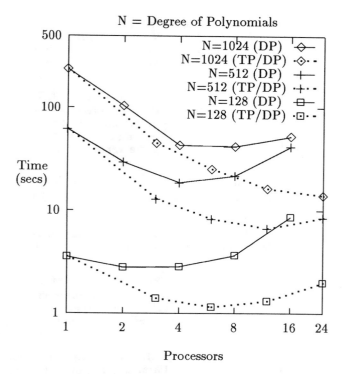

Figure 7: Performance of iterated polynomial multiplication using three FFT's on IBM SP2.

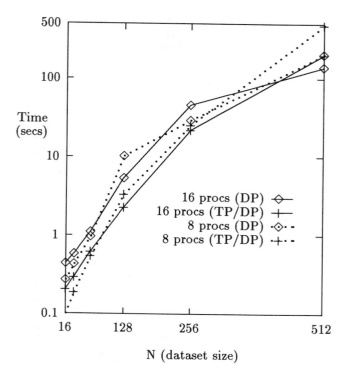

Figure 9: Performance of pattern recognition (computing moments) on 16 and 8 processors of IBM SP2.

between UC and existing data-parallel languages is its support for variable granularity of synchronization and its support of data and task parallelism.

SETL[15] is a high-level prototyping language that uses dynamic, heterogeneous sets as its primary data type. Dynamic typing and the fact that control-flow in SETL is specified using the standard sequential constructs may make parallelization less effective for SETL loops. Paralation Lisp[14] is a data-parallel language which is based on the notion of a group of related collections, called a *paralation*. A single communication mechanism is provided, which consists of two parts – setting up the communication pattern and

```
float ar[N],ai[N], br[N],bi[N], cr[N], ci[N], l;

int main(void) {
    for (l = 0; l <= L; l++) {
        fft(ar, ai);
        fft(br, bi);
        mult_and_ifft(cr, ci, ar, ai, br, bi);
    }
}
```

Figure 8: Purely data-parallel UC program to implement iterated polynomial multiplication using three FFT's.

actual data movement. In UC, the nature of the communication – neighbor communication, broadcast, reduction, etc. – is usually visible at compile-time by inspecting array subscripts or the parallel construct being used.

Recently some researchers have explored multi-paradigm integration. Most have focused on task-parallel coordination of data-parallel components. For example, Quinn et al.[16] propose a system for the Intel iWarp which is used to connect Dataparallel C computations. They extend the concept of C stream I/O to include inter-module communication channels. Their system requires each module to be written as a separate program instead of allowing computational phases to be different functions in the same program.

Subhlok et al.[17] propose a compile-time method for exploiting task and data parallelism also for the iWarp. The input program uses a FortranD-like notation with directives exposing dependencies, parallel sections, and data distributions. The compiler generates a task graph which is mapped onto the machine. However, tasks may only communicate at entry and exit points of functions. Our approach is more flexible, but also requires more programmer intervention.

Massingill[10] proposes a mechanism that uses PCN to coordinate SPMD programs written in message passing C. The SPMD programs may only communicate at entry and exit points, hence this approach is more restrictive than ours. Foster et al.[7] build on previous work in task-parallel and data-parallel Fortran compilers by using the task-parallel language Fortran M to coordinate data-parallel HPF tasks. Fortran M is designed to support dynamic task parallel programs where the identity of task is determined only at run-time and is run-time multithreaded. However data-parallel languages like HPF assume a dedicated machine of known size at compile-time. This raises the problem of integrating the run-time systems of the two compilers. Additionally, the Fortran M and HPF compiler may have different internal data representations. Data being transferred between modules may first have to be converted to representation used by the destination module.

5 CONCLUSION

The task-parallel and data-parallel paradigms are the two dominant parallel programming models in use today. The integration of the two can result in significant performance benefits for several applications. Also, some applications, such as pipelined computations and coupled simulations, can be designed more naturally in a combination of the two programming models. In this paper, we have proposed the integration of task and data-parallelism in an extension of C called UC. We have demonstrated the performance benefits of our approach for several applications on the IBM SP2.

References

[1] Bagrodia, R. and Mathur, S. Efficient implementation of high-level parallel programs. In *4th International Conference on Architectural Support for Programming Languages*, pages 142–153, April 1991.

[2] Bagrodia, R., Chandy, M., and Dhagat, M. UC: A Set-Based Language for Data-Parallel Programming. *Journal of Parallel and Distributed Computing*, 1994. To Appear.

[3] Chen, K. Efficient Parallel Algorithms for the Computation of Two-Dimensional Image Moments. *Pattern Recognition*, 23(1/2):109–119, 1990.

[4] Cormen, T.H., Leiserson, C.E., and Rivest, R.L. *Introduction to Algorithms*. MIT Press, Cambridge, MA, 1991.

[5] Dhagat, M., Bagrodia, R., and Chandy, M. Integrating Task and Data Parallelism in UC. Technical Report 940030, University of California, Los Angeles, CA 90024, September 1994.

[6] High Performance Fortran Forum. High Performance Fortran Language Specification, version 1.0. Technical Report CRPC-TR92225, Center for Research on Parallel Computation, Rice University, Huston, TX, January 1993.

[7] Foster, I., Xu, M., Avalani, B. and Choudhary, A. A Compilation System That Integrates High Performance Fortran and Fortran M. In *1994 Scalable High Performance Computing Conference*, May 1994.

[8] Hiranandani, S., Kennedy, K., Koelbel, C., Kremer, U., and Tseng, C.-W. An Overview of the Fortran D programming system. Technical Report CRPC-TR91121, Center for Research on Parallel Computation, Rice University, Huston, TX, March 1991.

[9] IBM, Kingston, NY 12401. *IBM AIX Parallel Environment Parallel Programming Reference*, 2.1 edition, December 1994.

[10] Massingill, B. Integrating Task and Data Parallelism. Master's thesis, Caltech, Pasadena, CA 91125, May 1993.

[11] Mehrotra, P. and Van Rosendale, J. Programming distributed memory architectures using Kali. Technical Report 90-69, Institute for Computer Application in Science and Engineering, Hampton, VA, 1990.

[12] Rose, J.R. and Steele Jr., G.L. C*: An Extended C Language for Data Parallel Programming. Technical Report PL-87.5, Thinking Machines Corporation, Cambridge, MA, March 1987.

[13] Rosing, M., Schnabel, R.B., and Weaver, R.B. The DINO parallel programming language. *Journal of Parallel and Distributed Computing*, 13:30–42, 1991.

[14] Sabot, G. *The Paralation Model: Architecture-Independent Parallel Programming*. The MIT Press, Cambridge, MA, 1988.

[15] Schwartz, J.T., Dewar, R.B.K, Dubinsky, E., and Schonberg, E. *Programming with Sets: An Introduction to SETL*. Springer-Verlag, New York, 1986.

[16] Seevers, B., Quinn, M. and Hatcher, P. A Parallel Programming Environment Supporting Multiple Data-Parallel Modules. In *Workshop on Languages, Compilers, and Run-Time Environments for Distributed Memory Multicomputers*, pages 44–47, Boulder, Colorado, September 1992.

[17] Subhlok, J., Stichnoth, J.M., O'Hallaron, D.R. and Gross, T. Exploiting Task and Data Parallelism on a Multicomputer. In *Fourth ACM SIGPLAN Symposium on Principles and Practice of Parallel Programming*, pages 13–22, San Diego, California, May 1993.

[18] Wegner, P. (Ed.). Special Issue on Programming Language Paradigms. *ACM Computing Surveys*, 21(3), September 1989.

DESIGN AND IMPLEMENTATION OF ABSTRACT MACHINE FOR PARALLEL LISP COMPILATION

M. D. Feng, W. F. Wong and C. K. Yuen

Department of Information Systems and Computer Science
National University of Singapore, Kent Ridge, Singapore 0511
email: {fengmd,wongwf,yuenck}@iscs.nus.sg

Abstract – *An abstract machine is presented as the front end for compilation of a parallel Lisp dialect, BaLinda Lisp, on multiprocessors. The abstract machine plays an important role in porting of the BaLinda Lisp compiler to parallel machines with different architecture. We will explain the abstract machine, its instruction set and its execution model based on group managing. The abstract machine has been realized on a network of Transputers and a SPARC-based shared memory multiprocessors. The experiment results show that the compiler, based on the abstract machine, generates efficient code for sequential BaLinda Lisp programs, and following upon this, parallel BaLinda Lisp programs could run much faster, and reasonable speedup could be achieved in most situations.*

1   INTRODUCTION

As more knowledge of effective algorithms for symbolic applications is developed, there is an increasing need for techniques for executing these algorithms efficiently. The higher expressiveness of parallel symbolic languages and the efficiency of the language implementation can be essential to such complex algorithms. This need for high-performance parallel symbolic processing systems was the initial motivation of the work. The result was BaLinda Lisp [24], a new member of the Lisp family of symbolic processing languages.

This paper describes the abstract machine used in the BaLinda Lisp compiler. Since it would be highly desirable to obtain a compiler which could be (relatively) easily portable to parallel machines, the abstract machine is defined at a higher level, with an instruction set suitable for execution of the compiled BaLinda Lisp programs. Our actual implementation shows the feasibility of the abstract machine as well as the language.

We will first discuss BaLinda Lisp in Section 2, then review the related work in Section 3. The compiling phases will be explained in Section 4. In Section 5, a description of the abstract machine, its instruction set, and its execution model are presented. In Section 6 and 7, we will discuss the ways to realize the abstract machine in concrete multiprocessor machines, namely a network of Transputer and SPARCserver 1000 respectively. Performance of our systems will be evaluated in Section 8. The last section will summarize the work.

2   BALINDA LISP

BaLinda Lisp is a parallel Lisp dialect designed for parallel execution on a multiprocessor system. Apart from constructs common to any sequential Lisp, BaLinda Lisp offers the following additional features:

- Parallel execution of sibling expressions within the same form by prefixing EXEC to a Lisp expression.

(... (EXEC $\langle$expression$\rangle$) $\langle$sibling expressions$\rangle$) $\langle$other expressions$\rangle$

causes $\langle$expression$\rangle$ to be executed in parallel with $\langle$sibling expressions$\rangle$[a]. When both complete, execution proceeds to $\langle$other expressions$\rangle$.

- Speculative processing to initiate parallel computation on the basis of speculation about usefulness of its result. In BaLinda Lisp, prioritized speculative processing is based on the Lisp COND construct. For example,

(COND (EXEC ($\langle$boolean$\rangle$) $\langle$action$\rangle$)
     $\langle$remainder$\rangle$)

The $\langle$remainder$\rangle$ is executed as a speculative task with its own local environment, in parallel with the main task which executes $\langle$boolean$\rangle$ and $\langle$action$\rangle$. The $\langle$remainder$\rangle$ task has a lower priority which will be increased if $\langle$boolean$\rangle$ evaluates to NIL or set to zero for termination if $\langle$boolean$\rangle$ evaluates to not NIL. Its local environment will also be incorporated into the main task environment or be discarded accordingly. EXEC may also be placed within (($\langle$boolean$\rangle$)) which makes $\langle$action$\rangle$ a speculative task.

Equally-prioritized speculative processing in terms of AND/OR parallelism is also supported by the UBOR (Un-Biased OR) construct:

(UBOR (($\langle$boolean$_1$$\rangle$) $\langle$action$_1$$\rangle$)
     (($\langle$boolean$_2$$\rangle$) $\langle$action$_2$$\rangle$)
     ...
     (($\langle$boolean$_n$$\rangle$) $\langle$action$_n$$\rangle$)))

where $\langle$boolean$_1$$\rangle$, $\langle$boolean$_2$$\rangle$ ... $\langle$boolean$_n$$\rangle$ execute in parallel with equal priorities in the UBOR environment, and as soon as one $\langle$boolean$_i$$\rangle$ returns T, $\langle$action$_i$$\rangle$, if it exists, executes and its result is returned as the result of UBOR. If $\langle$action$_i$$\rangle$ is omitted, then UBOR returns $\langle$boolean$_i$$\rangle$'s result. Other

---

[a] In this paper, a term which will be evaluated concurrently is underlined explicitly.

⟨boolean⟩'s are terminated, their side effects up to the point of the termination remain in the UBOR environment, their ⟨action⟩'s will no longer execute. If all ⟨boolean⟩'s return NIL, UBOR returns NIL also.

- Parallel tasks' communication and synchronization via the Linda *tuplespace* mechanism [2], a logically shared data space. The tuplespace contains multifield records (called *tuples*). A task puts a tuple into the tuplespace using an OUT command:

$$(\text{OUT } \langle exp_1 \rangle \ \langle exp_2 \rangle \ \dots \ \langle exp_n \rangle)$$

where each expression defines a value of any type, whether numerical, logical, character, string or even array/list. Tuples are not accessed by name or address, but by content, using the IN or RD commands:

$$(\text{IN } \langle exp_1 \rangle \ \dots \ \langle exp_m \rangle \ ? \ \langle var_1 \rangle \ \dots \ \langle var_{n-m} \rangle)$$

$$(\text{RD } \langle exp_1 \rangle \ \dots \ \langle exp_m \rangle \ ? \ \langle var_1 \rangle \ \dots \ \langle var_{n-m} \rangle)$$

These will retrieve from the tuplespace an n-field tuple whose first m fields match the result of the m expressions in the IN/RD, and then store the values of the last $n - m$ fields into the $n - m$ variables specified in the IN/RD. IN causes the tuple to be removed from the tuplespace, while RD leaves it for others to access. If no matching tuple can be found, a *template* will be left in the tuplespace for tuple matching in the future, and the task executing the IN/RD is suspended until another task OUTs the required tuple.

3 RELATED WORK

Some of the compilation techniques implemented are based on the previous work on compilation of Scheme. We are most influenced by the RABBIT compiler [20], the ORBIT compiler [14] and the HARE compiler [23]. The main differences between our work and the above include:

- the use of abstract machine as a target for code generation in order to achieve compiler portability.

- compilation of parallel constructs of BaLinda Lisp into abstract machine code.

- implementation of the abstract machine on a distributed memory machine with Transputers and a shared memory multiprocessor SUN SPARCserver 1000 in order to achieve realistic but low cost parallel operation.

The use of an abstract machine for compiling Lisp programs has been shown practical by the PSL compiler [7], though it was for a sequential Lisp, and obsolete. On parallel Lisp research, a number of parallel Lisp have been proposed, most of which have previously been intended for shared memory machines. These include Multilisp on the Concert machine [8], Mul-T on Encore Multimax [15], Qlisp on Alliant FX/8 [6], Spur Lisp on Spur multiprocessor workstation [25], MultiScheme and Butterfly Portable Standard Lisp on BBN Butterfly [17][22]. Since Lisp programs tend to employ a lot of pointers pointing to heap

cells, and share a common global space in passing information, parallel Lisp implementations tend to depend on shared memory architecture for efficiency and easy manipulation of Lisp pointers.

Our research is not confined to shared memory architectures, however. The Transputer system on which we make our experiments is based on a wholly distributed architecture. Thus, approaches that are independent of the shared memory techniques are required. Although there are several research activities on this issue [12] [9] [18], no common agreement has been reached on how to efficiently implement Lisp or other declarative languages on distributed machines. One aim of our work is to develop a method for implementing parallel Lisp languages for such architecture, giving performance comparable with imperative languages.

4 COMPILING TO ABSTRACT MACHINE

BaLinda Lisp programs are first compiled to an abstract machine. The idea of using an abstract machine is to allow for portability across a number of platforms.

The compiling of BaLinda Lisp consists of several passes, each of which performs relatively simple transformations on the Lisp form it receives.

- Macro expansion reduces the derived expressions, including parallel operations and tuplespace operations, into primitive expressions which are defined as abstract machine instructions.

- Continuation-passing style (CPS) transformation converts a macro expanded form to an equivalent CPS form. The advantage of the CPS transformation is that code becomes regularized and tail calls are made explicit.

- Optimization techniques are used to simplify the CPS form. These include the β-reduction rule of λ-calculus, boolean expression short-cutting, constant test evaluation and test result propagation, constant folding, and redundant subexpression elimination.

- Closure analysis determines the variable location in the stack or the heap, and decides what sort of information is to be stored in the environment.

- Abstract machine code generation utilizes all the analyzed results to produce the abstract machine code.

To these phases, one can attach a final code generation phase to generate actual machine code from the abstract machine code.

5 ABSTRACT MACHINE

Logical Processor

The abstract machine defined for BaLinda Lisp consists of a number of *logical processors* which have the same abstract machine instruction set. Each logical processor has its own stack and heap space; it is thus able to run one task independently. The communication among logical processors is fully described by abstract machine instructions. By clearly defining the semantics of all abstract machine instructions, it becomes straightforward to implement the logical processor on the real machine as we will show in the next two sections. For actual implementation, every logical processor will be mapped onto one physical processor. If the physical processor supports time-sliced multiprocessing, more than one logical processor could be mapped onto one physical processor.

The storage architecture of a logical processor includes a *stack* which holds the values of actuals being gathered for a combination, or temporary values; a *heap* which contains accessible data while user programs are being executed, and the following *registers* that maintain the current state of the computation:

pc The program counter pointing to the instruction currently being executed. Assuming the length of each abstract machine instruction is one unit, then $pc + 1$ points to the next instruction.

sp The stack pointer which holds a pointer to the element on the top of the stack. The stack grows from high addresses towards low addresses. Any element in the stack can be accessed or modified by giving the offset between its address in the stack and sp.

acc The accumulator which holds the value resulting from the most recent computation.

ep The environment pointer which points to the current environment object.

gp The global environment pointer which points to the global environment object.

nc The counter for the number of arguments, used to pass to a function the number of actuals with which it was invoked.

sb The subtask register which points to a set of child tasks of the current task. Upgrading or deletion of prioritized speculative tasks makes use of this register to propagate the required actions to all descendant tasks.

pr The priority register which holds the priority of the task running on this logical processor. If the task is still running, pr contains a value greater than 0, with bigger non-zero value indicating lower priority. The priority of a mandatory task is 1, while that of a speculative task could be greater than or equal to 1. If the task ends, pr becomes 0. We assume that the task could be killed if its pr is forced to 0.

tp The tuplespace register which points to the current visible tuplespace, either global or local. The global tuplespace is logically shared by any logical processors in the abstract machine. Whenever an %%IN or %%RD succeeds, fields of the matched tuple which should be assigned to variables are saved into a local dataspace τ which is only visible to this processor. Each prioritized speculative task has its own local subtuplespace visible only to itself and its non-prioritized child tasks.

up The upper tuplespace register which points to the tuplespace in the upper level to which the local subtuplespace could be merged (when the speculative task is confirmed).

Instruction Set

The instruction set is tailored to the execution of BaLinda Lisp. Table 1 shows all the instructions divided into several categories. An operational semantics for all the abstract machine instructions in the form of rewrite rules has been defined in [3]. Apart from these instructions, all the Lisp primitive operations are open coded. Depending on the physical machine instruction set, some operations, e.g., CAR and CDR, can be translated into physical machine instructions directly, while others, e.g., APPEND and MAP, will be translated into respective function-calls of the run-time library.

Category	Instructions
Function call/return	%%CALL, %%CALL-CL, %%CALL-ENV, %%JMP, %%JMP-CL, %%JMP-ENV, %%RETURN, %%RETURN-CL, %%CHK-NC
Task generation	%%EXEC, %%EXEC-CL, %%EXEC-ENV, %%SPEC-EXEC, %%SPEC-EXEC-CL, %%SPEC-EXEC-ENV,
Synchronization	%%SYNCHRONIZE, %%SPEC-SYNCHRONIZE, %%SELECT
Priority adjustment	%%UPGRADE, %%DISCARD
Tuplespace operation	%%OUT, %%IN, %%RD, %%TUPLE-READ
Global environment access	%%GET-GLOBAL, %%SET-GLOBAL
Jump	%%JMP-IF-TRUE, %%JMP-IF-FALSE
Stack operation	%%PUSH, %%DROP, %%MOVE
Heap operation	%%MAKE-CELL, %%GET-CELL, %%SET-CELL, %%ALLOC-ENV, %%MOVE-X, %%MAKE-CL
Lisp primitive	in-line or library function

Table 1: Abstract machine instructions.

Group Managing Model

The abstract machine adopts a group managing model to distribute the execution control mechanism. In this model, logical processors are clustered into groups, and each group is controlled by a group manager. A group manager runs a *kernel process* K (one-to-one relationship) and a logical processor runs an *application process* P (one-to-one relationship). The user application programs are executed by Ps. Each P is under control of a designated K (represented as K_P) and one K may control more than one Ps. The set of all Ps under a K's control is represented as $\mathcal{P}_K$. K handles all the communication for its Ps. These functions of K are:

- receive parallel (mandatory or speculative) tasks generated by $\mathcal{P}_K$.

- distribute the received tasks to $\mathcal{P}_K$ or other Ks according to certain task distribution strategies.

- receive evaluated tasks' results sent from $\mathcal{P}_K$ or transmitted from other Ks.

- support the upgrading or deletion of speculative tasks.

- managing the tuplespace and speculative subtuplespace.

As can be seen, K plays important roles in task management and tuplespace management. All Ks within an abstract machine coordinate with each other to provide the run-time support for parallel execution of BaLinda Lisp programs. P is tailored for different user applications, while K is common.

The number of Ks and Ps is not fixed, and could be changed by the user. In a multi-processor machine, the decision on how to divide processors into groups, and determine the proper size of each group should be made by the user because there is no fixed configuration versatile enough to cater for all different needs of user application programs. However, as a general guideline, there are pros and cons of the size of groups. If a group size is small, the work load of each K is alleviated compared with the one in a large group, but the total number of Ks would be large, which would incur extra overhead as Ks may compete with Ps on the same processor. If the group size is large, K would be slow in reacting the request of Ps in the group due to its heavy load, but few Ks are needed in a system and less overhead is incurred. The two extreme cases are a centralized control system where there is only one K in the system, and a totally distributed control system where each processor of the system runs a K. Depending on the different applications' requirements, it is necessary to make a balance between the number of groups and the size of each group.

In our abstract machine, users also have the freedom to determine the strategy of placement of Ks and Ps onto physical processors. There are few restrictions on the topology of physical processors, and users could choose a topology best suitable for their applications. The only assumptions we make are:

1. there is at least one route linking any two Ks, i.e., all Ks are connected, however, fully direct connection of Ks is not required.

2. there is at least one route linking any P to K_P.

All the connection information of a K can be expressed by two vectors ϕ and ψ:

ϕ_i contains the route information to send a message to the K identified by number i.

ψ_i contains the route information to forward a received message from the K numbered i. If there is no need to forward, then ψ_i is null.

Based on the different topology of user applications, only the ϕ and ψ vectors need to be specified. In fact, our implementation automatically calculates the values of these vectors for users, and later makes use of these two vectors to find the suitable route for message-passing.

The algorithm to compute ϕ and ψ could be applied to arbitrary processor network connection. We view the connection as a graph where each processor acts as a node and the link as an edge.

Assuming there are n nodes in the graph, the algorithm receives two input matrices r and d, containing the currently known shortest routes and their distances respectively. At the beginning, any two nodes i and j, if there is one edge between i and j, then $r_{i,j} = (j)$, $r_{j,i} = (i)$ and $d_{i,j} = d_{j,i} = 1$. If there is no edge, then $r_{i,j} = r_{j,i} = ()$ and $d_{i,j} = +\infty$. ϕ and ψ then could be calculated as shown in Figure 1.

Input:	r — a matrix of currently known shortest routes (list of nodes)
	d — a matrix of distance of the currently known shortest routes
Output:	ϕ — a vector of (neighbor) nodes leading to the shortest route
	ψ — a vector of (neighbor) nodes to forward a message

```
1    for ⌊log n⌋ times
2        for ∀i ∈ {1...n}
3            for ∀j ∈ {1...n}
4                for ∀k ∈ {1...n}
5                    if d_{i,k} + d_{k,j} < d_{i,j}
6                        d_{i,j} ← d_{i,k} + d_{k,j}
7                        r_{i,j} ← r_{i,k} ⋈ r_{k,j}
8    for ∀i ∈ {1...n}
9        f_i ← ∅
10       for ∀j ∈ {1...n}
11           if r_{i,j} ∉ f_i
12               f_i ← f_i ∪ r_{i,j}
13   for ∀m ∈ {1...n}
14       for ∀i ∈ {1...n}
15           φ_i ← j where r_{m,i} = (j ...)ᵃ
16           for ∀j ∈ f_i where j = (... m k ...)
17               ψ_i ← ψ_i ∪ kᵇ
```

ᵃNote that each node m has its own ϕ_i.
ᵇNote that each node m has its own ψ_i.

Figure 1: Algorithm of ϕ and ψ.

First, the Floyd-Warshall algorithm is applied to calculate the all-pairs shortest routes (line 1 to 7) [4]. The algorithm consists essentially of an iterated step. Within each iteration, the search for a possibly shorter route between each pair of nodes is extended to consider all possible routes utilizing up to twice the number of currently considered edges. Therefore, a logarithmic number of iterations $O(\log n)$ is sufficient where n is the number of processor nodes in a configuration graph. $\bowtie$ on line 7 means the concatenation of two lists. The shortest routes are stored in r. Line 8 to 12 combine all the shortest routes starting from i and removes the ones which are part of other shortest routes. Therefore f contains the shortest routes leading to all nodes. Line 13 to 17 calculate ϕ and ψ. Note that each node m has its own set of ϕ and ψ. ϕ_i is simply the first node in the shortest route $r_{m,i}$, while ψ_i could be obtained by searching all ks of the route pattern $(... m k ...)$ in f_i. This is because $(... m k ...)$ in f_i indicates the shortest route starting from i will go through m and k; therefore any message sent from i received by m should be forwarded to k.

6 REALIZATION ON TRANSPUTERS

In this section we will show how an abstract machine can be realized in a real target machine based on INMOS T805 Transputer running at 25 MHz [11]. The basic mapping can be illustrated as in Table 2.

Abstract machine	$\stackrel{map\ to}{\Longrightarrow}$	Transputer
kernel process K	$\Longrightarrow$	high-priority task
application process P	$\Longrightarrow$	task
stack	$\Longrightarrow$	workspace
heap	$\Longrightarrow$	memory
pc	$\Longrightarrow$	Iptr register
sp	$\Longrightarrow$	Wptr register
acc	$\Longrightarrow$	Areg register

Table 2: Mapping the abstract machine onto Transputer.

Both K and P are represented as Transputer tasks. Since K takes the responsibility of managing tasks and tuplespace, it runs at higher priority than P in our Transputer-based implementation.

The abstract machine makes heavy use of the stack to pass arguments; thus the efficiency of stack operations is crucial. Since each task in a Transputer has its own workspace, which allows us to use its workspace and workspace pointer (Wptr) as the stack and stack pointer (sp). This contributes much to the performance improvement of the Transputer code produced, as the access of the workspace is much faster than the one of external memory.

Since Transputer has limited registers, all the other registers defined in the abstract machine are mapped onto some locations in the memory. As Transputer has a built-in micro-code scheduler which supports multi-processing on one processor, more than one task can be placed on one Transputer, i.e., many to one relationship exists between Ps and Transputer. Thus, the number of Ps is not limited to the number of Transputers in the system.

Mapping abstract machine instructions onto Transputer instructions could be done after we clearly define the exact operational semantics of each abstract machine instruction. In the following, we show the mapping of some abstract machine instructions, with comments starting from "–". Details of translation of all abstract machine instructions could be found in [3]. A detailed description of Transputer instructions could be found in [10].

$(\text{\%\%CALL}\ \langle label \rangle\ \langle arg_1 \rangle\ \ldots\ \langle arg_n \rangle)$	
ajw $-\langle n+1 \rangle$	–adjust the stack pointer
{load $\langle arg_1 \rangle$}	–get the first argument
stl 0	–store the first argument
...	
{load $\langle arg_n \rangle$}	–get the $\langle n \rangle$th argument
stl $\langle n-1 \rangle$	–store the $\langle n \rangle$th argument
ldc $\langle label \rangle$-2	–get the offset to $\langle label \rangle$
ldpi	–get the address of $\langle label \rangle$
gcall	–jump to $\langle label \rangle$
rev	–shift the result to Areg

The %%CALL instruction is translated as follows: first, $\langle n+1 \rangle$ slots are allocated from the stack for passing $\langle n \rangle$ arguments

of %%CALL, and for the return address which will be stored into the stack by the called function $\langle label \rangle$ itself. Then $\langle n \rangle$ arguments are loaded into the respective slots separately. Different Transputer instructions would be generated for {load $\langle arg \rangle$} depending on whether the value of $\langle arg \rangle$ is in heap, stack, or acc register. The actual address of $\langle label \rangle$ will be loaded into Areg of the Transputer by ldpi, and be jumped to by gcall because gcall exchanges the values of Iptr and Areg. The return address of the call (i.e., the address after gcall) will be in Areg after gcall is made, which allows the called function to store it into the stack. When the function returns, the return result is stored in Breg while Areg holds the address after gcall. The result will then be shifted to Areg by rev which exchanges the values of Areg and Breg.

$(\text{\%\%EXEC}\ \langle label \rangle\ \langle arg_1 \rangle\ \ldots\ \langle arg_n \rangle)$	
ajw $-\langle n \rangle$	–adjust the stack pointer
{load $\langle arg_1 \rangle$}	–get the first argument
stl 0	–store the first argument
...	
{load $\langle arg_n \rangle$}	–get the $\langle n \rangle$th argument
stl $\langle n-1 \rangle$	–store the $\langle n \rangle$th argument
ldc $\langle funct\_table \rangle$-2	–get offset to $\langle funct\_table \rangle$
ldpi	–get address of $\langle funct\_table \rangle$
ldnlp $\langle offset \rangle$	–get $\langle label \rangle$
ldc $\langle gsb \rangle$-2	–get the offset to $\langle gsb \rangle$
ldpi	–get the address of $\langle gsb \rangle$
ldnl 0	–get $\langle gsb \rangle$
call \$SEND_TASK	–send the task to kernel
ajw $\langle n \rangle$	–restore the stack pointer

In essence, %%EXEC is translated to a call (SEND_TASK) of K by supplying the name of $\langle label \rangle$ and values of $\langle arg \rangle$s. K will take the responsibility of looking for a suitable P to evaluate the new task. In order to make a call of SEND_TASK, $\langle n \rangle$ slots are allocated from the stack for passing $\langle n \rangle$ arguments. The name of the function $\langle label \rangle$ stored in the function table is retrieved by using the compile-time determined values $\langle funct\_table \rangle$ and $\langle offset \rangle$. The pointer to the global static block $\langle gsb \rangle$ (for the purpose of resolving global symbol definitions of an external defined function such as SEND_TASK) is also fetched. SEND_TASK will be called with all these arguments, and a taskid will be returned in Areg after call. The stack pointer will then be adjusted to the original position before %%EXEC is executed.

$(\text{\%\%OUT}\ \langle n \rangle\ \langle field_1 \rangle\ \ldots\ \langle field_n \rangle)$	
ajw $-\langle n \rangle$	–adjust the stack pointer
{load $\langle arg_1 \rangle$}	–get the first argument
stl 0	–store the first argument
...	
{load $\langle arg_n \rangle$}	–get the $\langle n \rangle$th argument
stl $\langle n-1 \rangle$	–store the $\langle n \rangle$th argument
ldc $\langle n \rangle$	–load number of arguments
ldc $\langle gsb \rangle$-2	–get the offset to $\langle gsb \rangle$
ldpi	–get the address of $\langle gsb \rangle$
ldnl 0	–get $\langle gsb \rangle$
call \$OUT	–call OUT function
ajw $\langle n \rangle$	–restore the stack pointer

%%OUT is translated to a call (OUT) supported by K with length $\langle n \rangle$ and $\langle n \rangle$ arguments which will form a tuple in the tuplespace. %%IN and %%RD are handled similarly except calling different functions (IN and RD respectively). %%TUPLE-READ calls the function TUPLE_READ of K with ⟨gsb⟩ as the only argument.

7 REALIZATION ON SPARCSERVER

The abstract machine has also been implemented on a Sun SPARCserver 1000 with 8 CPUs running at 50 MHz, 512 MB of RAM and 20 GB of disk space under SUNOS 5.3 operating system. This is a shared memory multiprocessor which supports both the traditional UNIX interprocess communication constructs as well as multithreading with lightweight processes. The basic mapping can be illustrated as in Table 3.

Abstract machine	$\stackrel{map\ to}{\Longrightarrow}$	SPARCserver
kernel process K	$\Longrightarrow$	UNIX process
application process P	$\Longrightarrow$	UNIX process
stack	$\Longrightarrow$	register windows
heap	$\Longrightarrow$	memory
pc	$\Longrightarrow$	PC register
sp	$\Longrightarrow$	%sp register
acc	$\Longrightarrow$	%o0 register

Table 3: Mapping the abstract machine onto SPARC.

Both K and P are represented as UNIX processes. Up to 8 Ps are created, each bounded to a physical processor, while only one K, which is not frequently invoked, exists as an unbounded process free to run on any physical processor scheduled under the operating system. Inter-process communication is achieved by shared memory operations controlled by spin-locking using assembly level atomic operations, namely the SWAP instruction of the SPARC instruction set [19].

The stack is built onto the *ins* and *locals* of a register set (total 16 registers). If 16 stack slots are not enough for the current executing function, then the stack area (referenced by %sp) in the memory is used. As a result, the top 16 elements on the stack is referenced directly by the register name, while the rest are indexed by %sp. When mapping abstract machine code to SPARC code, the first SPARC instruction of each function is "save %sp, ⟨offset⟩, %sp", which moves the register window to the next register set, and adjusts the stack pointer %sp, therefore, the stack elements belonging to the parent function are preserved automatically. The last SPARC instruction of each function "restore" adjusts the register window and stack pointer properly to the parent call's position. Since the most frequently used part of a stack is situated in the register, the performance of the SPARC code generated is good although we did not employ any complicated register allocation algorithms.

Translation of abstract machine instructions to SPARC instructions is done in a way similar to that of the Transputer case. Our practice shows there is no difficulty in doing this as we complete it within three days. The program for the kernel process (written in C) is used without much change. In all, we successfully port the abstract machine to the shared memory SPARC systems and still achieve good performance of the generated code as we will show in the next section.

8 PERFORMANCE EVALUATION

In this section, we evaluate the performance of our implementation on a network of Transputers and SPARCserver 1000. For the experiments on Transputer, we use the Transputer clock to get the timing result. An important caveat for our benchmark numbers on SPARCserver 1000 is that the machine typically has over 100 users logged in. While we tried our best to take the timing measurements when the system workload is low, seldom is there less than 20 users logged in. The time reported are the real time returned by the UNIX gettimeofday system call. The lowest observed time is reported.

Table 4 shows the performance of the two versions of the compiler in compiling sequential programs. The benchmarks used include Takeuchi (with arguments 24, 16, 8), Fibonacci ($n = 30$) and Quicksort ($n = 4096$). For comparison, the equivalent C programs have also been tested. The aim is ascertain that the compiler generate good sequential code. The comparison is done with:

- the native INMOS C compiler producing code running on 25MHz T805 Transputer;

- the GNU CC compiler port to MS DOS (version 2.2.2) producing code running on 33MHz Intel 80386;

- the Scheme->C compiler [1] which translates Scheme programs to C with all optimization options on and compile with -O2 for GCC C optimization on SUN SPARCserver 1000;

- the native SPARCompiler C 2.0.1 with the -xO4 (highest) optimization option turned on.

The timing results show that BaLinda Lisp programs could run as fast as C programs on the same hardware, and sometimes even faster. This is mainly attributed to two factors:

- CPS transformation makes all tail-calls more explicitly, as a result, %%JMPs could be generated for these tail-calls rather than %%CALLs.

- Due to various optimization techniques applied during the optimization phase, the resulting code is more compact and efficient.

Our compiler has also been compared with other Lisp compilers on the widely used Gabriel benchmark programs [5], as shown in Table 5. The timing result of the ORBIT compiler comes from [14] and the Lucid Common Lisp compiler from [21] both of which run on Sun 3/160. We also give the performance result of a Transputer Common Lisp implementation from [18].

As shown, the performance figures of the BaLinda Lisp compiler for Transputer are much better than the Transputer Common Lisp in all the benchmarks being tested, with performance improvement from 1.8 to 8.1 times. The performance of our compiler on SPARCserver is better than the one of Scheme->C in most cases. For the other Lisp implementations, we should bear in mind that they are running on different machine architectures (e.g., different kind of processors, different memory space, with or without cache support, etc.). Allowing for these differences, we can see that the performance of our BaLinda Lisp compiler is comparable to those Lisp implementations, and sometimes better.

Test Programs	BaLinda Lisp (Transputer)	INMOS C (Transputer)	GNU CC (Intel 386)	BaLinda Lisp (SPARCserver)	Scheme->C (SPARCserver)	SPARCompiler C (SPARCserver)
takeuchi	6.889	7.828	6.207	0.761	0.819	0.751
fib	6.428	7.243	7.526	1.155	1.211	1.205
quicksort	0.199	0.210	0.165	0.017	0.073	0.015

Table 4: Execution time of sequential benchmark programs (in seconds).

Gabriel Benchmark	BaLinda Lisp (Transputer)	Common Lisp (Transputer)	BaLinda Lisp (SPARCserver)	Scheme->C (SPARCserver)	ORBIT (SUN 3/160)	Lucid Common Lisp (SUN 3/160)
tak	0.18	1.00	0.018	0.031	0.22	0.44
takl	1.58	-	0.137	0.111	1.63	-
idiv2	1.17	2.10	0.083	0.190	0.50	0.92
rdiv2	1.21	2.70	0.706	0.823	0.76	1.46
deriv	2.74	6.20	0.454	0.404	2.44	3.68
puzzle	3.82	21.00	0.406	0.422	2.40	7.70
triangle	43.85	356.00	5.158	5.551	79.18	131.72

Table 5: Execution time of Gabriel benchmark programs (in seconds).

Having established that our compiler produces quality sequential code, we now assess its performance on parallel code, in particular its ability to speed up execution of programs by harnessing parallelism.

Table 6 lists the timing results in seconds for three test programs running on Transputers. The first is the calculation of the Fibonacci number from its recursive definition. The next program is to find n queens' placement on an $n \times n$-chess board with the backtrack algorithm. The last is the shuffle of 15 square titles in a 4×4 board to get a shortest sequence of moves to the goal state by iterative deepening A* algorithm [13]. The execution time of the sequential versions running on a single Transputer is also listed for comparison. There is no time difference for the application process running the sequential versions with or without the kernel process. It means that the kernel process will not incur any timing overhead if no task distribution is required. The speedup, calculated as the sequential time divided by the parallel time, tells us how much real benefit could be obtained from the parallelism.

Overall, an effective speedup is achieved in all cases. With 18 processors, real speedup could reach above 15 for fibonacci problems, from 13.5 to 15.5 for queen problems. For the puzzle problems, the speedup exceeds 18 due to the abnormalities in search problems [16], that is, the branch leading to the goal would be searched "eagerly" in the parallel situations compared with the ordinary sequential search.

Table 7 shows the performance of same test programs on SPARCserver 1000. Note that the absolute performance is greatly improved due to the faster SPARC processor. Reasonable speedup is also achieved though it is not as good as we saw in Transputer. The reason is probably because the communication overhead among kernel process and application process takes bigger portion over computation. We will investigate this issue in more detail in the future research.

9 CONCLUSIONS

We present a design of the abstract machine for BaLinda Lisp, its logical processors, its instruction set, and its execution model. The architecture of the abstract machine is totally distributed as logical processors do not share any piece of memory. User application programs will be executed by application processes of logical processors while kernel processes take charge of the run-time support in terms of task management and tuplespace management.

Our abstract machine is portable to distributed systems with direct network because we impose few restrictions on the topology of physical processors within a multiprocessor system. Our implementation on Transputer shows its feasibility and usefulness. There is not much difficulty in realizing the abstract machine in shared memory systems as we exemplify in the SUN SPARCserver 1000. The flexibility of the group managing model allows users to choose a configuration to best suit their applications' need.

The abstract machine can be applied to other parallel programming languages when proper logical processors are developed for those languages. The group managing model with kernel processes could be adopted without any change. In consequence the run-time environment could be used in whole.

REFERENCES

[1] J. F. Bartlett. Scheme->C, a portable Scheme-to-C compiler. Research Report 89/1, Western Research Laboratory, Digital Equipment Corporation, January 1989.

[2] N. Carriero and D. Gelernter. Linda in context. *Communication of the ACM*, 32(4):444–458, 1989.

[3] M. D. Feng. *Compilation and Run-time Environment of Parallel Lisp on Distributed Systems*. PhD thesis, National University of Singapore, 1994.

[4] R. W. Floyd. Algorithm 97 (SHORTEST PATH). *Communication of the ACM*, 5(6):345, 1962.

[5] R. P. Gabriel. *Performance and Evaluation of Lisp Systems*. The MIT Press, 1985.

[6] R. Goldman and R. P. Gabriel. Qlisp: Parallel processing in Lisp. *IEEE Software*, pages 51–59, July 1989.

	Sequential (1 Transputer)		Parallel (5 Transputers)		Parallel (9 Transputers)		Parallel (18 Transputers)	
	time	speedup	time	speedup	time	speedup	time	speedup
fib(34)	44.78	1.00	9.86	4.54	5.58	8.02	2.93	15.30
fib(35)	72.46	1.00	15.85	4.57	8.93	8.12	4.70	15.40
fib(36)	117.24	1.00	25.44	4.61	14.41	8.14	7.47	15.70
queen(11)	19.75	1.00	4.44	4.45	2.58	7.67	1.46	13.52
queen(12)	109.16	1.00	23.53	4.64	13.30	8.21	7.04	15.51
puzzle	326.70	1.00	37.52	8.71	28.03	11.66	13.64	23.95

Table 6: Execution time of test programs with different arguments on Transputer (in seconds).

	Sequential (1 Processor)		Parallel (2 Processors)		Parallel (4 Processors)		Parallel (8 Processors)	
	time	speedup	time	speedup	time	speedup	time	speedup
fib(34)	7.92	1.00	5.36	1.47	2.70	2.94	1.34	5.91
fib(35)	12.78	1.00	8.66	1.47	4.36	2.93	2.26	5.65
fib(36)	20.57	1.00	13.75	1.50	6.92	2.97	3.34	6.17
queen(11)	1.45	1.00	0.98	1.48	0.46	3.15	0.29	5.00
queen(12)	7.90	1.00	5.40	1.46	2.82	2.80	1.33	5.93
puzzle	63.67	1.00	41.46	1.53	27.02	2.36	12.42	5.13

Table 7: Execution time of test programs with different arguments on SPARCserver (in seconds).

[7] M. L. Griss and A. C. Hearn. A portable Lisp compiler. *Software Practice and Experience*, 11:541–605, 1981.

[8] R. H. Halstead Jr. Multilisp: A language for concurrent symbolic computation. *ACM Transactions on Programming Language and Systems*, 7(4):501–538, 1985.

[9] C. Hammer and T. Henties. Parallel Lisp for a distributed memory machine. In *Proceedings of High Performance and Parallel Computing in Lisp*, Twickenham, London, UK, November 1990.

[10] Inmos Limited. *Transputer Instruction Set: A Compiler Writer's Guide*. Prentice-Hall, 1988.

[11] Inmos Limited. *Transputer Reference Manual*. Prentice-Hall, 1988.

[12] R. Kessler, H. Carr, L. Stoller, and M. Swanson. Implementing Concurrent Scheme for the Mayfly distributed parallel processing system. *Lisp and Symbolic Computation: An International Journal*, 5(1/2):73–93, May 1992.

[13] R. E. Korf. Depth-first iterative-deepening: an optimal admissible tree search. *Artificial Intelligence*, 27:97–109, 1985.

[14] D. A. Kranz. ORBIT: An optimizing compiler for Scheme. Technical Report YALEU/DCS/RR-632, Yale University, February 1988.

[15] D. A. Kranz, R. H. Halstead Jr., and E. Mohr. Mul-T: A high-performance parallel Lisp. In *Proceedings of the SIGPLAN'89 Conference on Programming Language Design and Implementation*, pages 81–90, Portland, Oregon, June 1989.

[16] T. H. Lai and S. Sahni. Abnormalities in parallel branch-and-bound algorithms. *Communication of the ACM*, 27(6):594–602, 1984.

[17] J. S. Miller. *MultiScheme, A Parallel Processing System Based on MIT Scheme*. PhD thesis, Department of Electrical Engineering and Computer Science, Massachusetts Institute of Technology, August 1987.

[18] B. Pages. Transputer Common-Lisp: A parallel symbolic language on Transputer. In M. Reeve and S. E. Zenith, editors, *Parallel Processing and Artificial Intelligence*, pages 149–174. John Wiley & Sons, 1989.

[19] SPARC International, Inc. *The SPARC Architecture Manual, Version 8*. Prentice Hall, 1992.

[20] G. L. Steele. RABBIT: A compiler for Scheme. Technical Report 474, MIT AI Lab, May 1978.

[21] Sun Microsystems, Inc. *Sun Common Lisp Performance Report*, July 1986.

[22] M. Swanson, R. Kessler, and G. Lindstrom. An implementation of portable standard Lisp on the BBN Butterfly. In *Proceedings of 1988 ACM Conference on Lisp and Functional Programming*, pages 132–142, 1988.

[23] D. Teodosiu. HARE: An optimizing portable compiler for Scheme. *ACM SIGPLAN Notices*, 26(1):109–120, 1991.

[24] C. K. Yuen, M. D. Feng, W. F. Wong, and J. J. Yee. *Parallel Lisp Systems: A Study of Languages and Architectures*. Chapman and Hall, 1993.

[25] B. Zorn, K. Ho, J. Larus, L. Semenzato, and P. Hilfinger. Multiprocessing extensions in Spur Lisp. *IEEE Software*, pages 41–49, July 1989.

A THREAD MODEL FOR SUPPORTING TASK AND DATA PARALLELISM IN OBJECT-ORIENTED PARALLEL LANGUAGES

Neelakantan Sundaresan
nsundare@cs.indiana.edu

Dennis Gannon
gannon@cs.indiana.edu

Computer Science Department
215 Lindley Hall
Indiana University
Bloomington, IN 47405

Abstract

Data- and task-parallelism are two important parallel programming models. Using object-oriented languages like C++, one can abstract data and control representations into a single active object. We propose a thread model of parallelism that addresses both data and task parallelism. Threads provide facilities to overlap computation and synchronization or communication. Thread objects depict control-parallelism. Thread objects are grouped, for collective computation and communication, into rope objects[16, 15] to abstract data-parallelism. Since rope objects are parallel objects, they can be customized, interestingly, in a serial or a parallel manner. We present results from a prototype system running on the SGI Challenge and the Intel Paragon.

keywords: *task-parallelism, data-parallelism, threads, object-oriented paradigm*

1 Introduction

User-level threads provide a good way of expressing parallelism at a level closer to the logic of the user algorithm than at the level of physical processors. Threads are to processes as processes are to physical processors. Even on a single processor system, threads provide for coroutine-style programming to express control separation. Since the cost per context-switch for these threads is just a few microseconds, a thread which cannot proceed because it has to wait for some condition to be satisfied(like data arrival, shared data consistency or control consistency) can enable another thread that is ready to execute, thus improving processor utilization.

Pthreads[8, 11] is an emerging standard of thread interface at C language level. A number of thread libraries are available[5, 4, 12, 7, 2]. A number of parallel object-oriented languages address task parallelism[13, 6, 9, 3].

Our model is an object-oriented thread library for both shared and distributed memory machines. Thread objects provide abstractions for control parallelism and rope objects abstract for data parallelism. They can be customized to specific application target by using object-oriented features like inheritance, polymorphism, dynamic dispatch and parameterized data types. This model was introduced for shared memory machines in [16] and later generalized in [15]. Its relevance to parallel extensions to C++ like pC++[10] is discussed in [1].

The paper is organized as follows. In the next section we introduce a portable machine and memory model. This is followed by a discussion on thread and rope objects and features for building a parallel object-oriented language. This is followed by performance results and conclusions.

2 Machine and Memory Model

In our model, a parallel machine is a three-level hierarchy of *processor-contexts*, *subdomains* and *domains*. A *processor-context* is a processing resource consisting of one or more physical processors each with a set of registers and a program counter. The processors share an address space in the context over which a thread migrates. A *subdomain* is a set of (one or more) processor-contexts that share an address space. It represents a tightly coupled architecture. Each processor-context has its own local space within this space, but processor-contexts share program globals and, may be, a heap within the same subdomain. A *domain* is a set of subdomains for which collective communication and synchronization operations are supported. It can be thought of as the set of address subspaces, processors and shared control hardware that is available to a user process when it is loaded.

Figure 1: Thread Class

```
class Thread
{
    // constructor to create a transient thread!
    Thread(FunctionType function, ArgType argument, const int immediate = 1);
    // constructor to create persistent thread!
    Thread(const int immediate = 1);
    // Assign a task to a persistent thread
    TaskId Run(Function function, ArgType argument);
    // Wait for the task to be done
    void Wait(TaskId tid);
    // An executing thread may call the following functions
    static void Block(); // Block myself(deschedule)
    static void Yield(); // Yield(not block) so that another thread may run
    static void Exit(void* result); // Exit. kill myself
    static Thread& Self(); // Query. who am I ?
    // Join: wait for a thread to finish executing
    void Join(void** ret_val);
    // Unblock a blocked thread
    void Unblock();
    // Cancel a thread (try to kill it)
    void Cancel();
    ...
};
```

Figure 2: Derived Thread Class for Bitonic Sort

```
class DataThread : public Thread
{
 public:
    DataThread()
      : Thread(0) {
      vector_ptr = new Pvector(BlkSize);
      Start(); // trigger the thread
    }
 private:
    Pvector* vector_ptr;
    void merge(int distance);
    void localSort(int flg);
    void grabFrom(int dist);
    void localMerge(int dist, int flg);
    void localBitonicMerge();
};
```

```
struct Pvector {
  public:
    Pvector(const int size);
    ~Pvector();
    E* part; // BlkSize;
    E* tmp; // BlkSize;
};
void DataThread::merge(int dist)
{
    int flg = 1;
    while (dist > 0) {
      grabFrom(dist);
      DataRope::Synchronize();
      localMerge(dist, flg);
      dist /= 2;
      flg = 0;
      DataRope::Synchronize();
    }
    localBitonicMerge();
}
```

3 Thread Objects

A *thread*[1] is a unit of control for parallel execution. A thread may be created at any point of the program where an independent control can be forked. The forking and the forked threads can run simultaneously subject to availability of processing resources. Thread objects may migrate between processors in a processor-context. Thread objects may be *transient* and *persistent*. Transient threads are created for one task(function) and die at the end of that task. Persistent threads are created to wait around to be assigned tasks. These threads die when the object goes out of scope or is *deleted*. Whether a thread is transient or persistent is defined by the constructor that it invokes (see figure 1). The constructors take an additional boolean parameter which is to delay actual thread triggering to after the derived class thread data is initialized. A persistent thread is assigned tasks through the *Run()* method which returns a task identifier and the calling thread can wait for this task to finish using the *Wait()* method. A transient thread may exit using the *Exit()* method and return a value to a thread waiting(through a *Join()* method) for it to finish. A running thread may yield or block using the *Yield()* or *Block()* methods respectively. A blocked thread may be unblocked invoking the *Unblock()* method on it. A *Cancel()* method on a thread is a is a hint to the thread that it is no longer required. Primitives for accessing shared memory data in a thread-specific way is available. Further, threads may communicate through *synchronous*(sends matched by receives) or asynchronous(remote action) styles of message passing.

---

[1] In this article threads stand for user-level non-preemptible threads.

Figure 3: A Rope Class

```
class Rope
{
    Rope(const int num_threads, Distribution& dist,
          Domain& rope_domain, Domain& curr_domain);
    // Return the size of the rope
    int Size() const;
    // domain over which the rope is defined
    Domain& RopeDomain() const;
    // Parallel execution
    TaskId Execute(FunctionType func, ArgType arg,Domain& curr_domain);
    void Wait(TaskId);
    // return the index of the current thread
    // in the rope called from a parallel function
    static int SelfIndex() const;
    // synchronization between threads in a rope
    static void Synchronize();
    // reduction
    static void Reduce(ReducerObj&);
    // broadcast
    static void Broadcast(BroadcastObj&, const int bcaster);
};
```

Customizing Threads

Threads can be customized to application-specific objects using the C++ inheritance facility to build data- and control- abstraction-rich active objects. Figure 2 gives an example of a customized persistent thread. Notice that activating the thread(through a *Start()* method) is delayed until after derived object initialization is done. The *DataThread* class is used in customizing a rope class for a bitonic merge-sort application to be discussed later.

4 Collections of Threads

A rope object is a group of threads working together in a cooperative manner. See figure 3 for a definition of the rope class. As a thread provides an abstraction for task parallelism, a rope, where each thread is assigned the same task with different data, provides an abstraction for data-parallelism. Threads in a rope are distributed over a domain object using a default or user-specified distribution scheme. They enter a parallel function together and barrier-synchronize at the end of the function. During the computation of the parallel function, the threads in the rope may enter a rope-specific barrier. A rope object also provides abstractions for global reduction and broadcast[14].

The rope method invocation mechanism uses a dynamic SPMD model which combines the simplicity of SPMD programming and efficiency of asynchronous scheduling. The information about the processors in the *caller* context(the set of processors which participate in invoking a data-parallel method) and the *callee* context (the processors whose threads participate in the data-parallel method) is used to minimize communication for exchanging information for invoking and returning values from the parallel method.

Customizing Rope Objects

Rope objects can be customized to data-parallel active objects in a *serial* or *parallel* sense. Serial inheritance is the standard **C++** inheritance where the derived data-rich ropes are created with abstractions added at the level of the rope class. In parallel inheritance, since a thread object can be customized to build data-rich threads, the rope object is built out of customized threads, adding abstractions at the thread-level. For this the rope class is defined as a template parameterized by derived thread class[2]. The simplest instantiation of this template is the template instantiation *RopeTemplate<Thread>*.

5 An Example: Bitonic Sort

Figure 4 shows a bitonic merge-sort program written in our model. Data of size N, is divided among the threads of a rope of size R. Each thread in the rope is an object of a data-rich thread class(see figure 2) and is responsible for $\frac{N}{R}$ elements. The class *DataRope* is a *serial-parallel* inherited *Rope* class, which defines two data-parallel methods, *ParSetKey()*,which initializes the data to random numbers, and *ParSort()*, which sorts it. In the parallel sort, each thread invokes the method *DataRope::Sort()*. In this routine, a thread locally sorts its data and then synchronizes with the other threads in the rope. Then, in a loop of size $\log(R)$, it calls *merge()*.

[2]C++ template syntax does not provide an explicit way to specify that the arguments to a template can be only classes derived from a particular base class. Our system internally ensures that only classes derived from the Thread classes form the components of the rope . Thus a program which tries to define a rope of non-thread objects will not compile.

Figure 4: Bitonic Sort Using Derived Rope

```
class DataRope : public
RopeTemplate<DataThread> {
  public:
    DataRope(const int size)
      :
    RopeTemplate<DataThread>(size) { }
    virtual ~DataRope() { }
    void ParSetKey() {
      TaskId tid =
    Execute(DataThread::SetKey);
      Wait(tid);
    }
    void ParSort() {
      TaskId tid = Execute(Sort);
      Wait(tid);
    }
  private:
    static void Sort(void* x);
};
```

```
main() {
  DataRope X(RopeSize);
  X.ParSetKey();
  X.ParSort();
}
void DataRope::Sort() {
  DataThread& this_thr =
Scheduler::Self();
  DataRope* this_rope =
Scheduler::SelfRope();
  Int size = this_rope→Size();
  this_thr.localSort(1);
  DataRope::Synchronize();
  for (Int i = 1; i < size; i *= 2)
    this_thr.merge(i);
}
```

Every thread taking part in the *merge()* routine, decrements - in a loop - the distance *dist* between the current thread and its exchange partner by factors of 2. The thread acquires the data from the *dist*-neighbor in the *grabFrom()* routine, and then performs a *localMerge()*, which it takes the lower or the upper half - as the case maybe - by merging the local and the grabbed bitonic sequence. Outside this loop is *bitonicMerge()* which merge-sorts the bitonic sequence of the local data. Assume a data size of N, rope size of R (each thread responsible for $\frac{N}{R}$ elements) distributed over a domain with P processors, where $R = kP$, where $k \geq 1$. Assuming a sequential average cost of $N \log N$, it can be shown that speed-up for bitonic sort is

$$S(N, P, k) = \frac{N \log N}{\frac{N}{P}(\log N + (\log kP)^2)} = \frac{P}{1 + \frac{(\log kP)^2}{\log N}}$$

For details of this derivation and study of the effect of thread distribution on communication is studied in [15].

Performance

We ran our tests on a 20-processor SGI-challenge symmetric multiprocessor machine and on the Intel Paragon distributed memory machine. As per our machine model, the SGI machine is a domain with 1 subdomain. The processors in the machine can be grouped so as to give different combinations of processor-context and subdomain sizes. A partition of size P on the Paragon is a domain with P subdomains, each with a processor-context of size 1. In our study, we concentrated on issues like scalability, overlapping independent rope operations and interference of threads (within or across ropes) on a processor.

Graph 5 shows that on both the machines the program scales well for increasing data sizes. The performance in better on the SGI due to better processor-speed and

Figure 5: **Scalability study for increasing data-sizes** 16-processor partition(16 processor-context on the SGI, 16 subdomains on the Paragon) rope size 16

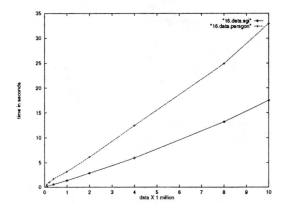

Figure 6: **Time taken for varying number of processors** data size = 1 million, rope size=# of processors

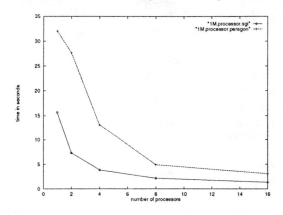

Figure 7: **Overheads of having multiple threads per processor:** For data size 1 million, #processors = 8, varying rope size

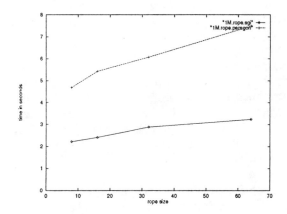

Figure 8: On the SGI multiple sorts scheduled simultaneously; **loadbalance.sgi1** (rope-size=8) corresponds to 8 processor-contexts with 1 processor each; **loadbalance.sgi2** (rope-size=8) and **loadbalance.sgi3** (rope-size=4) are for 4 processor-contexts with 2 processors each.

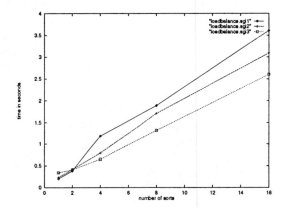

cheaper communication. Graph 6 shows that the speedup is linear for increasing number of processors for fixed data size and rope size.

Computation and communication in the bitonic sort algorithm is quite regular. The threads synchronize after every step of data exchange and every step of local merge computation. If all the processors are loaded about equally, having more than one thread per processor would mean just introducing the overhead of the context-switch and of synchronization between these threads. This effect can be seen in graph 7.

To study load balancing we divided up an an 8-processor partition on the SGI challenge in different ways and scheduled multiple sorts on different ropes at the same time. Having multiple processors per processor-context gives the threads more freedom for migration; at the same time, thread scheduling is more expensive because heavier-weight locks are required as multiple threads might be active in the context at the same time. Graph 8 shows 3 cases of partitioning. It can be seen that when the number of sorts is 1 or 2, case 1 (8 processor-contexts/1 processor each) shows better performance than the other two cases (4 processor-contexts/2 processors each). For fewer sorts, the processor-contexts with multiple processors are starved for work, but at the same time pay a penalty of higher context-switch times. As the number of sorts increases, the total number of threads increase too, thus processor-starvation is reduced in cases 2 and 3 and better load balancing is achieved. Since threads in the same rope are constrained by synchronizations, having fewer threads per rope and more independent ropes(more sorts) gives better performance. Hence case 3(1 thread per processor-context) posts better results than case 2(2 threads per processor-context).

6 Conclusions

In this paper, we proposed an object-oriented thread-model for parallelism for shared-memory and distributed-memory machines. We addressed both task and data parallelism issues through thread objects and aggregate thread objects called ropes. We showed, in the context of OO-paradigm, how these objects can be customized in an application-specific way. Our performance study shows that context-switching in the thread environment does not adversely affect the performance of regular data-parallel algorithms and can be used to advantage to schedule multiple data-parallel tasks.

We are working on supporting global active objects for a shared-memory programming paradigm, in addition to the message-passing paradigm for communication between threads. Also, using this model and library we plan to support nested parallelism and adaptive computation in **pC++**[1].

References

[1] Peter Beckman, Dennis Gannon, and Neelakantan Sundaresan. pC++ Meets Multi-Threaded Computation. In Jack Dongarra and Bernard Tourancheau, editors, *Proceedings of the second workshop on Environments and Tools for Parallel Scientific Computing*, Philadelphia, May 1994. SIAM.

[2] Brian Bershad, Edward Lazowska, and Henry Levy. Presto: A system for object-oriented parallel programming. *Software-Practice and Experience*, 18(8):713–732, August 1988.

[3] Mani Chandy and Carl Kesselman. *Compositional C++: A Declarative Concurrent Object-Oriented Programming Notation*. MIT Press, 1993.

[4] Eric C. Cooper and Richard P. Draves. *C Threads*. Technical Report CMU-CS-88-154, Carnegie Mellon University, June 1988.

[5] Edward Felten and Dylan McNamee. *NewThreads2.0 User's Guide*, August 1992. Williamsburg VA.

[6] Andrew Grimshaw. Easy-to-use Object-Oriented Parallel Processing with Mentat. *IEEE Computer*, May 1993.

[7] Dirk Grunwald. *A Users Guide to AWESIME: An Object-Oriented Parallel Programming and Simulation System*. Technical Report CU-CS-552-91, University of Colorado, Boulder, Colorado, November 1991.

[8] IEEE. *Thread Extensions for Portable Operating Systems (Draft 6)*, February 1992. P1003.4a/6.

[9] Laxmikant Kale and Sanjeev Krishnan. *CHARM++: A Portable Concurrent Object-Oriented System Based on C++*. Technical report, University of Illinois, Urbana-Champaign, March 1993.

[10] Allen Malony, Bernd Mohr, Peter Beckman, Dennis Gannon, Shelby Yang, François Bodin, and S Kesavan. Implementing a Parallel C++ Runtime System for Scalable Parallel Systems. *Proceedings, Supercomputing '93*, pages 588–597, November 1993.

[11] Frank Mueller. *Pthreads Library Interface*. Technical report, Florida State University, July 1993.

[12] Bodhisattwa Mukherjee, Greg Eisenhauer, and Kaushik Ghosh. *A Machine Independent Interface for LightWeight Threads*. Technical Report GIT-CC-93/53, College of Computing, Georgia Institute of Technology, June 1993.

[13] Stephen Murer, Jerome Feldman, and Chu-Cheow Lim. *pSather monitors: Design, Tutorial, Rationale and Implementation*. Technical Report TR-93-028, International Computer Science Institute, Berkeley, CA., June 1993.

[14] Neelakantan Sundaresan and Dennis Gannon. *Aggregate Thread Synchronization Operations on Shared and Distributed Memory Machines*. Technical Report 430, Indiana University, Computer Science Department, Bloomington, IN, 1995.

[15] Neelakantan Sundaresan and Dennis Gannon. *A Thread Model for Supporting Task and Data Parallelism in Object-Oriented Parallel Languages*. Technical Report 429, Indiana University, Computer Science Department, Bloomington, IN, 1995.

[16] Neelakantan Sundaresan and Linda Lee. An Object-Oriented Thread Model for Parallel Numerical Applications. In *Proceedings of the second annual Object-Oriented Numerics Conference*, April 1994. Sunriver, Oregon.

Static Message Combining in Task Graph Schedules

Stephen Shafer
Loral Federal Systems
1801 State Route 17c
Owego, NY 13827-3998
shafer@lfs.loral.com

Kanad Ghose
Department of Computer Science
State University of New York
Binghamton, NY 13902-6000
ghose@cs.binghamton.edu

Abstract: Inter-processor communication in a distributed memory multiprocessor can have a profound impact on the execution time of the program. We present and evaluate a technique that can reduce the overall execution time of a parallel program by combining two or more messages between a pair of processors into a single message. We first identify conditions under which such message combining is profitable and show how these conditions are used to guide the process of message combining. Our approach is static in nature and requires as input a schedule of the program on the processors. Finally, we evaluate the execution time savings realized using our technique on schedules of programs expressed as synthetic task graphs as well as task graphs corresponding to real applications.

Keywords: Distributed Memory Multiprocessors, Message Combining, Scheduling.

1. Introduction

Inter process communication time can significantly prolong the execution time of programs on distributed memory multiprocessors. A significant amount of work has thus been directed towards a reduction of the communication overhead of parallel applications for a given architecture by scheduling programs onto the processing nodes, taking into account the existence of inter-task communication. The schedulers presented in [ChAg 93], [ElLe 90], [GeYa 92], [GhMe 94], [KoSa 93], [ScJa 93] and [WuGa 90] represent examples of such an approach. Many of these schedulers, such as [ChAg 93], [ElLe 90], [GhMe 94], [KoSa 93] and [ScJa 93] address channel contention and attempt to minimize the additional delays due to such contention.

In [WaMe 91], Wang and Mehrotra proposed a technique for merging messages (that correspond to data dependencies) from the same processor to a common destination processor. Their approach is to reduce the communication between one pair of tasks at a time by combining a pair of messages from a source processor to the same destination processor into a single message that still satisfies all data dependencies involved. This message combining is done without an assessment of its impact on the overall execution time of the program. In [ShGh 94], we established that Wang and Mehrotra's approach does not necessarily lead to a reduction in the overall execution time. In fact, we showed that using their technique could lead to an increase in execution time and could also introduce deadlocking. A detailed critique of Wang and Mehrortra's scheme is found in [ShGh 94].

In [ShGh 94], we identified the reasons why message combining in general does not necessarily lead to a reduction of the overall execution time of a parallel program. Message combining is profitable only if it results in a reduction of the overall execution time of the program. The goal of this paper is to explain briefly our method of merging messages using the conditions introduced in [ShGh 94] for profitable message combining and to present an experimental evaluation of the resulting scheme for message combining.

2. The Message Combining Technique

This section describes our approach for message combining, based on the conditions for profitable message combining as introduced in [ShGh 94].

Figure 1 illustrates some of the notations and symbols used in this paper. It shows a sample (fictitious) program scheduled onto a 6-PE multiprocessor along with the messages needed to preserve the data dependences in the program.

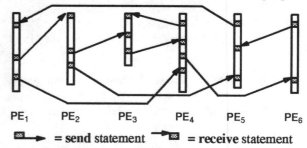

PE₁   PE₂   PE₃   PE₄   PE₅   PE₆

■→ = send statement   →■ = receive statement

Figure 1. An example program scheduled onto a 6-PE processor graph. The arrows indicate communications (messages) between tasks (code running on a PE). Each vertical bar represents processing time from top to bottom.

Only the source and sink of each message are shown. In some cases, the messages indicated need to be sent through an intermediate PE to reach the final destination PE.

*(Note that the notion of a task in a task graph is quite different from the way in which we have used the term "task" to describe the entire set of code running on a single PE. In the context of task graphs, a task, traditionally, refers to a sequence of computations that immediately follows one or more **receives** and which terminates with one or more **sends**.)*

2.1 The concept of merging

The merging process takes two messages in a task graph schedule, deletes one of them, and sends the data for both in the other. Thus after the merge, the data for both dependences are being sent in one message. Such a message, one that contains the data from more than one dependence, is called a *data dependence cluster* (DDC). By including messages that contain data from only *one* message in the definition, any message in the system can be considered a DDC. From this point on, we will use DDC to indicate any message in the system regardless of how many data dependences it is satisfying.

For two DDCs to be merged into one, they both must originate in one PE and terminate in another. When searching for candidate pairs of DDCs for merging, two major types of DDC pair emerge. Figure 2 shows an example of a non-

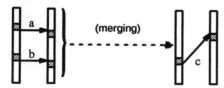

Figure 2. (a) non–crossed DDC pair; (b) after merging.

crossed pair, followed by the DDC that would remain after the pair was merged. Notice that the first DDC (a) was deleted, and the **receive** statement of the second DDC (b) is now where the **receive** statement of DDC a used to be. In this way, all data dependences are preserved. This can be viewed as deleting both of the original DDCs, and then inserting a third DDC (c) that contains both sets of data. Since a new communication is inserted into the system, it is possible that a cycle in the graph (i.e. a deadlock) has been introduced. To prevent that, non–crossed DDC pairs must meet the definition of mergeability. That is, there must be no path in the graph originating from between the pair's two **receive** statements and terminating between the pair's two **send** statements. If DDCs a and b were being considered

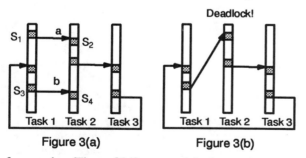

Figure 3(a)   Figure 3(b)

for merging (Figure 3(a)), any path in the graph that originates between S_2 and S_4 and terminating between S_1 and S_3 would create a deadlock if a and b are merged. Since such a path does exist , then merging the pair of messages would create a deadlock (Figure 3(b)). On the other hand, as long as no such path exists, no deadlock can be created by merg-

ing. As shown in [ShGh 94], Wang and Mehrotra's merging technique can introduce such deadlocks.

For a crossed DDC pair, as depicted in Figure 4(a), the situ-

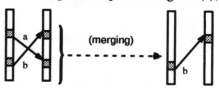

Figure 4. (a) crossed DDC pair; (b) after merging.

ation is slightly different. The first DDC (a) will be deleted, as in the previous scheme, but in this case the **receive** statement of the second DDC (b) actually appears before the **receive** of the first DDC in the code. This means that the second DDC doesn't need to be changed at all. No test for mergeability is required since no new communication is being inserted into the system. Thus, the merge of a crossed pair involves deleting the first DDC and sending its data in the second. No possibility of creating a deadlock exists since no new message is added. As in the previous case, all data dependences are preserved.

2.2 The notion of regions

DDCs entering a task define the *regions* in that task. Each **receive** statement defines the start of another region. The number of regions in a task is defined as one more than the number of entering communications. If the task starts with an incoming DDC, however, then the number of regions equals the number of entering communications (Figure 5).

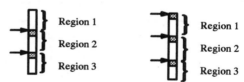

Figure 5. Task regions.

There are two basic types of regions in a task. If task execution arrives at a **receive** statement before the data it is expecting, the task waits. The region defined by this DDC is called a D–region, or DDC region, because the execution time through the DDC determines when execution enters the region. If task execution takes longer to arrive at the **receive** statement, the region below that statement is a T–region, or task region. The first region in every task is a D–region by definition.

2.3 Regional and DDC data

Two data values are associated with every region. $T_{strt}(x)$, or the start time for region x, indicates the time at which control arrives at the first statement in the region after the **receive** statement. $T_{reg}(x)$, or the regional time for region x, indicates the time it takes for execution to go from the beginning of the region to the end of the last statement in the region.

Associated with every DDC y is a value, $t_{arr}(y)$, that indicates the time it takes for the message to arrive at the sink task. Also associated with every DDC y is the time differential, or $t_{diff}(y)$. This value indicates the relationship between the data arrival time (t_{arr}) and the time that control reaches the **receive** statement in the task.

Finally, the merging of every DDC pair has an effect on the sink task at the point where the lone **receive** statement is after the merge. The value that indicates this effect is the change in arrival time of the message, or the delta t_{arr} value for the appropriate DDC. For a possible merge of two DDCs **a** and **b**, this value is indicated by $\Delta t_{arr}(a,b)$.

For detailed examples and equations for the regional and DDC related data, the reader is referred to [ShGh 94].

2.4 Task Sections and Merging Effects

Every pair of DDCs that enter a task partition the task into three sections. Section 1 is that part of the task before the **receive** statement of the first DDC, section 2 is between the two **receive** statements, and section 3 is after the **receive** statement of the second DDC. By examining how the start time of each of these sections is affected by a possible merge, it can be determined whether or not the merge can be performed without affecting the overall execution time of the program. The following tables define the effects of a merge on the start times of sections 2 and 3 (the start time of section 1 is not affected).

Table 1 shows the effect of a possible merge on section 2 of the sink task of the two non-crossed DDCs (**a** and **b**) being considered for merging. Thus, if $\Delta t_{arr}(a,b)$ is less than or

	Change in Section 2	Causal Condition
Case 1:	No change	$\Delta t_{arr}(a,b) \leq -t_{diff}(a)$
Case 2:	Increase	$\Delta t_{arr}(a,b) > -t_{diff}(a)$
Case 3:	Decrease	not possible

Table 1. Merge effects of non-crossed DDCs seen in section two of the sink task.

equal to the negative of $t_{diff}(a)$, then the merge will not change the start time of section 2. Table 2 shows the effect of a possible merge on section 3 of the sink task of the two DDCs (**a** and **b**) being considered for merging, given a specific effect on section 2. For example, Case 2b assumes that the start time of section 2 has already increased (see Case 2, Table 1), but the start time of section 3 will actually decrease if the indicated condition holds. Table 3 contains the different cases that are possible when a crossed DDC pair is being considered for a merge. Note that for a crossed pair merge, the effects seen by sections 2 and 3 of the task are the same. That is, if the start time of section 2 increases, so will the start time of section 3.

Tables 1, 2, and 3 show the effects of a possible merge only on the sink task of the DDC pair involved. As long as both sections 2 and 3 either have no change in start time, or a decrease in start time, then the merge will not increase the

	Section 2	Section 3	Causal Condition
Case 1a:	No change	No change	$t_{diff}(b) \leq 0$
Case 1b:	No change	Decrease	$t_{diff}(b) > 0$
Case 1c:	No change	Increase	not possible
Case 2a:	Increase	No change	$\Delta t_{arr}(a,b) + \tau = t_{diff}(b)$
Case 2b:	Increase	Decrease	$\Delta t_{arr}(a,b) + \tau < t_{diff}(b)$
Case 2c:	Increase	Increase	$\Delta t_{arr}(a,b) + \tau > t_{diff}(b)$

where $\tau = \min[0, t_{diff}(a)]$

Table 2. Merge effects of non-crossed DDCs seen in section three of the sink task given a specific change in section two.

	Sections 2 and 3	Causal Condition
Case 4a:	No change	$trans(M) = F$
Case 4b:	No change	$trans(M) < F$ and $t_{diff}(b) \leq 0$
Case 4c:	No change	$trans(M) > F$ and $trans(M) - F \leq -t_{diff}(b)$
Case 5:	Decrease	$trans(M) < F$ and $t_{diff}(b) > 0$
Case 6:	Increase	$trans(M) > F$ and $trans(M) - F > -t_{diff}(b)$

Table 3. Merge effects of crossed DDCs.

overall execution time of the system. This does not mean, however, that other cases will never produce a profitable merge, just that more information than that found in Tables 1, 2, and 3 is needed. For these cases, the merge needs to be performed and the total system execution time compared against the pre-merge time. Only when the new time is less will the merge be allowed.

3. Quantitative Assessments

To assess the benefits of our message merging scheme in a quantitative fashion, we implemented the merging technique to run as a post-scheduling tool for combining inter-PE messages in the schedule produced. We generated the schedule using the bottom-up scheduler described in [GhMe 94], [MeGh 94a], [MeGh 94b], using programs represented as task graphs as the input to the scheduler. This scheduler allocates tasks to PEs, taking into account actual communication costs and channel contention, and using heuristics for load balancing. The net result is the production of a highly optimized schedule [MeGh 94b], [Me 94]. *Any improvement in the execution time of the schedule produced by the bottom up scheduler through the use of our message merging technique will thus be quite useful.* The disadvantage of using a schedule produced by the scheduler as an input to our message merging tool is that the merging process had to assume that nothing it did would produce any channel conflicts, which is not necessarily true. An alternative approach will be to use our merging tool to merge messages based on the scheduler's output and then reallocate the channels, using the modified set of messages, to minimize channel contention.

It is expected that our message merging technique will be more beneficial for programs that are manually assigned to task nodes, without consideration of the channel contention or accurate estimates of the communication costs. Because of the subjective nature of such programs, the assessment of our merging technique was restricted to automatically produced and highly optimized schedules, such as the ones produced by the bottom up scheduler.

We ran our merging tool on two different types of schedules, first randomly generated (i.e., synthetic) task graphs and then task graphs corresponding to real applications.

Task Graph	# of PEs	# of Tasks	# of Inter-PE DDCs
Rand1	4	8	6
Rand2	4	8	12
Rand3	8	50	55
Rand4	8	50	57
Rand5	8	100	145
Rand6	8	100	168
Rand7	16	100	151
Rand8	16	100	171

Table 4. Summary – randomly generated test graphs

	Rand1	Rand2	Rand3	Rand4
Original Execution Time	39	419	538	420
Final Execution Time	39	418	537	413
Percent Decrease	0.0	0.2	0.2	1.7
Original # of DDCs	6	12	57	55
Final # of DDCs	6	11	56	43
Percent Decrease	0.0	8.3	1.8	21.8

	Rand5	Rand6	Rand7	Rand8
Original Execution Time	280	244	194	221
Final Execution Time	264	234	190	209
Percent Decrease	5.7	4.1	2.1	5.4
Original # of DDCs	168	145	151	171
Final # of DDCs	150	125	142	157
Percent Decrease	10.7	13.8	6.0	8.2

Table 5. The impact of message combining on the schedules for the random task graphs

3.1 Randomly generated task graphs

The merging tool was run on the schedule produced by the bottom up scheduler for 8 randomly-generated task graphs with varying characteristics, such as minimum and maximum computation times, minimum and maximum communication times, number of start nodes, and number of tasks. (Note again that a *task* in the context of a task graph refers to an indivisible unit of computation that commences with zero or more receives and terminates with one or more sends with no intervening communication primitives, whereas this paper has been referring to a task as that code

running on a single PE.) Table 4 summarizes the characteristics of these task graphs and also describes the number of PEs for which the graphs were scheduled (all of the systems considered were assumed to have a hypercube interconnection). For details on the generation of the random graphs see [Me 94].

These graphs listed in Table 4 were tested for two aspects of the merging process – a decrease in total execution time, and a decrease in the number of messages. A message-switched communication link with flat overhead of 1 time unit was used for these tests. Table 5 shows these results of message merging for the schedules of the randomly generated task graphs that were listed in Table 4.

The results of Table 5 show that, even with a highly optimized task graph schedule, improvements are possible by merging messages. As expected, the better results are seen with those graphs that have more DDCs to start with, but that is not the only factor. The number of initial DDCs also affects the effectiveness of this technique. For those graphs with less than 100 DDCs (Rand 1, 2, 3, and 4) the percent decrease in the execution time was minimal. For graphs with more than 100 DDCs, however, the results are slightly better, with an average of 3.8% decrease in execution time.

3.2 Task graphs for real applications

The merging process was also performed on 5 task graph schedules produced from real parallel program graphs. The task graphs of real parallel applications that were chosen are:

D&C: a divide-and-conquer application, and
SP: sum of products, both in [MKTM 94].
Atm: an atmospheric science application [LeEl 92].
Gauss: Gaussian elimination [WuGa 90].
Lap: the Gauss Seidel technique for solving the Laplace equations [WuGa 90].
For further details on these task graphs see [MeGh 94a], [Me 94].

As with the random graphs, the real graphs were scheduled by the bottom up scheduler and the resulting schedule was given to the merging tool for message combining. As before, the flat overhead of a **send** statement was assumed to be 1 time unit. The characteristics of the task graphs for the real applications are summarized in Table 6, which also shows the number of hypercube connected PEs for which the corresponding schedules were produced.

Task Graph	# of PEs	# of Tasks	# of Inter-PE DDCs
D&C	4	128	6
Atm	8	130	1045
Gauss	8	117	150
Lap	8	100	140
SP	8	128	93

Table 6. Some features of the real application graphs and their schedule

Table 7 shows the results of message combining for the schedules of the real task graphs. The results from the real graphs were much better in some cases than those from the random graphs. The percent decrease in execution time due to merging even reached 50.0% in one case, and the percent decrease in the number of DDCs was over 50.0% in two cases. At first glance, this may seem due to the large number of initial DDCs in some of these examples. The largest decreases, however, are not exclusively seen in those examples. The graph SP only had 93 initial DDCs, and yet its execution time was decreased by 50%.

	D&C	Atm	Gauss	Lap	SP
Original Execution Time	126	319	3442	2908	94
Final Execution Time	126	199	3400	2908	47
Percent Decrease	0.0	37.6	1.2	0.0	50.0
Original # of DDCs	6	1045	150	140	93
Final # of DDCs	6	352	107	138	43
Percent Decrease	0.0	66.3	28.7	1.4	53.8

Table 7. The impact of message combining on the schedules for the random task graphs

The number of each type of DDC pair checked for merging was collected for both the random and the real graphs. These results are very similar in both. By far, the largest number of DDC pairs checked (93%) were of type 2c, where the start times of both sections 2 and 3 in the sink task increased. This is precisely the reason why merging schemes that do not use DDC pair type information when choosing messages to merge are very likely to increase the execution time instead of decreasing it. For example, out of 9,893 attempts to merge type 2c DDC pairs, only 113 produced useable merges.

4. Conclusions

This paper presented an overview of a technique for combining messages from a PE to another with the explicit goal of reducing the execution time of the overall program. We also presented a quantitative assessment of the reduction in execution time brought about by our message combining technique, using as input schedules generated for real and synthetic task graphs. These schedules were already optimized by the scheduler of [MeGh 94b] that takes into account the exact inter-PE communication costs and channel contention, producing very optimized schedules. Any improvement obtained through message combining in these optimized schedules is thus quite useful. The performance improvements obtained from the use of our message combining technique range from a few percentage to more than 50%. The tests on the real graphs had better results than those on the random graphs, perhaps because of the larger number of DDCs present in the real graphs. Even with a smaller number of DDCs to start with, however, excellent results are still possible as can be seen in the case of the graph SP. This graph started with only 93 DDCs, and ended up with 43 producing a 50% decrease in the execution time. The reduction in execution time seems to depend on the structure of each graph itself, and not only on the number of

DDCs in the original graph. Further work will be focused on dealing with channel contention issues while merging.

References

[ChAg 93] Chaudhary, V., Aggarwal, J.K., "A Generalized Scheme for Mapping Parallel Algorithms", in IEEE Transactions on Parallel and Distributed Systems, March 1993, pp. 328-346

[ElLe 90] El-Rewini, H., Lewis, T.G., "Scheduling Parallel Program Tasks onto Arbitrary Target Machines", Journal of Parallel and Distributed Computing, 9, 1990, pp. 138-153.

[GeYa 92] Gerasoulis, A., Yang, T., "A Comparison of Clustering Heuristics for Scheduling Directed Acyclic Graphs on Multiprocessors", Journal of Parallel and Distributed Computing, 16, 1992, pp. 276-291.

[GhMe 94] Ghose, K. and Mehdiratta, N., "An Universal Approach for Task Scheduling in Distributed Memory Multiprocessors", in Proc. of the IEEE Sym. on Scalable and High Performance Computing (SHPCC '94), May 1994, pp. 577-584.

[KoSa 93] Kon'ya, S. and Satoh, T., "Task Scheduling on a Hypercube with Link Contentions," Proc. International Parallel Processing Symposium, pp. 363-368,1993.

[LeEl 92] Lewis, T.G. and El-Rewini, H., Introduction to Parallel Computing, Prentice Hall Inc. 1992.

[MTKM 94] McCreary, C.L., Khan, A.A., Thompson, J.J. and McArdle, M.E., "A Comparison of Heuristics for Scheduling DAGs on Multiprocessors", in Proc. Int'l. Sym. on Parallel Proc., pp. 446-451, 1994.

[Me 94] Mehdiratta, N., A General Strategy for Scheduling Parallel Programs on Distributed Memory Multiprocessors, Ph.D. dissertation. Depat. of Computer Science, SUNY, Binghamton, 1994.

[MeGh 94a] Mehdiratta, N. and Ghose, K., "Scheduling Task Graphs onto Distributed Memory Multiprocessors Under Realistic Constraints", in Proc. Parallel Architectures and Languages Europe (PARLE), July 1994, pp. 589-600.

[MeGh 94b] Mehdiratta, N. and Ghose, K., "A Bottom Up Task Approach to Task Scheduling on Distributed Memory Multiprocessors", in Proc. Int'l. Conf. on Parallel Proc., August 1994, Vol.-II, pp. 151-154.

[ShGh 94] Shafer, S., Ghose, K., "Improving Parallel Program Execution Time with Message Consolidation", Proc. 8th International Parallel Processing Symposium, April 1994, pp. 736-742.

[ScJa 93] Schwiebert, L. and Jayasimha, D.N., "Mapping to Reduce Contention in Multiprocessor Architectures", Proc. Intl. Parallel Processing Symposium pp. 248-253, 1993.

[WaMe 91][WaMe 91] Wang, K.Y., Mehrotra, P., "Optimizing Data Synchronizations On Distributed Memory Architectures, Proc. 1991 Intl. Conf. on Parallel Processing, Vol. II, pp. 76-82.

[WuGa 90] Wu, M-Y., and Gajski, D.D., "Hypertool: A Programming Aid for Message Passing Systems," IEEE Trans. Parallel and Distrib. Systems, vol. 1, no. 3, July 1990.

[YBN 91] J. Yang, L. Bic and A Nicolau, "A Mapping strategy for MIMD Computers," International Parallel Processing Conference, Vol. I, pp. 102-109, 1991.

Debugging Distributed Executions Using Language Recognition*

Özalp Babaoğlu
Dept. of Computer Science – University of Bologna
Piazza Porta S. Donato 5, 40127 Bologna (Italy)
E-mail: ozalp@cs.unibo.it

Eddy Fromentin, Michel Raynal
IRISA – Campus de Beaulieu
35042 Rennes Cedex (France)
E-mail: {fromentin,raynal}@irisa.fr

Abstract – *To a large extent, the dependability of complex distributed programs relies on our ability to effectively test and debug their executions. Such an activity requires that we be able to specify dynamic properties that the distributed computation must (or must not) exhibit, and that we be able to construct algorithms to detect these properties at run time. In this paper we formulate dynamic property specification and detection as instances of the language recognition problem. Considering boolean predicates on states of the computation as an alphabet, dynamic property specification is akin to defining a language over this alphabet. Detecting a property, on the other hand, is akin to recognizing at run time if the sentence produced by a distributed execution belongs to the language. This formal language-oriented view not only unifies a large body of work on distributed debugging and property detection, it also leads to simple and efficient detection algorithms. We give examples for the case of properties that can be specified as regular grammars through finite automata.*

1 Introduction

Our inability to formally prove the correctness of all but the most trivial distributed programs leaves testing and debugging as the only viable alternatives for arguing about their dependability. A fundamental step in program debugging is specifying which set of executions are considered desirable and which ones are considered erroneous. Informally, a desired (or undesired) temporal evolution of a distributed program's states is called a *dynamic property*. As such, a dynamic property (*property* for short) defines a subset of executions among all those that are possible for the program. Once the properties of interest for a distributed program have been defined, the act of debugging consists of verifying if its executions satisfy these properties.

There are three possible times at which one can prove properties of a distributed program: prior to any execution, during an execution or after an execution. The ability to prove a program property prior to executing it requires reasoning about the program itself as well as the distributed system. In other words, we need to characterize all possible executions of the program given a formal description of its actions (the code) and the environment in which it

is to be run. Model checking [5] is one such technique where the program, modeled as a finite-state transition system, is analyzed by traversing the trees representing all possible executions, checking to see if they satisfy properties expressed as temporal logic formulas. As an alternative to proving properties a priori, they can be checked concurrently with an actual execution of the program through *run-time property detection*. With this technique, conclusions that are drawn are not about all possible executions of the program, but about all possible *observations* [18, 1] of an actual execution. The third alternative — reasoning about program properties after an execution — is called *post-mortem* analysis and is similar to run-time property detection. The basic difference is that the analysis has to be based on data collected during the execution (traces) and is performed at program termination, thus making it unsuitable for reactive-architecture applications.

Ideally, one would like to prove as many properties as possible for a program prior to its execution. Unfortunately, techniques such as model checking may be infeasible or have prohibitive costs for complex programs. Furthermore, while these techniques are effective for proving program properties such as "is the program deadlock free?" they cannot address inherently run-time properties such as "has the program terminated?". Much of distributed debugging is an inherently run-time activity in that, short of having proven that the program is error free, we can only hope to detect desired or erroneous sequences of states that result during actual executions as soon as feasible. Applications that have a *reactive architecture* [12] (of which distributed debugging is an instance) are yet another example where the fundamental abstraction is run-time property detection. Clearly, run-time property detection and model checking can be seen as complementary techniques towards distributed debugging — the greater the number of properties that can be verified a priori, the fewer the number of properties that need to be detected at run time.

Early work in run-time property detection has concentrated on the detection of *stable properties* [4] including distributed termination [8] and deadlocks [3]. Informally, these properties are stable in the sense that once verified during an execution, they remain true thereafter. Efficient algorithms have been developed that can detect stable properties of distributed computations at run-time [4, 13]. Unfortunately, stable property detection has limited utility in the con-

*This work has been supported in part by the Commission of European Communities under ESPRIT Programme Basic Research Project 6360 (BROADCAST).

text of distributed debugging. Most of the properties that characterize desirable or erroneous executions for debugging purposes are transient whereby they may be verified for some prefix of the execution but cease to be satisfied later on. As such, the appropriate formalism for distributed debugging can be seen as *unstable property* detection. If erroneous behaviors are specified as unstable properties, then their detection during an execution reveals a fault in the corresponding program. Existence of many executions during which the unstable properties are not detected increases our confidence in the correctness of the corresponding program.

In this paper we consider the problem of specifying unstable properties and detecting them at run time. We formalize the problem as the design of a decision algorithm that, when superimposed on a distributed execution, will answer "yes" if and only if the distributed execution satisfies a property specified as a formula in some formal language. We develop a general framework for property specification and detection drawing on concepts from formal language theory. The framework, described in Section 3, is based on the labeling of a directed acyclic graph (DAG). Each node of the DAG is associated a set of labels from a given alphabet and each path of the DAG is associated a set of words constructed from the same alphabet. The set of all words corresponding to all paths terminating at a node defines a language. The language recognition problem is then defined in Section 3.3 as the decision procedure for determining if a given word belongs to this language. We instantiate this abstract problem for two possible models of distributed computations as a DAG — in Section 4 as the partially ordered set of local states and in Section 5 as the lattice of consistent global states. When the alphabet used for labeling these DAGs consists of boolean predicates, the formal language associated with the nodes of the graph effectively specifies dynamic properties of the corresponding distributed computation. In addition, run-time property detection reduces to the problem of language recognition. In Sections 4.3 and 5.3 we give examples of simple run-time detection algorithms that result for the case of properties that can be specified as regular grammars through finite automata. This framework unifies a large body of work on distributed debugging and property detection including: linked predicates [17], conjunction of local predicates [11], unstable predicates on global states [6], interval-constrained sequences [2], regular patterns [10] and regular properties [15].

2 Models for Distributed Computations

A distributed program is one that is executed by a collection of sequential processes, denoted $P_1, \ldots, P_n$ for some $n > 1$, that can communicate by exchanging messages. Processes have access to neither shared memory nor a global clock. Communication incurs finite but arbitrary delays. Without loss of generality, we assume that each process can reliably communicate with every other process.

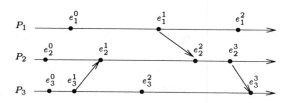

Figure 1: Distributed computation as a partially-ordered set $\mathcal{H}$ of events.

2.1 Partially-Ordered Sets

Execution of process P_i produces a sequence of events h_i called its *local history*. Each event of the local history may be either internal, causing only a local state change, or involve communication with another process through *send* or *receive* events. Let $h_i = e_i^0 e_i^1 \ldots$ be the local history of process P_i. Events of h_i are enumerated according to the total order in which they are executed by P_i and e_i^0 is a fictitious event introduced for initializing the local state of P_i.

Let H be the set of all events and let $\rightarrow$ be the binary relation denoting causal precedence [16] between events defined as follows:

$$e_i^a \rightarrow e_j^b \equiv \begin{cases} (i = j) \wedge (b = a + 1) \\ \text{or } (e_i^a = send(m)) \wedge (e_j^b = receive(m)) \\ \text{or } \exists e_k^c : (e_i^a \rightarrow e_k^c) \wedge (e_k^c \rightarrow e_j^b) \end{cases}$$

Formally, a distributed computation can then be modeled as the partially ordered set (poset) $\mathcal{H} = (H, \rightarrow)$. Figure 1 illustrates a distributed computation consisting of three processes using a graphical representation of the partial order between events known as a space-time diagram.

2.2 Directed Acyclic Graphs of Local States

Let σ_i^a be the local state of process P_i immediately after having executed event e_i^a, and let S be the set of all local states. Analogous to the causal precedence relation between pairs of events, we define the binary relation $\prec$ to denote *immediate causal precedence* between local states as follows:

$$\sigma_i^a \prec \sigma_j^b \equiv \begin{cases} (i = j) \wedge (b = a + 1) \\ \text{or } (e_i^{a+1} = send(m)) \wedge (e_j^b = receive(m)) \end{cases}$$

With respect to the set of local states, a distributed computation can be modeled as yet another DAG, denoted $\mathcal{S} = (S, \prec)$. Figure 2 depicts such a DAG of local states corresponding to the distributed computation of Figure 1.

Two local states of a distributed computation that are related through $\prec$ are said to be *adjacent* in that computation. A *control flow* associated with some local state σ of a computation $\mathcal{S}$ is a sequence of local states that are pairwise adjacent in $\mathcal{S}$ and the sequence begins at some initial state and terminates at σ.

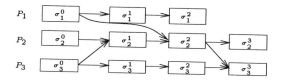

Figure 2: Distributed computation as a DAG $\mathcal{S}$ of local states.

2.3 Lattices of Global States

A global state $\Sigma = (\sigma_1, \ldots, \sigma_n)$ of a distributed computation is an n-tuple of local states, one for each process. Intuitively, a global state of a computation is said to be consistent if an omniscient external observer could actually observe the computation enter that state. More formally, a global state $\Sigma = (\sigma_1, \ldots, \sigma_n)$ is *consistent* if and only if for all pairs of its local states (σ_i, σ_j), neither $\sigma_i \stackrel{+}{\prec} \sigma_j$ nor $\sigma_j \stackrel{+}{\prec} \sigma_i$, where $\stackrel{+}{\prec}$ denotes the transitive closure of the immediate causal precedence relation between local states.

The set of all consistent global states for a distributed computation has a lattice structure whose minimal element corresponds to the initial global state $\Sigma^0 = (\sigma_1^0, \ldots, \sigma_n^0)$. Let $\mathcal{L}$ be this lattice. An edge exists from node $\Sigma = (\sigma_1, \ldots, \sigma_i^a, \ldots, \sigma_n)$ to node $\Sigma' = (\sigma_1, \ldots, \sigma_i^{a+1}, \ldots, \sigma_n)$ if and only if there exists an event e that can be executed by P_i in local state σ_i^a. Figure 3 depicts the lattice $\mathcal{L}$ of global states associated with the distributed computation of Figure 1. In the lattice, an n-tuple $(x_1, \ldots, x_n)$ of natural numbers is used as a shorthand to denote the global state $\Sigma = (\sigma_1^{x_1}, \ldots, \sigma_n^{x_n})$.

Informally, a *sequential observation* (*observation* for short) of a distributed computation is the sequence of its global states that could have been constructed by an omniscient external observer. Equivalently, an observation is a sequence of global states that would result if the distributed program were to be executed on a single sequential processor. More formally, a sequence of global states $\Sigma^0 \Sigma^1 \Sigma^2 \cdots \Sigma^{i-1} \Sigma^i \cdots$ is an observation if there exists a sequence of events $e^1 e^2 \cdots$ that is a linear extension of the partial order $\mathcal{H}$ (i.e., all events appear in an order consistent with the relation $\rightarrow$) such that Σ^i is the global state that results after executing event e^i in Σ^{i-1}.

By construction, each path of the lattice starting at the minimal element and proceeding upwards corresponds to an observation of the computation and each observation corresponds to a path in the lattice [1, 18]. In other words, the lattice of consistent global states represents all possible observations for the computation. Note that, internal to the computation, the actual sequence of global states that is produced cannot be known and this lattice represents the best information that is available.

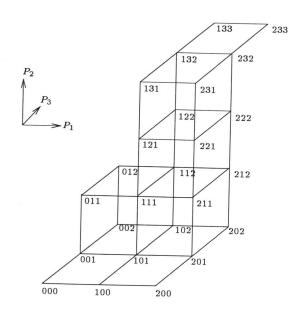

Figure 3: A distributed computation as a lattice $\mathcal{L}$ of global states.

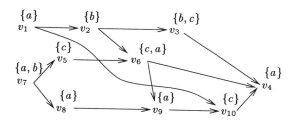

Figure 4: A directed acyclic graph and its labeling.

3 Properties as Languages over DAGs

In the previous section we have shown that a distributed computation $\mathcal{H}$ can be represented either as $\mathcal{S}$, the DAG of local states, or as $\mathcal{L}$, the DAG corresponding to the lattice of global states. In this section, we will develop a general framework for specifying dynamic properties based on labeling a generic DAG. In the following sections, we will instantiate this framework with the two specific DAGs $\mathcal{S}$ and $\mathcal{L}$, depending on whether we are interested in sequences of local or global properties, respectively. Development of this framework helps us understand and unify a large number of proposals for detection of properties on local and global states as instances of a single problem.

3.1 Graph Labeling and Languages

Let $G = (V, E)$ be a DAG and let A be a finite alphabet of symbols. We define a *labeling function* λ that maps nodes of G to non-empty sets of symbols drawn from A. For each node $v \in V$, the set $\lambda(v)$ is called the *label* of v (we assume that the empty symbol ϵ is implicitly in every alphabet and constitutes the implicit label of a node if the labeling function defines no other symbols). Figure 4 illustrates a DAG of ten nodes labeled from the alphabet $A = \{a, b, c\}$.

For each node v, let G_v be the subgraph obtained from G by retaining only node v and all of its predecessors in G. Clearly if G is a DAG, then so is G_v. A *directed path* of G_v is a sequence of nodes starting at a source node (i.e., one with no predecessors) and ending at node v. Let Π_v be the set of all such paths in G_v. We extend the notion of labeling to paths by associating with them words constructed from the same alphabet used to label nodes. Let A be a finite alphabet, λ be a labeling function and $\pi_v = u_0 u_1 \ldots u_k$ be a directed path of G_v. The label of path π_v is the set $\tilde{\lambda}(\pi_v)$ of all words $\omega = \omega_0 \omega_1 \ldots \omega_k$ such that $\omega_i \in \lambda(u_i)$ for each node u_i in path π_v.

Since each path label is a set of words, sets of path labels can be seen as defining a language. The language associated with node v of G under the labeling function λ, denoted $L^{\lambda}(v)$, is defined as the set of words that are in the labels of all directed paths of G_v. In other words,

$$L^{\lambda}(v) = \bigcup_{\pi_v \in \Pi_v} \tilde{\lambda}(\pi_v).$$

As an example, consider node v_{10} of the labeled DAG of Figure 4. The set of all directed paths for subgraph $G_{v_{10}}$ is $\Pi_{v_{10}} = \{v_1 v_{10}, v_1 v_2 v_6 v_9 v_{10}, v_7 v_5 v_6 v_9 v_{10}, v_7 v_8 v_9 v_{10}\}$. Under the labeling function λ that is illustrated, the language associated with node v_{10} is $L^{\lambda}(v_{10}) = \{ac, abaac, abcac, accac, bcaac, acaac, bccac, aaac, baac\}$.

3.2 Dynamic Properties

Informally, we would like dynamic properties to characterize the time evolution of states that occur during program execution. In our framework, a *dynamic property* (*property* for short) defines a set of words over some finite alphabet. Let $L(\Phi)$ be the set of words associated with property Φ. We find it convenient to distinguish the name of the property from the language that it defines.

Informally, we think of each word in language $L(\Phi)$ as satisfying property Φ. In defining properties for DAGs, we can extend this notion of satisfaction to the entire graph in two possible ways. Given an alphabet A, a directed acyclic graph G, a labeling function λ, a node of G and a property Φ over A, we define the following two satisfaction rules[1]:

Definition 1 *Given an alphabet A, a directed acyclic graph G, a labeling function λ, a node v of G and a property Φ over A, we say that node v satisfies* **SOME** Φ, *denoted* $v \models$ **SOME** Φ, *if and only if there*

---

[1] Satisfaction rules similar to these have been proposed in other contexts [6, 11, 15, 10, 9]. In particular, our definitions are in the same spirit as those of modal operators *POS* and *DEF* of [6], *strong* and *weak* of [11] and *POT, INEV, SOME* and *ALL* of more general transition systems [5].

exists some labeling of some directed path in G_v that defines a word in the language of Φ. In other words, $v \models$ **SOME** $\Phi \equiv L^{\lambda}(v) \cap L(\Phi) \neq \emptyset$.

Definition 2 *Given an alphabet A, a directed acyclic graph G, a labeling function λ, a node v of G and a property Φ over A, we say that node v satisfies* **ALL** Φ, *denoted* $v \models$ **ALL** Φ, *if and only if all labelings of all directed paths in G_v define words in the language of Φ. In other words,* $v \models$ **ALL** $\Phi \equiv L^{\lambda}(v) \subseteq L(\Phi)$.

As an example, consider the two properties Φ_1 and Φ_2 defined over the alphabet $A = \{a, b, c\}$ that define languages (without loss of generality, they are specified using regular expressions with the usual syntax rules) $L(\Phi_1) = aba^*c(ac)^*$ and $L(\Phi_2) = (b + c)^*a(a + b)^*c(a + b + c)^*$, respectively. With respect to the DAG of Figure 4, the following assertions hold (words $abaac$ and $abcac$ of $L^{\lambda}(v_{10})$ are in $L(\Phi_1)$):

$$v_{10} \models \textbf{SOME } \Phi_1$$

$$v_{10} \models \textbf{ALL } \Phi_2.$$

3.3 Detection of Dynamic Properties

For the sake of simplicity, in what follows we consider only properties that correspond to regular languages. As we shall see, many proposals including behavior patterns on local states [17, 10] and global states [6, 2] happen to be special cases of such properties. Moreover, these properties admit simple and efficient detection algorithms.

It is well known that regular grammars that specify regular languages are equivalent to deterministic finite-state automata. Formally, an *automaton* is a 5-tuple (Q, A, q_0, Q_F, δ) where Q is a finite set of states, q_0 an initial state, A a finite alphabet, Q_F a set of accepting states and δ a deterministic transition function.

Let Φ be a property such that $L(\Phi)$ is a regular language. A finite state automaton that recognizes $L(\Phi)$ is given the name Φ just as the property itself. Figure 5 illustrates the two automata recognizing the languages associated with the properties Φ_1 and Φ_2 of the above example. In the Figure, accepting states are shown as triangles.

Given a property Φ and DAG G, let $R^{\Phi}(v)$ denote the set of states of the automaton recognizing $L(\Phi)$ that are reached after analyzing all of the words in $L^{\lambda}(v)$. Recall that $L^{\lambda}(v)$ is the language associated with node v of G defined as the set of all words that are in the labels of all directed paths of G_v. For the automata of Figure 5 recognizing $L(\Phi_1)$ and $L(\Phi_2)$, and the DAG of Figure 4, we have $R^{\Phi_1}(v_{10}) = \{q_3, q_5\}$ and $R^{\Phi_2}(v_{10}) = \{q_2\}$. Note that $R^{\Phi_1}(v_{10}) \cap Q_F \neq \emptyset$ while $R^{\Phi_2}(v_{10}) \subseteq Q_F$. From the previous section, we know that for this example $v_{10} \models$ **SOME** Φ_1 and $v_{10} \models$ **ALL** Φ_2. In general, we can rewrite the satisfaction rules of Section 3.2 as follows as a consequence of the definitions:

$$v \models \textbf{SOME } \Phi \equiv R^{\Phi}(v) \cap Q_F \neq \emptyset$$

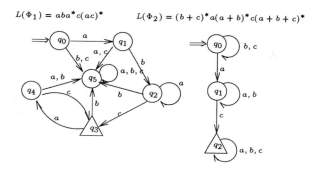

Figure 5: Two properties and their respective languages as finite automata.

$$v \models_{\textbf{ALL}} \Phi \equiv R^{\Phi}(v) \subseteq Q_F.$$

Expressing the satisfaction rules in terms of relationships between the accepting states of the automaton and the set of reachable states gives us effective decision procedures for computing them.

Given a property Φ, an automaton accepting $L(\Phi)$ and a labeled DAG G, the problem of detecting the property can be reduced to computing the sets $R^{\Phi}(v)$ for each node v of G. We proceed inductively in defining $R^{\Phi}(v)$.

Base case: Let G' be a DAG obtained from G by adding a fictitious node v_0 that is an immediate predecessor of all source nodes of G. The node labeling function is extended such that $\lambda(v_0) = \{\epsilon\}$. Then, by definition

$$R^{\Phi}(v_0) = \{q_0\}.$$

Inductive step: Let $pred(v)$ be the set of nodes that are immediate predecessors of node v in G. Let $R^{\Phi}_{pred}(v)$ be the set of reachable states of the automaton recognizing $L(\Phi)$ after analyzing all words associated with nodes in $pred(v)$. In other words,

$$R^{\Phi}_{pred(v)} = \bigcup_{u \in pred(v)} R^{\Phi}(u)$$

and by induction we have

$$R^{\Phi}(v) = \bigcup_{q \in R^{\Phi}_{pred(v)}, \alpha \in \lambda(v)} \delta(q, \alpha).$$

The above inductive definition can be easily transformed into a computation by doing a breadth-first traversal of G starting at node v_0 as shown in Figure 6.

4   Properties on Control Flows

In this section, we will instantiate the generic dynamic property detection algorithm of the previous section in order to detect properties on control flows.

```
R^Φ(v0) := {q0}; done := {v0}
while (∃v ∈ V : (v ∉ done) ∧ (∃u ∈ done : (u, v) ∈ E)
do
    foreach v ∉ done such that pred(v) ⊆ done
    do
        R^Φ(v) := {}
        foreach α ∈ λ(v), u ∈ pred(v), q ∈ R^Φ(u)
            do R^Φ(v) := R^Φ(v) ∪ {δ(q, α)} od
        done := done ∪ {v}
    od
od
```

Figure 6: A generic algorithm for detecting property Φ in directed acyclic graph $G = (V, E)$ with labeling function $\lambda()$.

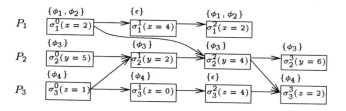

Figure 7: DAG of local states annotated with values of three variables and labels.

In other words, the directed acyclic graph of interest is the DAG of local states $\mathcal{S} = (S, \prec)$ defined in Section 2.2 and the alphabet of interest is a set of local predicates.

4.1   Local Predicates

A *local predicate* is a formula in propositional logic (boolean expression) naming only variables that are local to a single process. Let ϕ be such a local predicate. If the predicate holds in some local state σ of $\mathcal{S}$, we say that σ *satisfies* ϕ and write $\sigma \models \phi$. Let A be an alphabet consisting of a finite set of local predicates. We define a labeling function for the distributed computation $\mathcal{S} = (S, \prec)$ such that the labels of a local state are the local predicates it satisfies:

$$\forall \sigma \in S : \lambda(\sigma) = \{\phi \in A : \sigma \models \phi\}.$$

As an example, consider the distributed computation of Figure 7 where x, y and z are three variables local to processes P_1, P_2 and P_3, respectively. Each local state σ_i^k is characterized by the value of the corresponding local variable. In Figure 7, values of the variables are shown next to each of the local states (e.g., in local state σ_2^1 the local variable y has value 2). Let us consider the following four local predicates $\phi_1 \equiv (x < 3)$, $\phi_2 \equiv (x \text{ is prime})$, $\phi_3 \equiv (y \neq 0)$ and $\phi_4 \equiv (z < 4)$. For the computation of Figure 7, we have $\sigma_1^0 \models \phi_1$, $\sigma_1^0 \models \phi_2$, $\sigma_1^2 \models \phi_1$, $\sigma_1^2 \models \phi_2$,

$\sigma_2^0 \models \phi_3$, $\sigma_2^1 \models \phi_3$, $\sigma_2^2 \models \phi_3$, $\sigma_2^3 \models \phi_3$, $\sigma_3^0 \models \phi_4$, $\sigma_3^1 \models \phi_4$ and $\sigma_3^3 \models \phi_4$.

Thus $\mathcal{S}$ labeled with the alphabet $A = \{\phi_1, \phi_2, \phi_3, \phi_4\}$ is as shown in Figure 7.

4.2 Behavior Patterns on Local States

Consider the class of properties that can be specified as a set of local predicates that need to be satisfied in a particular order during a computation. Each such sequence is called a *behavior pattern*. Recall that in Section 2.2 we defined a control flow associated with some local state σ as a sequence of local states that are pairwise adjacent in $\mathcal{S}$ and the sequence begins at some initial state and terminates at σ. In other words, control flows of a computation correspond to directed paths of $\mathcal{S}$. A property Φ on control flows expresses behavior patterns that these control flows must exhibit as words of the language $L(\Phi)$ constructed out of local predicates.

The satisfaction rules introduced in Section 3.2 have the following meaning when interpreted in the context of distributed computations as a DAG of local states:

Definition 3 *Given a computation $\mathcal{S}$ and property Φ on control flows, a local state σ satisfies* SOME Φ *if and only if there exists a control flow π_σ terminating at local state σ such that $\widetilde{\lambda}(\pi_\sigma)$ includes at least one word of $L(\Phi)$. In other words,*

$$\sigma \models \text{SOME } \Phi \equiv L^\lambda(\sigma) \cap L(\Phi) \neq \emptyset.$$

Definition 4 *Given a computation $\mathcal{S}$ and property Φ on control flows, a local state σ satisfies* ALL Φ *if and only if for each control flow π_σ terminating at local state σ, every word of $\widetilde{\lambda}(\pi_\sigma)$ is included in $L(\Phi)$. In other words,*

$$\sigma \models \text{ALL } \Phi \equiv L^\lambda(\sigma) \subseteq L(\Phi).$$

For example, consider the properties $\Phi_1 = \phi_1 \phi_3^+$, $\Phi_2 = \phi_1 \phi_3^+ \phi_4$, $\Phi_3 = \phi_1 \phi_4$ and $\Phi_4 = (\phi_1 + \phi_2 + \phi_4)^* \phi_3^+$ on control flows specified as regular expressions from the alphabet $A = \{\phi_1, \phi_2, \phi_3, \phi_4\}$ as defined before. For the computation of Figure 7, the following results hold: $\sigma_2^2 \models$ SOME Φ_1, $\sigma_3^3 \models$ SOME Φ_2, $\neg(\sigma_3^3 \models$ SOME $\Phi_3)$, $\sigma_2^3 \models$ ALL Φ_4 and $\neg(\sigma_2^3 \models$ ALL $\Phi_1)$.

4.3 Run-Time Detection

As discussed in Section 3.3, regular languages can be recognized using deterministic finite state automata. In the case of properties on control flows, we can instantiate the generic algorithm of Figure 6 such that detection can be done at run time without introducing any delays and without adding any control messages to the distributed computation. All that is necessary is to piggyback control information on the existing messages of the computation.

Let Φ be a property such that $L(\Phi)$ can be recognized by the finite state automaton $\Phi = (Q, A, q_0, Q_F, \delta)$. A *controller* is superimposed on each process P_i of the computation that maintains an array $A_i[Q]$ of boolean values with the following

```
when P_i enters a new local state σ_i:
    foreach α ∈ λ(σ_i) do
        A^α[Q] := (false, ···, false)
        foreach q ∈ Q such that A_i[q] do
            foreach r ∈ δ(q, α) do A^α[r] := true od
        od
    od
    foreach q ∈ Q do A_i[q] := ⋁_{α∈λ(σ_i)} A^α[q] od
```

```
when P_i sends a message m:
    piggyback A_i[Q] on m
```

```
when P_i receives m containing A_m:
    foreach q ∈ Q do A_i[q] := A_i[q] ∨ A_m[q] od
```

Figure 8: Algorithm executed by the controller of process P_i in order to detect properties on control flows.

semantics: For each $q \in Q$, element $A_i[q] = true$ if and only if there exists a control flow π_σ terminating at the current local state σ of P_i such that at least one word in $\widetilde{\lambda}(\pi_\sigma)$ places automaton Φ in state q. Initially, only $A_i[q_0]$ is defined to be true. The algorithm executed by each controller can be easily derived from the generic algorithm and shown in Figure 8.

Let A_σ denote the value of A at local state σ as maintained by the algorithm of Figure 8. Then, the satisfaction rules for SOME and ALL can be computed through the following simple definitions:

$$\sigma \models \text{SOME } \Phi \equiv \exists q \in Q : ((A_\sigma[q] = true) \land q \in Q_F)$$

$$\sigma \models \text{ALL } \Phi \equiv \forall q \in Q : ((A_\sigma[q] = true) \Rightarrow q \in Q_F)$$

5 Properties on Observations

Here properties are on sequences of global states and consequently the lattice $\mathcal{L}$ of global states introduced in Section 2.3 instantiates the DAG G of Section 3 and a behavior pattern is a word on an alphabet of global predicates.

5.1 Global Predicates

A *global predicate* is a general boolean expression defined over a consistent global state; such an expression may reference any variable of any process. Satisfaction of global predicate φ by a global state Σ is denoted $\Sigma \models \varphi$.

The lattice introduced in Section 2.3 constitutes the basic model where properties related to global states are interpreted. Given a finite alphabet A of global predicates, a labeling function λ is defined on global states in a way analogous to that for local states:

$$\forall \Sigma : \lambda(\Sigma) = \{\varphi \in A : \Sigma \models \varphi\}.$$

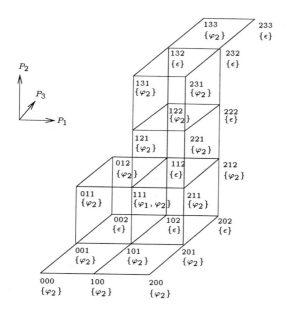

Figure 9: Lattice $\mathcal{L}$ labeled with predicates.

Consider for example the lattice of Figure 9 corresponding to the distributed computation of Figure 7 and the two global predicates:

$$\varphi_1 \equiv ((x > y) \wedge (y > z))$$
$$\varphi_2 \equiv ((x \neq y) \Rightarrow (x > z)).$$

We can see that global state $\Sigma_1 = (\sigma_1^1, \sigma_2^1, \sigma_3^1)$ (indicated as 111 in Figure 9) satisfies both φ_1 and φ_2, obtaining the label $\{\varphi_1, \varphi_2\}$. The labeling of the lattice with the alphabet $A = \{\varphi_1, \varphi_2\}$ is shown in Figure 9.

5.2 Behavior Patterns on Observations

A property Φ on global states is expressed as a set $L(\Phi)$ of sequences of global predicates: each sequence of global predicates defines a particular pattern. A sequence of global predicates defining a pattern must be satisfied by a sequence of global states in order for the corresponding pattern be recognized. In other words, such a property Φ describes behavior patterns on observations. By definition, a distributed computation is said to satisfy a property Φ if its final global state Σ^{last} satisfies it. Informally, the satisfaction rules introduced in Section 3.2 express the following properties:

Definition 5 $\Sigma^{last} \models$**SOME** Φ *is true if and only if there exists a labeled observation of the computation that belongs to* $L(\Phi)$.

Definition 6 $\Sigma^{last} \models$**ALL** Φ *is true if and only if all possible labelings of all observations of the computation belong to* $L(\Phi)$.

$previous := \left\{ \Sigma^{-1} \right\}$
$current := \left\{ \Sigma^0 \right\}$
while $(current \neq \emptyset)$ **do**
    foreach $\Sigma \in current$ **do**
      foreach $q \in Q$
        do $A_{pred}[q] := \bigvee_{\Sigma' \in pred(\Sigma)} A_{\Sigma'}[q]$ **od**
      foreach $\alpha \in \lambda(\Sigma)$ **do**
        $A^\alpha[Q] := (false, \cdots, false)$
        foreach $q \in Q$ **such that** $A_{pred}[q] = true$
        do
          foreach $r \in \delta(q, \alpha)$ **do** $A^\alpha[r] := true$ **od**
        od
      od
      foreach $q \in Q$ **do** $A_\Sigma[q] := \bigvee_{\alpha \in \lambda(\Sigma)} A^\alpha[q]$ **od**
    od
    $previous := current;$
    $current := \{global\ states\ directly\ reachable$
      $from\ those\ in\ previous\}$
od

Figure 10: Algorithm executed by the monitor in order to detect properties on observations.

For the example of Figure 9, we have $\Sigma^{last} \models$**SOME** $(\varphi_2^+ \varphi_1 \varphi_2^+)$ with alphabet $A = \{\varphi_1, \varphi_2\}$ while $\Sigma^{last} \models$**ALL** φ_2^+ with alphabet $A = \{\varphi_2\}$.

5.3 Run-Time Detection

A basic prerequisite for detecting properties on global states is to build these global states in one way or another in order to check if they satisfy some global predicates. In the most general case, the entire lattice of global states, representing all observation has to be traversed. Each process P_i of the computation is augmented with a controller that sends local states produced by P_i to a monitor. The monitor pieces together local states received from the processes in order to construct consistent global states and to incrementally build the lattice. Several algorithm are known of doing such constructions [6] that use vector clocks to ensure consistency of the global states they compute.

Construction of the lattice and checking of properties is done at run time (i.e., concurrently with the distributed computation) by the monitor through the algorithm shown in Figure 10. As indicated in Section 3.3 we consider only the detection of properties corresponding to regular languages (and so each property Φ can be recognized by a finite state automaton $\Phi = (Q, A, q_0, Q_F, \delta)$). With each global state Σ of the lattice, we associate a boolean array $A_\Sigma[Q]$. For each $q \in Q$, element $A_\Sigma[q]$ is true if and only if there exists an observation terminating at Σ whose labeling puts the automaton in state q. A fictitious global state Σ^{-1} is added to the lattice as the unique predecessor of Σ^0 and $A_{\Sigma-1}[q_0]$ is the only element of $A_{\Sigma-1}$ that is initially true. Function $pred(\Sigma)$ returns the set of global states that immediately precede Σ in the lattice. The algorithm needs to consider only the global states of two adjacents levels of the lattice maintained in sets **previous** and **current** (the immediate predecessors of a global state in **current** belongs to the set **previous**).

The algorithm computes values of $A_\Sigma[Q]$ for each state Σ of **current** from the values associated with its predecessors and the labeling of Σ. Then, the satisfaction rules for **SOME** and **ALL** can be computed through the following simple definitions:

$$\Sigma \models \textbf{SOME}\ \Phi \equiv \exists q \in Q : ((A_\Sigma[q] = true) \land q \in Q_F)$$

$$\Sigma \models \textbf{ALL}\ \Phi \equiv \forall q \in Q : ((A_\Sigma[q] = true) \Rightarrow q \in Q_F)$$

This general detection algorithm for behavior patterns on observations can be simplified according to specifications of a particular behavior pattern. As indicated earlier, it includes as special cases proposals described in [6, 2].

6 Conclusion

This paper has first presented a general framework which allows the expression of properties (behavior patterns described as words on an alphabet of predicates). It has been shown that if the predicates are on local states the patterns are on the control flows of the execution; if the predicates are on global states, the patterns are on observations of the distributed executions (remember an observation is a sequence of global states that could be produced by executing the program on a monoprocessor). A control flow satisfies a pattern if the sequence of local states defining this control flow satisfies the sequence of local predicates defining the pattern. Similarly, an observation satisfies a pattern if the sequence of global states defining this observation satisfies the sequence of global predicates defining the pattern.

It has been shown that the detection of regular patterns on control flows can be done at run time and without requiring additional messages; only piggybacking of an array of bits (one for each state of the finite state automaton) is necessary. So the detection of these properties is not expensive. Detection of regular patterns on observations, on the other hand, in general requires an additional monitor process whose aim is to build all possible observations of the computation. These detection algorithms can be seen as on-the-fly model checkers that work on only a small part of a directed acyclic graph of local or of global states.

From a practical point of view, run-time detection of dynamic properties (properties expressed here as regular patterns) is a fundamental point when one is interested in analyzing or debugging distributed executions. The work presented in this paper originated from a project whose aim is to design and implement a debugging facility for distributed programs [14]. The algorithms which have been described are currently implemented in this debugging facility.

References

[1] Ö. Babaoğlu and K. Marzullo. *Consistent global states of distributed systems: fundamental concepts and mechanisms, in Distributed Systems*, chapter 4, pages 55–93. *ACM Press, Frontier Series*, (S.J. Mullender Ed.), 1993.

[2] Ö Babaoğlu and M. Raynal. Specification and verification of dynamic properties in distributed computations. *Journal of Parallel and Distributed Computing*, 1995.

[3] Gabriel Bracha and Sam Toueg. Distributed deadlock detection. *Distributed Computing*, 2(3):127–138, 1987.

[4] K. M. Chandy and L. Lamport. Distributed snapshots : determining global states of distributed systems. *ACM TOCS*, 3(1):63–75, Feb. 1985.

[5] E.M. Clarke, E.A. Emerson, and A.P. Sistla. Automatic verification of finite state concurrent systems using temporal logic specifications. *ACM Toplas*, 8(2):244–263, 1986.

[6] R. Cooper and K. Marzullo. Consistent detection of global predicates. In *Proc. ACM/ONR Workshop on Parallel and Distributed Debugging*, pages 167–174, Santa Cruz, California, May 1991.

[7] C. Diehl, C. Jard, and J. X. Rampon. Reachability analysis on distributed executions. In *Theory and Practice of Software Development*, pages 629–643, Springer Verlag, LNCS 668, April 1993.

[8] N. Francez. Distributed termination. *ACM TOPLAS*, 2(1):42–55, January 1980.

[9] E. Fromentin, C. Jard, G.V. Jourdan, and M. Raynal. *On the fly analysis of distributed computations*. To appear in In. Proc. Letters.

[10] E. Fromentin, M. Raynal, V.K. Garg, and A.I. Tomlinson. On the fly testing of regular patterns in distributed computations. In *Proc. of the 23rd International Conference on Parallel Processing*, pp. 73–76, 1994.

[11] V. K. Garg and B. Waldecker. Detection of unstable predicates in distributed programs. In *12th Int. Conf. on Foundations of Software Tech. and Theor. Comp. Science*, pp. 253–264, Springer Verlag, LNCS 625, Dec. 1992.

[12] David Harel and Amir Pnueli. On the development of reactive systems. In *Logics and Models of Concurrent Systems*, pages 477–498, Springer-Verlag, 1985.

[13] J.-M. Hélary, C. Jard, N. Plouzeau, and M. Raynal. Detection of stable properties in distributed applications. 6th *ACM SIGACT-SIGOPS, Symp. Principles of Distributed Computing*, Vancouver, Canada, 125–136, 1987.

[14] M. Hurfin, N. Plouzeau, and M. Raynal. A debugging tool for distributed Estelle programs. *Journal of Computer Communications*, 16(5):328–333, May 1993.

[15] C. Jard, T. Jeron, G.V. Jourdan, and J.X. Rampon. A general approach to trace-checking in distributed computing systems. In *Proc. 14th IEEE Int. Conf. on DCS*, pages 396–403, Poznan, Poland, June 1994.

[16] L. Lamport. Time, clocks and the ordering of events in a distributed system. *Com. ACM*, 21(7):558–565, 1978.

[17] B.P. Miller and J. Choi. Breakpoints and halting in distributed programs. In *Proc. 8th IEEE Int. Conf. on Dist. Comp. Syst.*, San Jose, pp. 316–323, 1988.

[18] R. Schwarz and F. Mattern. Detecting causal relationships in distributed computations : in search of the holy grail. *Distributed Computing*, 7(3):149–174, 1994.

Impact of Load Imbalance on the Design of Software Barriers

Alexandre E. Eichenberger

Santosh G. Abraham

Advanced Computer Architecture Laboratory
EECS Department, University of Michigan
Ann Arbor, MI 48109-2122
alexe@eecs.umich.edu

Hewlett Packard Laboratories
1501 Page Mill Road
Palo Alto, CA 94303
abraham@hpl.hp.com

Abstract

Software barriers have been designed and evaluated for barrier synchronization in large-scale shared-memory multiprocessors, under the assumption that all processors reach the synchronization point simultaneously. When relaxing this assumption, we demonstrate that the optimum degree of combining trees is not four as previously thought but increases from four to as much as 128 in a 4K system as the load imbalance increases. The optimum degree calculated using our analytic model yields a performance that is within 7% of the optimum obtained by exhaustive simulation with a range of degrees. We also investigate a dynamic placement barrier where slow processors migrate toward the root of the software combining tree. We show that through dynamic placement the synchronization delay can be reduced by a factor close to the depth of the tree, when sufficient slack is available. By choosing a suitable tree degree and using dynamic placement, software barriers that are scalable to large numbers of processors can be constructed. We demonstrate the applicability of our results by performing measurements on a small SOR relaxation program running on a 56-processor KSR1.

KEYWORDS: Synchronization barrier, fuzzy barriers, combining tree, shared-memory multiprocessors, parallel processing

1 Introduction

Synchronization barrier constructs are used by programmers to ensure that all processors have reached a particular point in a computation before any processors are allowed to advance beyond that point. Parallel supercomputer applications typically use data-parallel programming techniques where large data structures are updated in parallel by all the processors. Barriers are used in such programs to separate the phases of the computation and to ensure that all processors have finished updating a data structure in step t before any processor uses the updated values as input in step $t+1$. The simplest implementation of a barrier uses a counter protected by a lock. The overhead of such barriers increases linearly with the number of processors and can dominate overall execution time.

In response to the potential overhead of software barrier synchronization schemes, several hardware schemes have been proposed. The NYU Ultracomputer [1] and the IBM RP3 [2] employ combining networks which combine accesses to the same memory location, thus alleviating contention on the counters used to implement synchronization barriers. Other machines such as the Sequent, SGI, and Alliant have provided special synchronization buses. Vector supercomputers such as the Cray and the Convex provide a set of communication registers which are used for fast barrier synchronization. The Cray T3D multiprocessor has a fast synchronization network for barrier synchronization [3]. The hardware support for efficient synchronization is indicative of the potential impact of synchronization on overall performance.

Software barrier synchronization schemes that can approach the performance of hardware techniques are extremely attractive because hardware schemes have several disadvantages. Hardware synchronization schemes are expensive; for instance, the combining network is at least six times as expensive as a noncombining network [4]. Also, hardware schemes employ special-purpose logic that has a large design cost, especially significant for low-volume parallel systems. Thirdly, software barriers are more flexible and can be adapted to suit the application or to exploit advances in synchronization techniques such as fuzzy barriers. Finally, software barrier schemes are more easily portable to different platforms.

The disadvantages of hardware synchronization schemes have motivated the study of software barriers using software combining trees. In some cases, these studies have demonstrated that the synchronization performance of software schemes approach that of hardware synchronization [5] [6]. In a software combining tree, a tree of counters is used for synchronization. Processors are divided into groups and a group is assigned to each leaf of the combining tree. Each processor updates the counter and if it finds it is the last one to reach the counter it proceeds to the parent of the counter. The last processor to reach the counter at the root of the tree releases all the processors by updating a shared variable.

In previous work, the performance of synchronization barriers is evaluated and optimized for the case where all processors arrive at the barrier simultaneously. This assumption is motivated by the perception that synchronization mechanisms and load imbalance are two different issues that can be solved independently. As a result, synchronization mechanisms are developed assuming the case of zero load imbalance and consequently maximal contention.

However, experiments on parallel machines have demonstrated that processors typically fail to reach a synchronization point at the same time for several reasons. Firstly, the workload may be unevenly partitioned among the processors. As a result, certain processors consistently arrive late at a synchronization point thus idling other processors and resulting in *systemic load imbalance*. Though, systemic imbalance can be handled by effectively partitioning the workload, sufficient information may not be available to perform the best partitioning. In *non-deterministic* imbalance, processors fail to reach a synchronization point at the same time but typically the processor arriving last changes on each iteration. Non-deterministic load imbalance is generated by several factors: the workload associated with a processor may change from cycle to cycle; the interprocessor communication may incur random delays due to contention; or there may be contention for hardware or software resources.

Theoretically, the earliest a processor can leave a barrier is when the last processor arrives at the barrier. In practical implementations such as combining trees, processors have to wait till this last processor updates the root counter. Thus, we define *synchronization delay* of combining trees as the difference between the release time (time when last processor updates root counter) and the arrival time of the last processor. There are two components to this delay: *update delay* of updating counters and *contention delay* for locks that govern access to the counters. The first component is determined by the tree depth and the second by the number of processors simultaneously attempting to gain access to a particular counter.

The first contribution of this paper is to investigate the degree of a software combining tree that minimizes the synchronization delay as a function of the load imbalance. On the one hand, if all processors arrive simultaneously, contention delay is maximum and a deep tree is desirable. On the other hand, if one of the processors arrives significantly later than the others, the main issue is to reduce the update delay and a wide tree is desirable. Thus the requirements for the two cases, viz. deep tree (small degree) and wide tree (large degree), are conflicting and the best degree of the combining tree is a function of the amount of load imbalance. In this paper, we develop an approximate analytic model for estimating synchronization delay as a function of load imbalance and combining tree degree. Simulation results show that the performance of the combining tree obtained using the analytic model is within 7% of the performance of the optimum combining tree obtained through exhaustive simulation, for normally distributed processor execution times.

The second contribution of this paper is a novel software barrier where late arriving processors migrate toward the root of the combining tree to minimize the synchronization delay. In the normal software combining tree, late arriving processors tend to update more counters on average. In presence of systemic load imbalance, some processors tend to be consistently late and are consistently delayed further by the synchronization barrier. We explore the possibility of giving less synchronization work to these processors by modifying a tree structure first proposed in Mellor-Crummey and Scott [7]. In their tree structure, one processor is statically attached to each non-leaf counter in the software combining tree. The rest of the processors are split into groups and assigned to the leaf counters. A processor updates the counter it is attached to as well as its parent if it is the last processor to arrive at the counter. In contrast to the static assignment in Mellor-Crummey and Scott, we attach processors dynamically to the non-leaf counters. Processors that tend to arrive late are attached to counters closer to the root to reduce the latency delay of syn-

chronization. We compare the performance of the dynamic placement scheme with the static placement scheme used by Mellor-Crummey and Scott and obtain significant performance improvement.

This paper is organized as follows. First, we present a summary of the related work in Section 2. We present an analytical model for approximating the synchronization delays in Section 3. Using this model, we estimate the optimal combining tree degree of a barrier in Section 4. We investigate a dynamic placement barrier in Section 5. In Section 6, we present a quantitative comparison of the simulation results of the two previous sections. Section 7 presents our measurements on a parallel machine (KSR1). Finally, we conclude in Section 8.

2 Related Work

Performance degradations due to busy wait synchronization are widely regarded as a serious performance problem. Pfister and Norton [4] showed that the presence of hot spots can severely degrade performance for all traffic in multistage interconnection networks. Agarwal and Cherian [8] investigated the impact of synchronization on overall program performance and showed that cache line invalidations due to synchronization references can account for more than half of all invalidations.

In response to performance concerns, hardware support has been designed and implemented in several parallel machines. Combining networks [1] [2], that combine concurrent accesses to the same memory location, have been advocated as a technique that significantly reduces the impact of busy waiting. Similarly, special purpose cache protocols [9] [10] have been designed to include synchronization primitives that reduce communication due to synchronization.

New software synchronization mechanisms have been developed to approach the synchronization performance of dedicated hardware at lower cost. Yew, Tzeng and Lawrie [6] investigated the use of software combining trees to distribute hot spots in large scale multiprocessors. Their analysis indicates that combining trees effectively decrease memory contention. Furthermore, they showed that the optimal degree of a combining tree (fan in) is around four. Mellor-Crummey and Scott [7] refined this technique further, presenting an algorithm that generates the theoretical minimum number of communications on machines without broadcast. Michael and Scott [5] showed that a software implemented exclusion mechanism could outperform naive hardware locks, even under heavy contention.

Alternatives to the usual synchronization barriers have also been investigated. Gupta [11] developed and investigated Fuzzy Barriers. He measured significant performance improvements with software implemented Fuzzy Barriers on a four processor Encore Multimax. He presents techniques [11] [12] that detect and increase the number of independent operations, and hence the slack time. Eichenberger and Abraham [13] characterized the performance improvements due to fuzzy barriers and showed that the expected idle time at a fuzzy barrier is inversely proportional to the slack time. Finally, Nguyen [14] investigated compiler techniques that transform synchronization barriers into point to point synchronizations, showing encouraging performance improvements.

The source and extent of variation of thread (processor) execution times have been investigated in a few studies. Adve and Vernon [15] have measured the fluctuations of parallel execution times for a large number of applications and observed that the empirical execution time distribution very closely tracks the normal distribution. Dubois and Briggs [16] obtained an analytical formula describing the expected number of cycles and its variance for memory references in tightly coupled systems. Sarkar [17] provided a framework to estimate the execution time and its variance based on the program's internal structure and control dependence graph. Finally, Eichenberger and Abraham [13] analyzed the fluctuation of processor execution time due to random replacement caches and communication contention. We derived an analytical formula describing the expected variance for programs with simple memory and communication access patterns. In all four papers, the variation in execution times was found to approximate a normal distribution and we assume that thread execution times are normally distributed in this paper.

The effects of load imbalance on idle times, assuming a perfect barrier with zero synchronization delay, have been investigated in several articles. Kruskal and Weiss [18] have investigated the total execution time required to complete k tasks for various distributions. The performance of parallel algorithms that have regular control structures and non-deterministic task execution times is quantified by Madala and Sinclair [19]. Durand *et al* provide experimental measurements on the impact of memory contention in NUMA parallel machines [20].

Axelrod [21] has considered both the effects of load imbalance and synchronization costs and derived an analytical result that takes both load imbalance and synchronization costs into consideration. However, while considering the synchronization costs, he assumed that processors arrive simultaneously at the synchronization point thus overestimating the effects of contention. We determine synchronization delays as a function of both the particular synchronization structure used and the load imbalance.

Beckmann and Polychronopoulos [22] have investigated the effects of barrier synchronization and dynamic loop dispatch overhead. They classify loops as synchronization bound or arrival-time bound, depending on the spread of processor arrival time at the barrier. They derived an analytical result for these two cases for shared-bus multiprocessors and present theoretical and simulated speedup curves.

Definitions

In this article, we distinguish two synchronization phases: the release and the enforce phase. During the *release* phase, a processor signals its arrival at the synchronization point by incrementing counters in a combining, or synchronization, tree. A synchronization tree consists of L levels of counters, where each counter is connected to at most d other counters, where d is the tree degree. During the *enforce* phase, a processor checks if all processors have completed their release phase.

We furthermore define the *arrival* time as the time at which a processor arrives at the release phase and the *release* time as the time at which a processor completes the release phase. The time needed by a processor to update one counter is defined as t_c. This time includes the communication time to fetch the counter and to execute an atomic operation.

3 Analytic Model for Estimating Synchronization Delays

In this section, we will first derive the synchronization delay assuming simultaneous processor arrivals and then estimate this delay for general processor distributions.

When assuming simultaneous processor arrivals, the resulting synchronization delay for a combining tree of degree d with L full levels is obtained as follows. At the lowest level of the combining tree, d processors will simultaneously attempt to increment their counter. Since only one processor can update its counter at a time, the d processors are serialized. The last processor completes incrementing of its counter after $d \cdot t_c$. Since all processors arrive simultaneously, the same process occurs at each level of the combining tree. Therefore, the total synchronization delay for a combining tree with L full levels is defined as follows:

$$T_{sync,0}(L) = L\,d\,t_c \qquad (1)$$

Given that the number of levels in a combining tree for p processors is defined as $L = \log_d p$ (p chosen such that it results in full levels), Equation (1) yields a minimum synchronization delay for a combining tree of degree $d = e \simeq 2.71$.

Two problems arise when extending the previous model to processors that do not arrive simultaneously. The first problem is that only the slowest processors propagate upward in the tree, requiring the use of order statistics [23] at each level of the tree. The second problem is that contention at one level changes the distribution of the processors that propagate to the next level in the combining tree. As a result, a direct solution of the synchronization delay in the presence of load imbalance would require expensive numerical computations. Therefore, we introduce three assumptions and find an approximate analytical solution that takes the load imbalance into account.

First, we partition the processors into subsets and assume that all processors of a subset arrive simultaneously. Second, we assume a specific ordering of the arrival time of each subset of processors: the closer a subset is to the last processor, the later it arrives. Third, we relate the arrival and release times of each subset to its respective position in the combining tree.

In Figure 1a, we selected processor p8 to be the last one. Along its path to the root, it will experience contention with each of the three subsets S0 through S2. The first assumption of our approximation states that all processors in subsets S0, S1, and S2 arrive respectively at time t0, t1, and, t2. The ordering of these arrival times is illustrated in Figure 1b. Figure 1c and 1d illustrate how subset arrival times and contention delays are merged together. In Figure 1c, the distribution is wide enough to prevent the last processor from being slowed down by the contention of earlier processors. In Figure 1d, however, the distribution is narrower and contention from previous processors affects the last processor.

We compute the subset arrival time by first defining the processor subsets in a combining tree, then computing the percentage of processors that arrive earlier, and finally applying the density function of the processor arrival time.

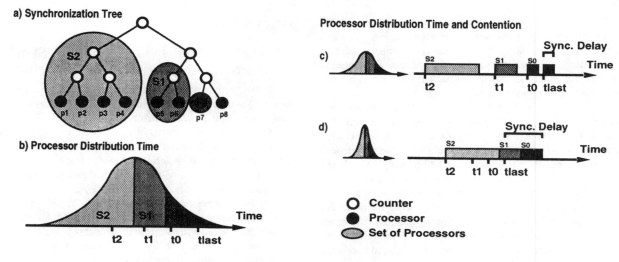

Figure 1: Estimating the synchronization delay.

We define a subset S_l as the subset that includes all the subtrees of exactly depth l that are connected to the counters along the path from the last processor to the root of the combining tree. For example, the subset S1 in Figure 1a contains all the subtrees of exactly depth 1, and therefore consists of processors $p5$ and $p6$. In general, we see that subset S_l consists of $d - 1$ subtrees of depth l and therefore contains $(d - 1)d^l$ processors. Furthermore, we know from our second assumption that all the processors in subsets $S_{l+1} \ldots S_{L-1}$ arrived before the ones in subset S_l. Therefore, the expected fraction of processors that arrived before the processors in subset S_l is defined as follows:

$$P_{before}(S_l) = 1 - P_{in/after}(S_l) = 1 - d^{l-L+1} \qquad (2)$$

Now that we have obtained the expected fraction of processors to arrive before the processors of subset S_l, we are able to determine the expected arrival time of each subset:

$$T_{arr}(S_l) = F^{-1}(P_{before}(S_l)) \qquad (3)$$

where F^{-1} is the inverse of the distribution function of the processors. If the processors are normally distributed with parameters μ and σ, the expected arrival time of subset S_l is

$$T_{arr}(S_l) = \sigma \, \Phi^{-1}(P_{before}(S_l)) \qquad (4)$$

where Φ^{-1} is the inverse of the normal distribution function[1]. Since we consider here only the arrival time relative to the mean, we omitted the μ term in the preceding equation. Finally, we can asymptotically estimate the arrival time of the last processor with the help of order statistics [23]:

$$T_{arr}(last) = \sigma \left(\sqrt{2 \log p} - \frac{\log \log p + \log 4\pi}{2\sqrt{2 \log p}} \right) \qquad (5)$$

Furthermore, we compute the release times as follows. The release time (T_{rel}) of a subset S_l corresponds to the sum of its arrival time, contention delays, and propagation time from the subset root counter to the combining tree root counter,

$$T_{rel}(S_l) = T_{arr}(S_l) + T_{sync,0}(S_l) + (L - l)t_c \qquad (6)$$

---
[1]Since $P_{before}(S_{L-1}) = 0$ and $\Phi^{-1}(0) = -\infty$ we approximate $P_{before}(S_{L-1})$ as $P_{before}(S_{L-2})/2$.

The release time of the last processor corresponds to the sum of its arrival time and its propagation time to the root counter through all levels,

$$T_{rel}(last) = T_{arr}(last) + Lt_c \qquad (7)$$

Since the last processor experiences contention with the processors of each subset, its release time corresponds to the maximum of all release times. The synchronization delay is defined as the difference between its release and arrival times:

$$T_{sync,\sigma} = \max_{s=S_0 \ldots S_{L-1}} (T_{rel}(last), T_{rel}(s)) - T_{arr}(last) \qquad (8)$$

The steps required to compute Equation (8) are summarized in Algorithm 1. This approximation is useful in estimating the optimal degree of a combining tree, since we can approximate the synchronization delay associated with a given number of processors, processor distribution, and degree of its combining tree.

Algorithm 1 *Given a combining tree with L levels and degree d, the synchronization delay is computed as follows.*

1. *The release time of each subset $S_0 \ldots S_{L-1}$, with respective levels $0 \ldots L - 1$, are computed by using Equations (1), (2), (4) and (6).*

2. *The release time of the last processor is computed by using Equations (5) and (7).*

3. *The synchronization delay is computed by using Equation (8).*

To characterize the accuracy of Equation (8), we compared its results against simulation results. We assumed the processors to be normally distributed and obtained data for various processor numbers, combining tree degrees and standard deviations of processor distributions. The time to update a counter, t_c, was experimentally measured on a KSR1 and that value[2] was used in our set of simulations.

---
[2]On the KSR1, it takes on average $20\mu s$ to acquire a subpage in atomic mode and to increment its value.

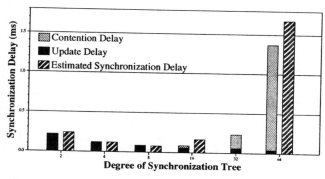

Figure 2: Synchronization delay for various degrees. (4K Processors, $\sigma = 250$ ms, $t_c = 20 \mu s$)

Figure 2 illustrates this comparison for 4K processors and for six different degrees of combining trees. Degrees 2, 4, 8, 16, 32, and 64 resulted in tree depths of 12, 6, 4, 3, 3, and 2, respectively. In each pair, the right bar represents the approximated synchronization delay. Since we assumed full trees when deriving Equation (8), there is no data for the degree 32. The left bar represents the simulated results and consists of two components: update delays and contention delays. The update delay is directly proportional to t_c and the tree depth. The contention delay increases dramatically after a threshold tree degree (16 in this figure).

Despite the strong assumptions used to obtain Equation (8), namely the simultaneous arrivals within subsets of processors and the ordering of the subset arrival times, we see that this approximation still captures the behavior of synchronization under workload imbalance.

4 Optimal Degree of Combining Tree

In this section, we investigate the optimal combining tree degree of a synchronization barrier for various numbers of processors and processor distributions. We assumed the processors to be normally distributed, an assumption supported by measurements in [13] and [15]. First, we will determine the optimal degree experimentally. Then, we will use the approximation presented in Section 3 to estimate this degree and compare the performance improvement between the experimental and estimated optimal degree.

Figure 3 presents the optimal combining tree degree for synchronizing 64, 256, and 4K processors for various standard deviations defined in units of t_c. The optimal degree corresponds to the degree that resulted in the smallest synchronization delay, as defined in Section 1. The horizontal axis lists a range of standard deviations for the distribution of processor execution times. The vertical axis lists the number of processors simulated. The first number in each entry is the optimal degree obtained by exhaustive simulation. The second number, in parenthesis, is the synchronization speedup obtained when using a combining tree with optimal degree, expressed as a ratio of the synchronization delay for a combining tree of degree four to that of the optimal degree. The performance of a combining tree of optimal degree is compared to a combining tree of degree four because degree four was previously considered as optimal [6] [7].

To obtain these optimal degrees we used a conventional event driven simulator. The time for updating a counter (t_c) was set to

Processors	Optimal Degree (Sync. Speedup)		
	$\sigma = 0t_c$	$\sigma = 25t_c$	$\sigma = 500t_c$
64	4 (1.00)	64 (2.87)	64 (2.99)
256	4 (1.00)	32 (1.97)	256 (4.00)
4K	4 (1.00)	32 (1.98)	128 (2.99)

Figure 3: Simulated optimal degree of combining trees.

$20 \mu s$ and the contention for updating the counters was accounted for in the simulation. The optimal degree was determined by carrying a simulation for all feasible degrees and choosing the one that yields the smallest synchronization delay.

There are several conclusions that can be drawn from this experiment. First, this experiment confirms the fact that combining trees of degree four are optimal when processors arrive simultaneously, or when the distribution of processors is small compared to t_c. Second, this experiment confirms our assertion that with a wide distribution of processors, large degrees yield smaller synchronization delays. For example, when 64 processors are distributed with a standard deviation of $25t_c$, a single counter yields the smallest synchronization delay. Finally, the synchronization speedup gained by synchronizing processors with a combining tree of optimal degree compared to a combining tree of degree of four ranges from 30 percent faster with a degree of eight up to 300 percent faster with a degree of 256.

One could argue that if there is substantial load imbalance relative to the counter update time t_c, that the overall parallel application is inefficient and any improvements in synchronization performance will have only a small impact on the overall performance. However, when fuzzy barriers are employed, load imbalance does not necessarily translate into idle times [13] and an application could have substantial load imbalance and still be efficient, provided synchronization delays are not excessive.

We now investigate the use of the approximated synchronization delay of Equation (8) to estimate the optimal degree of a combining tree. Figure 4 presents the estimated optimal degree for synchronizing 64, 256, and 4K processors for various standard deviations defined in units of t_c. These estimated optimal degrees are found in the rows labeled "est". For comparison, the optimal degree obtained by exhaustive simulation are shown in the rows labeled "opt". Results in bold indicate the cases where the estimated degree differs from the simulated optimum. As in the previous figure, we present the synchronization speedup gained by synchronizing processors with combining tree of optimal degree compared to a degree of four. The difference between the speedup associated with the optimal and the estimated optimal degree is particularly interesting, because it is a metric of how accurate our approximation is.

As indicated in Figure 4, the approximated synchronization delay is useful in determining the optimal degree. Moreover, when the estimation fails to identify the optimal degree, the speedup associated with the estimated optimal degree is usually not significantly smaller than the optimal speedup. Indeed the optimal degree combining trees are only 7% faster on average than the estimated degrees.

We also investigated the combining trees proposed by Mellor-Crummey and Scott [7], as described in Section 1. We simulated

Processors		Optimal Combining Tree Degree (Synchronization Speedup)					
		$\sigma = 0t_c$	$\sigma = 6.2t_c$	$\sigma = 12.5t_c$	$\sigma = 25t_c$	$\sigma = 50t_c$	$\sigma = 500t_c$
64	opt.	4 (1.00)	8 (1.31)	16 (1.47)	64 (2.87)	64 (2.94)	64 (2.99)
	est.	4 (1.00)	8 (1.31)	8 (1.46)	8 (1.48)	64 (2.94)	64 (2.99)
256	opt.	4 (1.00)	8 (1.26)	32 (1.89)	32 (1.97)	64 (1.99)	256 (4.00)
	est.	4 (1.00)	4 (1.00)	16 (1.78)	16 (1.96)	16 (1.99)	256 (4.00)
4K	opt.	4 (1.00)	4 (1.00)	8 (1.43)	32 (1.98)	128 (2.97)	128 (2.99)
	est.	4 (1.00)	4 (1.00)	8 (1.43)	16 (1.96)	64 (2.94)	64 (2.99)

Figure 4: Simulated and estimated optimal degree of combining trees.

these trees and obtained results similar to the one of Figure 4. Comparing these results with the ones of Figure 4, we noticed performance improvements of 5%, on average, for all combining trees with an optimal degree of four. However, this performance improvement vanishes when the optimal degree is larger than four. For degree four, this performance improvement is due to the fact that the average depth seen by the processors is smaller, since some of the processors are attached at higher levels of the combining tree; however, for larger degrees this improvement decreases as the proportion of processors at the higher levels also decreases.

We note that if one could guess which of the processors will arrive late at a synchronization point, one could place those processors near the top of the combining tree, and therefore reduce the synchronization delay. The next section will investigate the feasibility of this technique.

5 Dynamic Placement in Combining Trees

In this section, we use the combining tree presented by Mellor-Crummey and Scott [7]. As mentioned in Section 1, this technique uses combining trees of degree d where each counter in the tree is connected to at least one processor, and where leaf counters are connected statically to at most $d+1$ processors. We will investigate the feasibility of a scheme that positions late processors near the top of the tree, thus reducing the synchronization delay of the slower processors. This technique can be viewed as the shifting of the synchronization cost from the slower to the faster processors. We use a prediction scheme based on recent history that is expected to work well in two situations: with systemic workload imbalance and with fuzzy barriers.

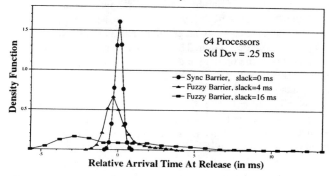

Figure 5: Distribution of processor arrival time for fuzzy barriers. ($\sigma = 0.25$ ms, $p = 64$)

With systemic workload imbalance, the workload is unevenly partitioned among processors, and therefore processors that were attributed a larger amount of work will systematically arrive late at synchronization points. A similar situation arises with evolving workload imbalance, where the workload slowly fluctuates from iteration to iteration. In both cases, recent history is a good indication of future processor arrival order.

With fuzzy barriers [11], independent operations are inserted between the release and the enforce phase, thus significantly reducing the expected idle time due to non-deterministic workload imbalance. The independent operation execution time corresponds to the *slack* of a fuzzy barrier. In [13], we showed that the expected idle time is inversely proportional to slack.

Here, we will consider another interesting property of fuzzy barriers. With increasing slack, some processors can be significantly slower than others without requiring faster processors to wait for them. As a consequence, processors that arrive late at a synchronization point are likely to arrive late in the near future.

Figure 5 illustrates the processor arrival time distributions for 64 processors and a range of slacks after 500 iterations. In this figure, the arrival times are relative to the mean: negative times correspond to processors faster than the average and vice versa. The curve labeled "Sync Barrier" corresponds exactly to the processor arrival time distribution after one single iteration since all processors were strongly synchronized at the previous synchronization barrier. We see that the larger the slack is, the wider the distribution is. This effect is explained by the fact that with large slack, processors are allowed to desynchronize themselves and are eventually distributed over the entire slack.

Several conclusions can be drawn from this experiment concerning the processor distribution synchronized with fuzzy barriers. First, we see that the processor distributions become wider with increasing slack. Second, we see that the distributions are not symmetric on their mean, resulting in an increasing number of fast processors and a decreasing number of slow processors. Finally, we see that the slowest processors are far away from the mean and are likely to remain far away during the next few iterations. Since the curve labeled "Sync Barrier" indicates the distribution after a single iteration, we know from the graph that it is unlikely that a processor changes its relative position by more than a half millisecond in this graph. As a result, processors that are ten milliseconds or more away from the mean are very likely to remain significantly slower for the next 20 iterations. This means, in turn, that a dynamic placement scheme is feasible with fuzzy barriers when the slack is larger than the distribution of processors after one iteration.

a) Synchronization Tree

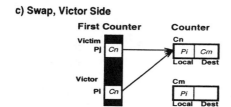

Victim →

Victor →

○ Counter
● Processor

b) Synchronization Data Structure

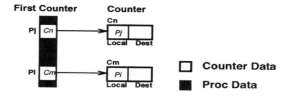

□ Counter Data
■ Proc Data

c) Swap, Victor Side

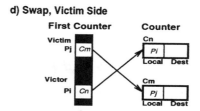

d) Swap, Victim Side

Figure 6: Dynamic placement barrier mechanism.

5.1 Dynamic Placement Barrier Algorithm

This section presents a *Dynamic Placement Barrier* algorithm that predicts the ordering of the processor arrival time based on previous history and places slower processors near the top of the combining tree. This algorithm is based on the combining tree presented in [7]. However, we expect to reduce the critical path from $O(\log p)$ to $O(1)$ when the prediction is successful.

The dynamic placement barrier proceeds as follows. When a processor propagates toward the top of the tree, it positions itself at the highest level counter where it arrived last. Figure 6a illustrates this scheme: processor Pi starts incrementing up the tree from the counter Cm, its initial counter, up to the counter Cn+1. We can deduce from this fact that processor Pi arrived last in the whole subtree attached to the counter Cn. Our dynamic placement scheme swaps processor Pi with the processor associated with Cn, namely Pj. For the remainder of this section, we name the processor that is swapped to a higher level in the combining tree the *victor* processor, and the one that is swapped to a lower level the *victim* processor.

This scheme requires two data structures, as illustrated in Figure 6b. The first one, First-Counter, is private to each processor and provides a pointer to the first counter associated with each processor. The second data structure, Counter, is associated with each counter and consists of two entries, Local and Destination. Its first entry allows us to locate the processor that is currently attached to a counter and the second entry provides an index to the new initial counter of the victim.

The swapping of a pair of processors occurs in two phases: the first phase occurs on the victor side and is followed by a second phase on the victim side. During the first phase, shown in Figure 6c, the victor processor checks if the conditions are fulfilled for a swap. If a swap is indicated, it updates its First-Counter pointer to the counter of the victim and modifies the two entries of the victim's initial Counter as follows: it writes its processor ID in the Local entry and its old First-Counter value in the Destination entry. During the next synchronization phase, the victim processor detects that it has been swapped by inspecting its initial counter Local field

```
TYPE Counter = RECORD
        count, init: INTEGER;    (* #children  *)
        local: [0..P-1];         (* current id *)
        dest, parent: *Counter;
     END;
VAR sense: BOOLEAN;              (* shared var *)
    first_counter: *Counter;     (* private var *)
    private_sense: BOOLEAN;      (* private var *)

PROCEDURE Release(id: [0..P-1]);
  private_sense := NOT private_sense;
  IF (first_counter->local != id) AND
     NOT Leaf(first_counter) THEN
     (* swap, victim size *)
     first_counter := first_counter->dest;
     first_counter->local := id;
  END;
  curr := first_counter; prev := NULL;
  LOOP (* through combining tree levels *)
     c := FetchAndDecrement(curr->count);
     IF (c != 0) THEN EXIT END;
     (* last at this level: reinit counter *)
     curr->count := curr->init;
     prev := curr; curr := curr->parent;
     if (curr = NULL) THEN EXIT END;
     (* last in last level: exit *)
  END;
  IF not Leaf(prev) AND (prev->local != id) THEN
     prev->local := id;    (* swap, victor size *)
     prev->dest := first_counter;
     first_counter := prev;
  END;
  (* last in last level reverse sense *)
  IF (curr = NULL) THEN sense := NOT sense END;
END

PROCEDURE Enforce()
  (* wait for last processor *)
  WHILE (private_sense != sense) DO END;
END
```

Figure 7: Dynamic placement barrier algorithm.

and by noticing that it is no longer local to its initial counter. As illustrated in Figure 6d, the victim uses the Destination entry of its initial counter to find out its new initial counter and updates its First-Counter entry accordingly. Figure 7 presents the detailed operations of this algorithm.

Assuming that the cache line is large enough to accommodate

a counter, local, and destination field, the communication overhead of the dynamic placement algorithm is one communication per swap, since one additional communication occurs on the victim side to acquire its new initial counter. Fortunately, this overhead occurs on the victim side, which was the faster of the two processors. Since there is at most one such swap among a counter and its direct children, the communication overhead is bounded by $1/(d+1)$ additional communications per processor. As a result, we can limit the upper bound communication overhead of this algorithm by choosing an appropriate degree of the combining tree.

	Slack				
	0ms	1ms	2ms	4ms	16ms
Degree: 4					
Last Proc Depth	5.85	3.34	1.88	1.44	1.24
Sync. Speedup	1.00	1.73	3.07	3.98	4.71
Comm. Overhead	1.09	1.08	1.07	1.04	1.01
Degree: 16					
Last Proc Depth	2.99	2.16	1.59	1.36	1.21
Sync. Speedup	0.99	1.34	1.85	2.21	2.45
Comm. Overhead	1.04	1.03	1.02	1.01	1.00

Figure 8: Performance of the dynamic placement barriers.

Figure 8 presents the performance improvement obtained with the dynamic placement barrier for 4K processors, normally distributed with a standard deviation of 0.25 ms for various slacks. Each set of measurements presents the average depth of the combining tree seen by the last processor releasing the barrier, the synchronization speedup of the dynamic placement scheme relative to the static placement scheme, and the fractional increase in communication occurred in the dynamic placement scheme.

First, we notice that as the slack increases the average depth of the last processor is effectively reduced from 5.85 to 1.24 and from 2.99 to 1.21 for barriers of degree 4 and 16, respectively. Second, we see that the synchronization speedup, relative to the performance of a static placement scheme, increases from 1 to 4.71 and from 0.99 to 2.45 for barriers of tree degree 4 and 16, respectively. It is interesting to notice that dynamic placement does not improve the performance with a slack of zero, in which case a static placement is as relevant a placement as the one of the previous iteration.

6 Quantitative Comparison

The performance improvements due to individual techniques were presented in earlier section. This section attempts to combine these techniques together and presents the cumulative performance effect.

Figure 9 presents the performance improvements due to synchronization with an optimal degree combining tree. First, we see that, for the curves corresponding to combining trees of degree four, the synchronization delay exactly corresponds to the depth of the tree, indicating that there is no contention. Therefore, the smallest standard deviation presented is sufficient to remove all contention problems for combining trees of degree four. Second, we see the performance improvements due to combining trees of optimal degree. Their synchronization delay is consistently less

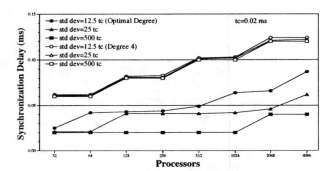

Figure 9: Degree four versus optimal degree.

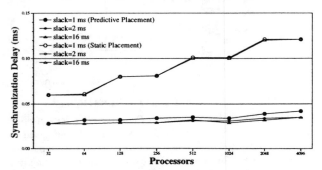

Figure 10: Static versus dynamic placements (degree 4).

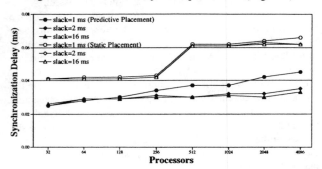

Figure 11: Static versus dynamic placements (degree 16).

than combining trees of degree four. Furthermore, the slope of these curves decreases with increasing standard deviation. Thus, the synchronization delay is relatively insensitive to the system size when load imbalance is sufficiently large.

Figure 10 illustrates the benefits of dynamic placement for execution times that are distributed with a very small standard deviation ($3.1t_c$). First, we see that the curves that correspond to a synchronization degree of four are similar to the curves of the previous figure, again indicating that there is no contention problem. Second, we see the performance improvement due to dynamic placement. The dynamic placement scheme almost neutralizes the tree depth in larger systems, and the synchronization delay is nearly constant.

Finally, Figure 11 illustrates the combined effects of higher tree degrees and dynamic placement. As in the previous figures, the curves that correspond to a static placement with a combining tree degree of 16 result in curves increasing stepwise with the number of processors, again indicating that there is no con-

Optimal Tree Degree (Synchronization Speedup)			
$d_y = 52$ $\sigma = 42\mu s$	$d_y = 210$ $\sigma = 110\mu s$	$d_y = 420$ $\sigma = 220\mu s$	$d_y = 840$ $\sigma = 421\mu s$
4 (1.00)	8 (1.04)	32 (1.22)	32 (1.23)

Figure 12: Measured optimal degree on the KSR1 for 56 processors.

tention problem. The curves associated with the dynamic placement present a synchronization delay that is smaller and increasing at a slower rate. We note that the curves associated with the dynamic placement scheme are more sensitive to the variations in slack than the corresponding curves of Figure 10. This effect is due to the fact that flatter trees have proportionately fewer processors near the top of the tree, making the choice of the slower processors more critical.

7   Measurements

In order to test the performance improvements due to optimal tree degree and dynamic placements on the KSR1 [24], we used a program that has a well defined computation and communication pattern. We used a relaxation algorithm (SOR) where each element is averaged with its four neighbors. The relaxation is performed in two alternating arrays, thus avoiding additional communication due to race conditions among processes. The two-dimensional data of size (d_x, d_y) is partitioned along the x-dimension, resulting in $4\lceil d_y/16 \rceil$ communication events per processors[3]. By varying the y-dimension, we change the total number of communications and therefore the variation of execution time [13]. All measurements consist of 200 relaxations on 56 processors[4] with d_x sets to 60 data points per processor.

In the first set of measurements, we investigated the optimal combining tree degree for various numbers of data along the y-dimension. Figure 12 shows the optimal combining tree degree as well as the synchronization speedup achieved by using an optimal degree as opposed to a degree of four. As the data size along the y-dimension increases, the variance of the execution time also increases. The experimentally determined standard deviation σ, is given for each data size. As the standard deviation increases, the optimal degree increases from 4 to 32 and the resulting speedup increases from zero to 23 percent.

In the second set of measurements, we evaluate the performance improvements due to the dynamic placement barriers. We used the same SOR program with 56 processors and 210 data points along the y-dimension, resulting in an execution time of 9.5 ms and a statistical standard deviation of $110\mu s$. Figure 13 presents the achieved performance improvements for various slacks and for a tree degree of 2, 4, and 16. We also present a combining tree of degree of two to test our dynamic placement barriers with as deep a combining tree as possible. Each set of measurements presents the combining tree depth seen by the last processor releasing the barrier[5] and the synchronization speedup achieved by using dynamic placement instead of static placement.

---

[3]The denominator, 16, corresponds to the cache sub-line size on the KSR1.

[4]We used 56 processors out of 64 to avoid using dedicated nodes (IOs) that would perturb the precision of the measurements.

[5]On the KSR1, the processors are organized in rings of 32 processors.

	Slack				
	0ms	1ms	5ms	10ms	15ms
Degree: 2 Last Proc Depth Sync. Speedup	 4.38 0.89	 3.93 0.98	 2.23 1.11	 1.755 1.44	 1.67 1.73
Degree: 4 Last Proc Depth Sync. Speedup	 3.24 0.84	 2.88 0.96	 1.63 1.15	 1.75 1.23	 1.44 1.25
Degree: 16 Last Proc Depth Speedup	 2.88 0.82	 2.7 1.09	 1.72 1.53	 1.57 1.34	 1.24 1.32

Figure 13: Performance of dynamic placement barriers on the KSR1.

First, we see that the combining tree depth for the last processor is effectively decreased from 4.38 to 1.67 and from 2.88 to 1.24 for trees of degree 2 and 16 respectively. Second, we see that dynamic placements result in a slower performance up to approximately a slack of 1 ms, and a performance improvement up to 1.73 and 1.32 for trees of degree 2 and 16 respectively.

8   Conclusion

Prior to our work, software synchronization barriers have been developed, evaluated and optimized under the assumption that all processors arrive simultaneously at the barrier. But, in most parallel systems there is some load imbalance. Therefore, we evaluate and reconfigure software combining trees for varying amounts of load imbalance.

Through analytic models and simulation results we show that the synchronization delay is minimized by choosing an optimum combining tree degree that is a function of the load imbalance. Thus, software barriers configured under the common assumption of simultaneous arrival are not merely over-designed for the probably more common case of distributed arrivals but have higher synchronization delays. Our analytic model can be used by a compiler to estimate the optimum degree and this estimate yields a performance that is within 7% of the actual optimum obtained by exhaustive simulation. This finding also indicates the feasibility of barriers that would adapt their degree at run time to minimize their synchronization delay.

Fuzzy barriers are effective in reducing the impact of non-deterministic load imbalance on overall performance. These barrier constructs also tend to distribute the arrival times of processors at a barrier over the slack interval. As a result, higher degree combining trees perform better when fuzzy barriers are used. Furthermore, processor arrival times are asymmetrically distributed with a few processors being much slower than average.

We show that dynamic placement schemes are very effective when fuzzy barriers are used. The average depth of the last processor to arrive at the barrier is indicative of the update delay component of the overall synchronization delay. The average depth reduces from close to L to close to 1.2 as the slack increases.

To preserve the ring locality, our dynamic placement scheme does not cross ring boundaries. As a result, the number of tree levels corresponds to two subtrees of 32 processors merged by an additional level. This explains why a tree degree of 16 results in an initial tree depth of three.

Since the contention delay component is small once the slack is sufficiently large, a speedup of close to L/1.2 is obtained over the static placement scheme for large slack.

The two techniques are combined and evaluated: a suitable tree degree is chosen based on load imbalance considerations and a dynamic placement scheme is used to exploit the last processor predictability under fuzzy barriers. The resulting synchronization delay is relatively insensitive to the number of processors when sufficient slack is present. These experiments demonstrate that software barriers implemented using simple hardware locks are scalable to large numbers of processors provided slack is available.

Acknowledgements

This work was supported in part by the Office of Naval Research under grant number N00014-93-1-0163 and by Hewlett-Packard. The University of Michigan's Center for Parallel Processing, site of the KSR1, is partially funded by NSF Grant CDA-92-14296.

References

[1] A. Gottlieb et al., "The NYU ultracomputer-designing an MIMD shared memory parallel computer," *IEEE Transactions on Computers*, vol. 32, no. 2, pp. 175–189, February 1983.

[2] G. Pfister et al., "The IBM research parallel processor prototype (RP3): Introduction and architecture," *Proceedings of the International Conference on Parallel Processing*, pp. 764–771, August 1985.

[3] *Cray T3D System Architecture Overview*, Cray Research, Inc, revision 1.c edition, September 1993.

[4] G. Pfister and V. A. Norton, "Hot spot contention and combining in multistage interconnection networks," *IEEE Transactions on Computers*, vol. C-34, no. 4, pp. 943–948, October 1985.

[5] M. M. Michael and M. L. Scott, "Fast mutual exclusion, even with contention," Technical Report TR-460, University of Rochester, 1993.

[6] P.-C. Yew, N.-F. Tzeng, and D. H. Lawrie, "Distributing hot-spots addressing in large-scale multiprocessors," *IEEE Transactions on Computers*, vol. 36, no. 4, pp. 388–395, April 1987.

[7] J. M. Mellor-Crummey and M. L. Scott, "Algorithms for scalable synchronization on shared memory multiprocessors," *ACM Transactions on Computer Systems*, vol. 9, no. 1, pp. 21–65, February 1991.

[8] A. Agrawal and M. Cherian, "Adaptive backoff synchronization techniques," *Proceedings of the Sixteenth Annual International Symposium on Computer Architecture*, pp. 396–406, May 1989.

[9] J. R. Goodman, M. K. Vernon, and P. J. Woest, "Efficient synchronization primitives for large-scale cache-coherent multiprocessors," *3rd International Conference on Architectural Support for Programming Languages and Operating Systems*, pp. 64–75, apr 1989.

[10] J. Lee and U. Ramachandran, "Synchronization with multiprocessor cache," *Proceedings of the Seventeenth Annual International Symposium on Computer Architecture*, pp. 27–37, May 1990.

[11] R. Gupta, "The fuzzy barrier: A mechanism for high speed synchronization of processors," *3rd International Conference on Architectural Support for Programming Languages and Operating Systems*, pp. 54–63, 1989.

[12] R. Gupta, "Loop displacement: An approach for transforming and scheduling loops for parallel execution," *Proceedings of Supercomputing '90*, pp. 388–397, 1990.

[13] A. E. Eichenberger and S. G. Abraham, "Modeling load imbalance and fuzzy barriers for scalable shared-memory multiprocessors," *Proceeding of the 28th Hawaii International Conference on System Sciences*, vol. I, pp. 262–271, January 1995.

[14] J. Nguyen, *Compiler Analysis to Implement Point-to-Point Synchronization in Parallel Programs*, PhD thesis, MIT, August 1993.

[15] V. S. Adve and M. K. Vernon, "The influence of random delays on parallel execution times," *ACM SIGMETRICS Conference on Measurement and Modeling of Computer Systems*, pp. 61–73, 1993.

[16] M. Dubois and F. A. Briggs, "Performance of synchronized iterative processes in multiprocessor systems," *IEEE Transactions on Software Engineering*, vol. SE-8, no. 4, pp. 419–431, July 1982.

[17] V. Sarkar, "Determining average program execution times and their variance," *Proceedings of the ACM SIGPLAN'89 Conference on Programming Language Design and Implementation*, vol. 24, no. 7, pp. 298–312, 1989.

[18] C. P. Kruskal and A. Weiss, "Allocating independent subtasks on parallel processors," *IEEE Transactions on Software Engineering*, vol. SE-11, no. 10, pp. 1001–1016, October 1985.

[19] S. Madala and J. B. Sinclair, "Performance of synchronous parallel algorithms with regular structure," *IEEE Transactions on Parallel and Distributed Systems*, vol. 2, no. 1, pp. 105–116, January 1991.

[20] M. D. Durand, T. Montaut, L. Kervella, and W. Jalby, "Impact of memory contention on dynamic scheduling on numa multiprocessors," *Proceedings of the International Conference on Parallel Processing*, vol. 1, pp. 258–267, 1993.

[21] T. S. Axelrod, "Effects of synchronization barriers on multiprocessor performance," *Parallel Computing*, vol. 3, pp. 129–140, 1986.

[22] C. J. Beckmann and C. D. Polychronopoulos, "The effect of barrier synchronization and scheduling overhead on parallel loops," *Proceedings of the International Conference on Parallel Processing*, vol. II, pp. 200–204, August 1989.

[23] A. H.-S. Ang and W. H. Tang, *Probability Concepts In Engineering Planning And Design*, volume 2, New York : Wiley, 1984.

[24] *KSR1 Principles of Operation*, Kendall Square Research Corporation, 1991.

Location Consistency: Stepping Beyond the Memory Coherence Barrier

Guang R. Gao
School of Computer Science
McGill University
3480 University Street
Montreal, Canada H3A 2A7

Vivek Sarkar
Application Development Technology Institute
IBM Software Solutions Division
555 Bailey Avenue
San Jose, California 95141

Abstract

In this paper, we introduce a new memory consistency model called Location Consistency (LC). The LC model uses a novel approach for defining memory consistency. The state of a memory location is modeled as a partially ordered multiset (pomset) of write operations and synchronization operations. The partial orders are determined solely by the ordering constraints imposed by the program being executed. We illustrate how the LC model can enable more compiler and hardware performance optimizations to be applied, compared to other memory consistency models which rely on the memory coherence assumption.

1 Introduction

The hardware memory consistency model that has been most commonly used as a basis for past work is *sequential consistency* (SC) [7]. It has been observed that sequential consistency limits performance by preventing the use of common uniprocessor hardware optimizations such as store buffers and out-of-order memory operations [6, 1].

The main approach taken in recent work on memory consistency models is to allow performance optimizations to be applied, while guaranteeing that sequential consistency is retained for a restricted class of programs — mainly programs that do not exhibit data races [2]. Therefore, we refer to these weaker memory consistency models as *SC-derived* models. Recently proposed SC-derived models include *weak ordering* (WO) [3] , *release consistency* (RC) [6], *data-race-free-0* (DRF0) [1], and *data-race-free-1* (DRF1) [2].

A central assumption in the definitions of all SC-derived memory consistency models is the *memory coherence* assumption, which can be stated as follows [6]: "all writes to the same location are serialized in some order and are performed in that order with respect to any processor". Memory coherence is a less restrictive form of serializability — it enforces serializability of operations performed on the same location, rather than serializability of all memory operations. However, it is still an additional restriction on allowable memory orderings, compared to the basic ordering constraints imposed by the program.

In contrast to the SC-derived models, we model the state of a memory location as a partially ordered multiset (pomset). Each element of the pomset corresponds to either a write operation to the location or a synchronization operation. The partial orders in the pomsets are determined entirely by the ordering constraints imposed by the program being executed. The LC model has thus stepped beyond the barrier of memory coherence. The LC model should be viewed as an extension of existing models by relaxing the ordering constraints imposed by the memory coherence assumption and by extending the definition of memory consistency to programs that may exhibit data races.

The rest of this paper is organized as follows. In section 2, we illustrate the restrictions on performance optimizations imposed by the SC-derived models through some motivating examples. In section 3, we summarize the program model assumed in this paper. In section 4, we define the pomset abstraction of memory systems used in our work. The LC model is defined in section 5, and its performance potential and implementation implications are discussed in section 6. and compare them to the LC model. Our conclusions are presented in section 7.

2 Motivating Examples

Figure 1 contains an example to illustrate the impact of the memory coherence assumption on register and cache locality. The program in Figure 1 contains

```
    Processor 1      Processor 2    PARTIAL ORDER
    -------------    -------------  ---------------
    w1: L := val1;   w3: L := val3; w1        w3
        . . .            . . .       |        /
    w2: L := val2;       . . .      w2        /
        . . .            . . .        \      /
    sync(P1,P2)<-->sync(P1,P2)       sync
        . . .            . . .         /    \
    r1: S1 := L;     r2: S2 := L;  r1        r2
        . . .            . . .
```

Figure 1: Example to illustrate register and cache locality

write-write data races on location L. The program partial order arising from the barrier synchronization ensures that the writes to L ($w1$, $w2$, $w3$) are completed before the reads ($r1$, $r2$) are initiated. The program partial order does not require that the write operations be atomic or that they be observed in the same order by both processors. If a write operation is meant to be performed atomically, then extra synchronization operations (e.g. lock/unlock, barriers) should be introduced in the program so that the resulting program partial order ensures that the write is performed in mutual exclusion.

In contrast, the SC-derived models require that write operations $w1$, $w2$, and $w3$ be observed in the same order by both processors; in particular, the observed order must be one of $< w1, w2, w3 >$, $< w1, w3, w2 >$, $< w3, w1, w2 >$. The memory coherence assumption prohibits an execution sequence in which (say) read operation $r1$ returns $val2$, and read operation $r2$ returns $val3$, because both processors are required to observe the same ordering of write operations to location L.

In fact, that the scenario that is prohibited by the memory coherence assumption is likely to be the most efficient way of satisfying the memory requests because of register locality. It is most efficient for location L to be allocated to a register across operations $w1$, $w2$, and $r1$ in processor P_1, and to a separate register across operations $w3$ and $r2$ in processor P_2. In this case, read operation $r1$ on processor P_1 will pick up the value $val2$ for location L, and read operation $r2$ on processor P_2 will pick the value, $val3$. This value assignment is consistent with the concurrency defined by the program partial order. There is no reason why

this scenario should be prohibited by a memory consistency model.

Note that compilation of processes one at a time, with only local knowledge, is the default compilation mode in use today for most concurrent programs. The previous discussion illustrates that the memory coherence assumption will force register allocation (and most other optimizations) to be disabled for all shared variables and all potentially shared variables, if a process is be executed correctly in a concurrent context. This is a serious restriction. We will revisit this example in later discussions in Sections 4 and 6.

3 Program Model

The memory and synchronization operations assumed by the LC model that are relevant to defining memory consistency are as follows:

- *Memory write* — if processor P_i wants to write value v in location L, it performs a $write(P_i, v, L)$ operation, which we represent by the notation, $L := v$, in processor P_i's instruction sequence.

- *Memory read* — if processor P_i wants to read a value from location L, it performs a $read(P_i, L)$ operation, which we represent by a read reference to L in processor P_i's instruction sequence.

- *Undirected control synchronization* — if processors $P_1, \ldots, P_k$ all need to synchronize among each other, the synchronization is accomplished by each processor performing a $sync(\{P_1, \ldots, P_k\})$ operation. This operation can be efficiently implemented by a counting semaphore [8]. A *sync* operation performed on the entire set of processors is equivalent to a barrier synchronization. The matching of *sync* operations from different processors is done at run-time, in general.

For a complete list of memory operations, readers are referred to [4].

4 Abstraction of Memory System

In this section, we define the memory abstraction used by the LC model. Section 4.1 defines the pomset *abstraction*. Section 4.2 specifies how the state (pomset) of a memory location is *updated* by a write operation or a synchronization operation. Section 4.3 specifies how the state of a memory location is *observed* by a read operation.

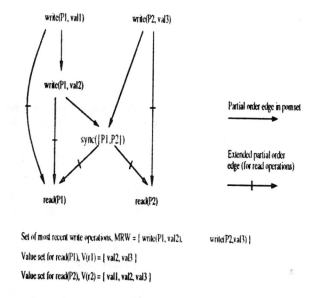

write(P1, val1)   write(P2, val3)

write(P1, val2)

sync({P1,P2})

read(P1)   read(P2)

Partial order edge in pomset

Extended partial order edge (for read operations)

Set of most recent write operations, MRW = { write(P1, val2),   write(P2,val3) }

Value set for read(P1), V(r1) = { val2, val3 }

Value set for read(P2), V(r2) = { val1, val2, val3 }

Figure 2: Value sets for read operations r1 and r2 in figure 1

- **Condition 2:** $e' = write(P_j, v)$ and $e' \not\prec' e$
  If e' is a write operation that does not precede e in $\prec'$, then it is automatically included in V (even though it may not be a most recent write operation).

As an example, Figure 2 illustrates the value sets defined for read operations $r1$ and $r2$ from the sample program in Figure 1. Consider any execution of this program. Figure 2 shows the pomset[1] for location L that is obtained after the write operations and synchronization operation have completed (note that the same pomset will be obtained, regardless of the order in which the write operations are completed). From Figure 2, we see that the set of most recent write operations is $MRW = \{w2, w3\}$.

The pomset is then extended with the edge $(w2, r1)$ for read operation $r1$, and (separately) with edge $(w3, r2)$ for read operation $r2$ (for convenience, we show both extended edges together in Figure 2). The value set for read operation $r1$ is given by $V(r1) = \{val2, val3\}$; $val2$ is included because write operation $w2$ satisfies Condition 1, and $val3$ is included because write operation $w3$ satisfies Condition 2 with respect to read operation $r1$. The value set for read operation $r2$ is given by $V(r2) = \{val1, val2, val3\}$; $val3$ is included because write operation $w3$ satisfies Condition 1, and $val1$ and $val2$ are included because write

---

[1]To avoid clutter, we do not show all transitive partial order edges in Figure 2.

operations $w1$ and $w2$ satisfy Condition 2 with respect to read operation $r2$.

5   Definition of Location Consistency (LC) model

In this section, we define the Location Consistency (LC) model for memory consistency with the help of an abstract interpreter. The execution model for the abstract interpreter maintains the state of each memory location as a pomset. The initial state of the memory location is assumed to be the empty set. The abstract interpreter mimics the execution of the memory and synchronization operations encountered in the concurrent program, and updates the states of memory locations according to the rules specified in section 4.2. For each read operation r on location L, the abstract interpreter computes the value set $V(r)$ from the pomset for location L and (arbitrarily) returns a value from the set $V(r)$ as the result of the read operation[2]. As in other definitions of memory consistency models [6], we retain the assumption that each processor in the abstract interpreter sequentially executes its assigned instruction sequence, subject to all uniprocessor control and data dependences being satisfied.

Analogous to the notion of a *sequential consistent execution* defined in [2], we introduce the notion of a *location consistent execution* as any execution of a program by the abstract interpreter discussed above. Note that the execution model of the abstract interpreter makes no assumption on the timing of events in the program execution. Therefore, there may be many location consistent executions for a given set of program inputs due to the nondeterminism that may be inherent in the program.

The Location Consistency (LC) model is now defined as follows:

Definition 5.1 *A multiprocessor system is location consistent if for any execution of a program on the system, 1) the operations of the execution are the same as those for some location consistent execution of the program, and 2) for each read operation R (with target location L) that is executed on the multiprocessor, the result returned by R belongs to the value-set, V, specified by the state of the memory location L as main-*

---

[2]If $V(r) = \emptyset$, the empty set, it means that the memory location has not been initialized, and the abstract interpreter should take whatever action is specified in the program model for a read access to an uninitialized location e.g. return an undefined value, or raise an exception.

tained by the abstract interpreter in the corresponding location consistent execution.

6 Performance Potential and Implementation Issues

As mentioned earlier, the LC model does not rely on the memory coherence assumption presented in [6] and other SC-derived models. Implementation difficulties that arise when trying to accommodate the memory coherence assumption in conjunction with such hardware optimizations are discussed in [5]. Since the LC model does not depend on the memory coherence assumption, such complications in implementing hardware optimizations will not exist, and hence the potential for higher performance.

The implementation requirements imposed by the LC model are weaker than the requirements imposed by the SC-derived models. For example, the LC model permits multiple writes from different processors to the same location to be performed concurrently. Consider the sample concurrent program in Figure 1. In the LC model, there is no need for either processor P_1 or P_2 to gain unique ownership of location L before it commits a write. Both processors can process the write operations concurrently, so long as each subsequent read operation will return some value from its value set defined by the LC model. We believe that more aggressive cache consistency protocols can be developed for the LC model, along these lines.

7 Conclusions and Future Work

The LC model provides a new framework for memory consistency models beyond the SC-derived models. As future work, we plan to exploit the power of the LC model by designing and extending compiler optimizations for shared-memory multiprocessors as permitted by the LC model and by designing new, efficient, and perhaps more aggressive, implementations of memory systems (including caches) in shared memory multiprocessors.

8 Acknowledgments

We thank K. Gharachorloo and M. Dubois for discussions on this work. We also thank V.C. Sreedhar for commenting on an earlier draft of the paper. Nasser Elmasri helped with the typesetting of this manuscript. Finally, we thank the National Sciences and Engineering Research Council (NSERC) and IBM Corporation for their continued support.

References

[1] S. V. Adve and M. D. Hill. Weak ordering—a new definition. In *Proceedings of the 17th ACM Symposium on Computer Architecture*, pages 2–14, May 1990.

[2] S. V. Adve and M. D. Hill. A unified formalization of four shared-memory models. *IEEE Transaction on Parallel and Distributed Systems*, pages 613–624, August 1993.

[3] M. Dubois, C. Scheurich, and F. A. Briggs. Memory access buffering in multiprocessors. In *Proceedings of the 13th ACM International Symposium on Computer Architecture*, pages 434–442, June 1986.

[4] G. R. Gao and V. Sarkar. Location Consistency: Stepping Beyond the Barriers of Memory Coherence and Serializability. ACAPS Tech. Memo 78 (Preliminary Version), Sch. of Comp. Sci., McGill U., Montréal, Qué., Dec. 1994.

[5] K. Gharachorloo, A. Gupta, and J. Hennessy. Revision to 'memory consistency and event ordering in scalable shared-memory multiprocessors'. Technical report, Computer System Laboratory, Stanford University, April 1993. Technical Report No: CSL-TR-93-568.

[6] K. Gharachorloo, D. Lenoski, J. Laudon, P. Gibbons, A. Gupta, and J. Hennessy. Memory consistency and event ordering in scalable shared-memory multiprocessors. In *Proceedings of the 17th ACM International Symposium on Computer Architecture*, pages 15–27, May 1990.

[7] L. Lamport. How to make a multiprocessor that correctly executes multiprocess programs. *IEEE Transactions on Computers*, 28(9):690–691, September 1979.

[8] Vivek Sarkar. Synchronization using counting semaphores. In *Conference Proceedings, 1988 International Conference on Supercomputing*, pages 627–637, St. Malo, France, July 4–8, 1988. ACM.

LOCAL ITERATION SET COMPUTATION FOR BLOCK-CYCLIC DISTRIBUTIONS

Samuel P. Midkiff
I.B.M. T.J. Watson Research Center
P.O. Box 704
Yorktown Heights, NY, 10598
midkiff@watson.ibm.com

Abstract – *The block-cyclic (cyclic(bs), where bs is a block size > 1) distribution is thought to be difficult to compile. The apparent difficulty arises from the inability to describe the array element to processor mapping as an affine function in one variable. In this paper, we show how an affine function in two variables can be used to describe the mapping. This mapping leads to an efficient method of computing the local iteration sets for cyclic(bs) distributed variables. We also describe how to precisely intersect, and conservatively union, these local iteration sets.*

INTRODUCTION

Large scale distributed memory computers pose special problems for compilers. Because the penalty for accessing off-processor memory is high, the distribution of data onto processors has a pronounced effect on performance. Languages motivated by the rise of distributed memory computing – High Performance Fortran (HPF)[9], Vienna Fortran[3] and Fortran-D[5] – recognize this by allowing the programmer to specify the distribution of data onto processors.

In the common compilation model for these languages, the data distribution implies both the communication between processors and the scheduling of computation onto processors. These in turn imply the realizable parallelism in the program.

Given a data distribution and a reference, the compiler must be able to determine what elements of an array that reside on the processor are accessed by the reference. This is typically done by finding the *local iteration set (LIS)* for a processor, i.e. the set of loop iterations for which a reference accesses elements owned by the processor. With the LIS, the loop iterations that execute on the processor, the elements of the array that are accessed, and the elements of the array that a processor is responsible for communicating can be determined.

For cyclic(1) and block distributions, efficient and straightforward methods of computing the LIS are known[10, 12]. LIS computation for the cyclic(bs) distribution is still problematic. In this paper we describe an efficient method for describing cyclic(bs) LIS's. This method has several useful properties:

1. The method is efficient. Most of the work of selecting elements of the LIS at run-time is performed by loop iteration. Unlike other methods, *gcd*'s and solutions of diophantine equations are not computed at run-time.

2. The local iteration sets can be intersected. This allows efficient communication to be generated, and iteration scheduling (based on the *owner computes* rule[1]) to be performed in the presence of references with *coupled* subscripts[2].

3. Local iteration sets can be unioned. This allows efficient scheduling of iterations, based on the owner computes rule, to be performed on loops with multiple left hand sides.

4. The efficiency of the method is not reduced if the loop bounds, array bounds, block size or number of processors is unknown at compile time.

NOTATION AND DEFINITIONS

The operations of multiplication and concatenation are used in this paper. Multiplication (of arrays and scalars) will be denoted using ".", e.g. $\mathbf{I} \cdot \mathbf{A}$. Concatenation of arrays is denoted by juxtaposing the array names, e.g. $\mathbf{IA}$.

The upper (lower) bound of an object o is U_o (L_o). Thus, L_a is the lower bound of a dimension of the array a, and U_i is the upper bound of the index variable i. The type of the object (array, index variable, etc.) will be clear from the context.

The subscript function for a dimension of a is $\sigma_a(i)$. The inverse of the subscript function is $\sigma_a^{-1}(i)$. The dimension of a will either be clear from the context or be irrelevant to the discussion.

Triplet notation will be used to specify collections of integers. Therefore

$$b \in [1:9:2]$$

says that b takes on the values $1, 3, 5, 7$ and 9.

We assume processors are organized as in HPF, i.e. they logically form a grid. Each processor has a number $[p_0, p_1, \ldots, p_n]$, where n is the rank of the processor grid. The processor grid numbers are zero origined.

Data is mapped onto the processor grid using *distribution* and *alignment*. At most one dimension of

---

[1] The owner computes rule says that the processor that owns the data element being assigned a value performs the computation.

[2] Coupled subscripts contain references to a common loop index variable. An example is $a(i, i)$.

the array is mapped onto a dimension of the processor grid, and an array dimension is cyclic(bs) distributed onto at most one dimension of the processor grid. Because of this, the technique we describe can be independently applied to each array/processor grid dimension. The dimension under discussion will either be clear from the context or irrelevant to the discussion. The processor number for a dimension will be denoted p, and the number of processors in that dimension will be denoted $\#p$.

An array and its cyclic(bs) distribution onto a one dimensional processor grid is shown in Figure 1. Adjacent elements of the array are grouped into blocks which are then distributed in a cyclic fashion onto the processor grid. Four elements make up each block, giving a block size (bs) of 4.

Alignment has little effect on the difficulty of computing LIS's. We are considering subscripts of the form $\sigma(i) = c \cdot i + k$. HPF alignment functions have the form $\alpha(i) = a \cdot i + b$. Computing the LIS of an aligned array with subscript σ and alignment function α is the same as computing the LIS of an array with subscript $\alpha(\sigma(i))$ Since affine functions in one variable are closed under composition, the resulting subscript is an affine function in one variable.

BACKGROUND

In this section we present some results concerning the solution of diophantine equations[3]. In later sections, solutions to diophantine equations will be used in the derivation of the local iteration set. We present these results without proof. The reader should consult [2] or another reference for details.

Let the diophantine equation we are attempting to solve be:

$$c = a_1 \cdot i_1 + a_2 \cdot i_2 + \ldots a_n \cdot i_n \qquad (1)$$

A is the coefficient matrix of the system. The j'th row of **A** contains coefficients of the j'th variable in the system. Therefore:

$$\mathbf{A} = \begin{bmatrix} a_1 \\ a_2 \\ \ldots \\ a_n \end{bmatrix}$$

C is the left hand side, or solution vector for the system. Since we are solving a single equation:

$$\mathbf{C} = [c]$$

To solve the system of Equation 1, a matrix is formed by concatenating **I**, the identity matrix, with **A** to get the matrix **IA**:

$$\begin{bmatrix} 1 & 0 & 0 & \ldots & 0 & a_1 \\ 0 & 1 & 0 & \ldots & 0 & a_2 \\ \ldots \\ 0 & 0 & 0 & \ldots & 1 & a_n \end{bmatrix}$$

---

[3] A diophantine equation is one whose coefficients and constant terms are all integers, and whose solutions are restricted to the integers.

An elimination algorithm can be applied to this matrix, transforming it to the matrix **UD**:

$$\begin{bmatrix} u_{1,1} & u_{2,1} & \ldots & u_{n,1} & d_1 \\ u_{1,2} & u_{2,2} & \ldots & u_{n,2} & 0 \\ \ldots \\ u_{1,n} & u_{2,n} & \ldots & u_{n,n} & 0 \end{bmatrix}$$

The matrix **U** is an $n \times n$ matrix and the matrix **D** contains the gcd of the the coefficients as its first element, and zeros for all other elements.

Let $\mathbf{T} = [t_1, t_2, \ldots, t_n]$. Then $\mathbf{T} \cdot \mathbf{D} = \mathbf{C}$, i.e.

$$c = d_1 \cdot t_1$$

The system of Equation 1 has integer solutions if and only if d_1 evenly divides c. Since d_1 is the gcd of the equation coefficients, this is simply the gcd test. If solutions exist to the equation,

$$\mathbf{T}[1] = \frac{d_1}{c}$$

and

$$\mathbf{T} = \left[t_1 = \frac{c}{d_1}, t_2, \ldots, t_n \right]. \qquad (2)$$

Also, if solutions exist, the relation $[i_1, i_2, \ldots, i_n] = \mathbf{T} \cdot \mathbf{U}$ holds. Less tersely:

$$\begin{aligned} i_1 &= t_1 \cdot u_{1,1} + t_2 \cdot u_{1,2} + \ldots + t_n \cdot u_{1,n} \\ i_2 &= t_1 \cdot u_{2,1} + t_2 \cdot u_{2,2} + \ldots + t_n \cdot u_{2,n} \quad (3) \\ \ldots \\ i_n &= t_1 \cdot u_{n,1} + t_2 \cdot u_{n,2} + \ldots + t_n \cdot u_{n,n} \end{aligned}$$

By substituting the solution for t_1 and arbitrary integer values for $1 < t_k \le n$ into the system of Equation 3, all solutions to the system of Equation 2 can be generated.

LIS COMPUTATION

This section describes a method for computing the local iteration set for cyclic(bs) distributed data. The LIS computed is represented as expressions emitted by the compile-time and evaluated at run-time. To find these expressions:

- represent the set of data resident on a processor ($\mathcal{P}$) as an affine function in two bounded variables;

- represent the set of data accessed by a reference ($\mathcal{R}$) as an affine function in one bounded variable;

- find the intersection of $\mathcal{P}$ and $\mathcal{R}$;

- generate the values of the index variable i that cause the reference to access members of $\mathcal{P} \cap \mathcal{R}$. These values of i form the LIS.

The expressions found by these steps are valid for all cyclic(bs) distributions and references whose subscript is an affine function in one variable. Therefore, these steps need not be performed at compile-time – the compiler only needs to emit the expression.

```
dim A(60)
!HPF$ processors p(4)
!HPF$ distribute a(cyclic(4)) onto p
do i = 0, 19
    a(3 · i + 1) = ...
end do
```

(a) A cyclic(bs) distributed array

$p = 0$				$p = 1$				$p = 2$				$p = 3$			
1	2	**3**	4	5	**6**	7	8	9	**10**	11	12	**13**	14	15	**16**
17	18	**19**	20	21	**22**	23	24	**25**	26	27	**28**	29	30	**31**	32
33	**34**	35	36	**37**	38	39	**40**	41	42	**43**	44	45	**46**	47	48
49	50	51	**52**	53	54	**55**	56	57	**58**	59	60				

(b) The distribution of the array a (referenced elements are shown in bold)

Figure 1: An example of a cyclic(bs) distribution

Representing cyclic(bs) Distributions

The portion of a cyclic(bs) distributed array that is owned by a processor cannot be represented as an affine function in a single variable. This is because the stride between adjacent elements is not constant. The stride between elements within a block is 1, and the stride between the last element in a block and the first element in the next block is $bs \cdot (\#p - 1) + 1$. Since the stride of an affine function in one variable is the value of the constant coefficient of the variable, the distribution cannot be represented by an affine function in one variable.

It can, however, be represented as an affine function in two variables, if the values of the variables are properly bounded. Let the blocks that are owned by (i.e. reside on) processor p be

$$\beta = [bs \cdot p + 1 : U_a : bs \cdot \#p]$$

and let the elements in each block be

$$\epsilon = [0 : bs - 1 : 1]$$

Thus, the region of an array owned by a processor p is $\mathcal{P}$:

$$\mathcal{P} = \{\beta + \epsilon \mid \forall \beta, \epsilon\}$$

Normalizing[4] the bounds for the blocks owned by processor p gives

$$L_b = 0; \quad U_b = \left\lfloor \frac{U_a - bs \cdot p - 1}{bs \cdot \#p} \right\rfloor .$$

Define b to be bounded by L_b and U_b, i.e.

$$L_b \leq b \leq U_b \qquad (4)$$

Normalizing the bounds on the elements in a block p gives the original bounds:

$$L_e = 0; \quad U_e = bs - 1.$$

---
[4] Normalization adjusts the variable bounds so that the induction sequence producing the values of the variable runs from 0 to some upper bound with a step size of 1.

and e is defined to be bounded by L_e and U_e, i.e.

$$L_e \leq e \leq U_e \qquad (5)$$

Then $\mathcal{P}$ becomes:

$$\mathcal{P} = \{bs \cdot \#p \cdot b + bs \cdot p + e + 1 \mid \forall b, e\} \qquad (6)$$

In the example, the array elements owned by processor $p = 1$ are:

$$\mathcal{P} = \{16 \cdot b + e + 5 \mid 0 \leq b \leq 3, 0 \leq e \leq 3\}$$

Representing Subscripts as an Affine Function

The region of an array accessed by a subscript function in a loop nest can be represented in the same way as $\mathcal{P}$ above. Let $\sigma(i) = c \cdot i + k$ be the reference subscript ($c = 3$, $k = 1$ in the example). The upper and lower bounds of the i loop are U_i and L_i (0 and 19 in the example). The stride of the loop is S_i (1 in the example). Normalizing the loop gives the set of accessed elements:

$$\mathcal{R} = \{S_i \cdot c \cdot i + c \cdot L_i + k \mid \forall i\} \qquad (7)$$

where i takes on the values:

$$0 \leq i \leq \left\lfloor \frac{U_i - L_i}{S_i} \right\rfloor \qquad (8)$$

The set of elements accessed in the example is:

$$\mathcal{R} = \{3 \cdot i + 1 \mid 0 \leq i \leq 19\}$$

Intersecting the $\mathcal{P}$ and $\mathcal{R}$ Regions

The LIS for a block-distributed reference is the intersection of the set of array elements accessed by the reference ($\mathcal{R}$) and the set of array elements that resides on the processor ($\mathcal{P}$).

Ignoring the bounds on the variables in $\mathcal{P}$ and $\mathcal{R}$, the intersection of the regions described in Equations 6 and 7 are solutions to the equation

$$S_i \cdot c \cdot i + c \cdot L_i + k = bs \cdot \#p \cdot b + bs \cdot p + e + 1$$

Moving loop terms not involving the variables of the equation (b, e, i) to the left hand side gives:

$$c \cdot L_i + k - bs \cdot p - 1 = bs \cdot \#p \cdot b + e - S_i \cdot c \cdot i \quad (9)$$

We now solve this system using results from the background section. The left hand side of the equation are placed in the matrix $\mathbf{C}$:

$$\mathbf{C} = [c \cdot L_i + k - bs \cdot p - 1]$$

The coefficients of the equation are placed in the matrix $\mathbf{A}$:

$$\mathbf{A} = \begin{bmatrix} bs \cdot \#p, \\ 1, \\ -S_i \cdot c \end{bmatrix}$$

We now form the 3×4 matrix $\mathbf{IA}$: and use the elimination procedure to transform the matrix into the matrix $\mathbf{UD}$:

$$\begin{bmatrix} 1 & 0 & 0 & bs \cdot \#p \\ 0 & 1 & 0 & 1 \\ 0 & 0 & 1 & -S_i \cdot c \end{bmatrix} \implies$$

$$\begin{bmatrix} 0 & 1 & 0 & 1 \\ 0 & S_i \cdot c & 1 & 0 \\ 1 & -bs \cdot \#p & 0 & 0 \end{bmatrix}$$

The $\mathbf{U}$ and $\mathbf{D}$ matrices are:

$$\mathbf{U} = \begin{bmatrix} 0 & 1 & 0 \\ 0 & S_i \cdot c & 1 \\ 1 & -bs \cdot \#p & 0 \end{bmatrix}, \mathbf{D} = \begin{bmatrix} 1 \\ 0 \\ 0 \end{bmatrix}$$

Since $\mathbf{T} \cdot \mathbf{D} = \mathbf{C}$, $t_1 = c \cdot L_i + k - bs \cdot p - 1$ and $\mathbf{T}$ can be written:

$$\mathbf{T} = [c \cdot L_i + k - bs \cdot p - 1, t_2, t_3].$$

$\mathbf{T} \cdot \mathbf{U}$ gives the parametric equations for b, e and i:

$$b = t_3 \quad (10)$$
$$e = c \cdot L_i + k - bs \cdot p - 1 + \quad (11)$$
$$\quad S_i \cdot c \cdot t_2 - bs \cdot \#p \cdot t_3$$
$$i = t_2 \quad (12)$$

Note that we have formed the parametric equations without any knowledge of the values of bs, $\#p$, S_i and c. Therefore the equation does not need to be solved at run-time or at compile-time. The equation can be solved without knowing the actual values of the left hand sides or coefficients for b and i because the coefficient of e is always one, and therefore the gcd of the coefficients of Equation 9 is always one. Because one evenly divides any integer, the equation always has solutions. This allows us to express the output of the elimination procedure symbolically in terms of the other elements of the $\mathbf{A}$ matrix.

Computing the LIS

In order to determine the elements owned by processor p that are accessed by the reference, values for the parameters cannot be any integer, but only those integers that yield values for b, e and i that lie within the bounds given in Equations 4, 5 and 8.

Since $b = t_3$ (Equation 10) the bounds on t_3 are the bounds of b:

$$0 \leq t_3 \leq \left\lfloor \frac{U_a - bs \cdot p - 1}{bs \cdot \#p} \right\rfloor$$

The bounds of t_2 can be found by combining the bounds for e in Equation 5 with the parametric equation for e (Equation 11)

$$0 \leq$$
$$c \cdot L_i + k - bs \cdot p - 1 + S_i \cdot c \cdot t_2 - bs \cdot \#p \cdot t_3$$
$$\leq bs - 1$$

Subtracting all terms that do not contain t_2 from the lower bound, and dividing by the coefficient of t_2 gives a lower bound on t_2 of

$$L_{t_2} = \frac{-c \cdot L_i - k + bs \cdot p + 1 + bs \cdot \#p \cdot t_3}{S_i \cdot c}$$

This inequality is in the rational numbers. Since t_2 is an integer, the lower bound of t_2 should be the smallest integer greater than the rational lower bound, i.e. the ceiling of the rational lower bound. Doing this gives a lower bound for t_2 of

$$L_{t_2} = \left\lceil \frac{-c \cdot L_i - k + bs \cdot p + 1 + bs \cdot \#p \cdot t_3}{S_i \cdot c} \right\rceil \quad (13)$$

Subtracting all terms that do not contain t_2 from the upper bound, and dividing by the coefficient of t_2 gives an upper bound on t_2 of

$$U_{t_2} = \frac{-c \cdot L_i - k + bs \cdot p + bs \cdot \#p \cdot t_3 + bs}{S_i \cdot c}$$

The upper bound should be the largest integer less than the rational upper bound, i.e. the floor of the rational upper bound. Doing this gives an upper bound on t_2 of

$$U_{t_2} = \left\lfloor \frac{-c \cdot L_i - k + bs \cdot p + bs \cdot \#p \cdot t_3 + bs}{S_i \cdot c} \right\rfloor$$

Combining these gives bounds for t_2 of

$$L_{t_2} \leq t_2 \leq U_{t_2} \quad (14)$$

Note that by Equation 12, $i = t_2$. Therefore, Equation 14 describes the local iteration set for the index variable.

Figure 2 gives an algorithm for computing the LIS. The algorithm extends the previous discussion in the following way. It is likely that the n array elements owned by a processor are packed into a zero origined buffer of length n. Steps two and three

Assume as input the subscript expression $\sigma(i) = c_i \cdot i + k$, and $\#p$, p, bs, L_i, U_i, L_a and U_a as defined previously.

1. form the equations:

$$L_b = 0$$
$$U_b = \left\lfloor \frac{U_a - bs \cdot p - 1}{bs \cdot \#p} \right\rfloor$$
$$L_i' = L_{t_2}$$
$$U_i' = U_{t_2}$$

2. replace the do i loop with:

```
do b = L_b, U_b
    L_{A,b} = p · bs + L_A + b · #p · bs
    e_f = σ(L_i') − σ(L_{A,b}) − c
    do i = max(L_i, L_i'), min(U_i, U_i'), s
        e_f = e_f + c
        ...loop body ...
    end do
end do
```

3. replace the reference $a(\sigma(i))$ with:

$$\dots buffer(b, e_f) \dots$$

Figure 2: Algorithm for cyclic(bs) local iteration set computation

give the code necessary for converting a value of the subscript expression evaluated at some member of the LIS into an offset into this buffer. If the buffer is not zero origined, the calculation can be adjusted accordingly.

Using this algorithm, the loop in the example can be rewritten:

```
dim A(60)
!HPF$ processors p(4)
!HPF$ distribute a(cyclic(4)) onto p

p = processor_number(1)
do b = 0, 3
    L_i' = ⌈(16·b+4·p)/3⌉
    L_{A,b} = 4 · p + 1 + 16 · b − 3
    e_f = 3 · L_1' − 3 · L_{A,b} − 3
    do i = max(0, L_i'), min(19, ⌊(16·b+4·p+3)/3⌋)
        e_f = e_f + 3
        buffer(b, e_f) = ...
    end do
end do
```

The algorithm of Figure 2 assumes that, in the normalized loop, that array elements are accessed in ascending order. i.e. that $S_i \cdot c > 0$. If $S_i \cdot c < 0$ the technique still works if the upper and lower bounds on b and i are reversed, and e_f is made to run from high to low.

Finally, we note that a caching scheme in the spirit of that used in in [8] can be used to minimize the amount of arithmetic performed in the generated loop nest.

LIS Computation for Aligned Variables

When arrays are aligned with other arrays or templates, the basic technique still applies. Some changes, however, needs to be made in forming the value of e_f. Let $\mathcal{T}$ be the array, or template, that array **a** is ultimately aligned to. Let $\sigma_\alpha = c_\alpha \cdot i + k_\alpha$ be the alignment function, i.e. element i of array **a** is aligned with element $\sigma_\alpha(i)$ of $\mathcal{T}$. Let $\sigma_a = c_a \cdot i + k_a$ be the subscript function for the dimension of **a** under consideration.

We assume that blocks of **a** are stored as adjacent elements, as before. The length of these blocks is

$$\left\lfloor \frac{c_\alpha}{c_a} \right\rfloor + 1$$

The value of i for the first element of $\mathcal{T}$ in a block, $L_{\mathcal{T},b}$ is given by:

$$L_{\mathcal{T},b} = bs \cdot p + bs \cdot \#p \cdot b + L_{\mathcal{T}}$$

Applying the inverse of σ_α to $L_{\mathcal{T},b}$, and taking the ceiling, gives the first element of **a** in the block, i.e.

$$L_{a,b} = \left\lceil \sigma_\alpha^{-1}(L_{\mathcal{T},b}) \right\rceil$$

Let the *effective subscript* for **a** be the composition of the alignment and subscript functions for **a**, i.e.

$$\sigma_e(i) = \sigma_\alpha(\sigma_a(i)),$$
$$\sigma_e(i) = c_\alpha \cdot c_a + c_\alpha \cdot k_a + k_\alpha$$

Then the value of i that indexes first element of **a** in the block accessed by the subscript, L_i', can be found by applying Equation 13 to σ_e and L_i.

With these two values e_f, the first position in the buffer of an accessed element of **a**, is given by:

$$e_f = c_a \cdot L_{i,b} + k_a - L_{a,b} - c_a$$

Incrementing e by c_a within the inner loop (as in the algorithm of Figure 2) will access the desired elements of **a** in the buffer.

The bounds of b are computed as before using $L_{\mathcal{T}}$ for the distributed object lower bound and $\sigma_\alpha(U_a)$ for the distributed object upper bound, and the block size of the $\mathcal{T}$ for the block size. The bounds for i are computed using $\sigma_e(i)$.

Figure 3 gives an example of LIS computation for a reference to an aligned variable.

LIS UNION AND INTERSECTION

In this section we describe how to form the union and intersection of LIS's resulting from the algorithm of Figure 2. When forming unions and intersections we will be dealing with multiple references or multiple dimensions of a reference. Therefore the variables bs, $\#p$, p, c, k, e, b and i will be written bs_j, $\#p_j$, p_j, c_j, k_j, e_j, b_j and i_j to indicate that the variable is associated with the j'th reference (or dimension of a reference.)

Finding the Union of LIS's

Finding an approximate union of LIS's is straightforward. An approximate union of n references is shown in Figure 4. The union is computed by taking the union of the b and i bounds of each reference.

```
REAL a(25)
!HPF$ processor P(3)
!HPF$ template T(48)
!HPF$ distribute T(cyclic(5)) onto P
!HPF$ align a(i) onto T(2 · i − 1)
do i = 0, 8
    a(3 · i + 1) = . . .
end do
```

$$\sigma_\alpha(i) = 2 \cdot i - 1$$
$$\sigma_a(i) = 3 \cdot i + 1$$
$$\sigma_e(i) = 6 \cdot i + 1$$

(a) A program   (b) The σ_α, σ_a and σ_e functions

		$p=0$					$p=1$					$p=2$		
T	1	2	3	4	5	6	7	8	9	10	11	12	13	14	15
a	1		2		3		4		5		6		7		8
T	16	17	18	19	20	21	22	23	24	25	26	27	28	29	30
a		9		10		11		12		13		13		15	
T	31	32	33	34	35	36	37	38	39	40	41	42	43	44	45
a	16		17		18		19		20		21		22		23
T	46	47	48	49											
a		24		25											

(c) The distribution of T and $\mathbf{a}$

$$\text{do } b = 0, \left\lfloor \frac{50 - 5 \cdot p - 1}{bs \cdot \#p} \right\rfloor$$
$$L_{T,b} = bs \cdot p + bs \cdot \#p \cdot b + L_T$$
$$L_{a,b} = \left\lceil \frac{L_{T,b}+1}{2} \right\rceil$$
$$L'_i = \max\left(0, \left\lceil \frac{5 \cdot p + bs \cdot \#p \cdot b}{6} \right\rceil\right)$$
$$e_f = 3 \cdot L'_i + 1 - L_{a,b} - 3$$
$$\text{do } i = \max\left(0, L'_i\right), \min\left(8, \left\lfloor \frac{5 \cdot p + bs \cdot \#p \cdot b + 4}{6} \right\rfloor\right)$$
$$e_f = e_f + 3$$
$$buffer(b, e_f) = \ldots$$
$$\text{end do}$$
$$\text{end do}$$

(d) The resulting code

Figure 3: An example of computing the LIS for an aligned array

This union can be made more exact. The union of the b bounds is done as before. Let β be a value of b in the union. When determining the union of i bounds, add the bounds of i_j only if β is within the bounds of b_j, i.e. $L_{i_j} \leq \beta \leq U_{i_j}$. This can be done with a series of if statements.

Finding the Intersection of LIS's

The LIS for two references r_1 and r_2 can be found by solving equations like Equation 9. This gives two sets, $\mathcal{L}_1$ and $\mathcal{L}_2$. What we now wish to compute is a third set, $\mathcal{L} = \{i \mid i \in \mathcal{L}_1 \wedge i \in \mathcal{L}_2\}$. Figure 5 shows code to compute this intersection.

The basic idea behind our method is as follows. Each block b_1 of the first reference's distribution is visited. During each visit to b_1, each block b_2 of the second reference's distribution is visited. The bounds on i_1 and i_2 are computed, and if the ranges of i_1 and i_2 overlap, the common values of i are members of the intersection.

$$\text{do } b = \min(L_{b_1}, L_{b_2}, \ldots, L_{b_n}),$$
$$\max(U_{b_1}, U_{b_2}, \ldots, U_{b_n})$$
$$\text{do } i = \max(L_i, \min(L_{i_1}, L_{i_2}, \ldots, L_{i_n})),$$
$$\min(U_i, \max(U_{i_1}, U_{i_2}, \ldots, U_{i_n}))$$
$$\ldots$$
$$\text{if } b \in \left[L_{b_j}, U_{b_j}\right] \text{ and } i \in \left[L_{i_j}, U_{i_j}\right] \text{ then}$$
$$r_j$$
$$\ldots$$
$$\text{end do}$$
$$\text{end do}$$

Figure 4: Code to compute an approximate union of LIS's

In Figure 5 we do better than this. By the Algorithm of Figure 2, the lower bound of i_2 is given by:

$$L_{i_2} = \left\lceil \frac{-c \cdot L_i - k + bs \cdot p + 1 + bs \cdot \#p \cdot t_3}{S_i \cdot c_2} \right\rceil$$

```
do b_1 = L_{b_1}, U_{b_1}
  calculate L_{i_1}, U_{i_1}
  do b_2 = max(L_{b_2}, low_{b_2}), min(U_{b_2}, up_{b_2})
    calculate L_{i_2}, U_{i_2}
    do i = max(L_i, L_{i_1}, L_{i_2}), min(U_i, U_{i_1}, U_{i_2})
      code requiring the intersection
    end do
  end do
end do
```

Figure 5: Code to generate the intersection of two LIS's

i.e. by the t_2 lower bound function applied to i_2. Assuming that integer division truncates, the ceiling function can be dropped by adding $S_i \cdot c_2 - 1$ to the numerator, giving:

$$L_{i_2} = \frac{-k_2 - c_2 \cdot L_i + bs_2 \cdot p_2 + bs_2 \cdot \#p_2 \cdot b_2 + S_i \cdot c_2}{S_i \cdot c_2}$$

For there to be common values of i_1, it is necessary that $L_{i_2} \leq L_{i_1}$, i.e. that

$$\frac{-k_2 - c_2 \cdot L_i + bs_2 \cdot p_2 + bs_2 \cdot \#p_2 \cdot b_2 + S_i \cdot c_2}{S_i \cdot c_2} \leq L_{i_1}$$

Assume that $S_i \cdot c_2 \geq 0$. Solving for b_2 yields:

$$b_2 \leq \frac{U_{i_1} \cdot S_i \cdot c_2 + k_2 + c_2 \cdot L_i - bs_2 \cdot p_2 - S_i \cdot c_2}{bs_2 \cdot \#p_2}$$

We call the function that computes the upper bound on b_2 up_{b_2}[5]. A similar function, which we call low_{b_2}, can be found to compute the lower bound for b_2, i.e. $U_{i_2} \geq L_{i_1}$ and therefore

$$low_{b_2} = \frac{L_{i_1} \cdot S_i \cdot c + k_2 + c_2 \cdot L_i - bs_2 \cdot p_2 - bs_2}{bs_2 \cdot \#p_2}$$

Now, instead of searching through all blocks associated with i_2, we only search through the blocks that may have solutions, i.e the region $max(L_{b_2}, low_{b_2}) \leq b_2 \leq min(U_{b_2}, up_{b_2})$

At this point, the lower and upper bounds on i_1 and i_2 can be computed, and overlapping values enumerated as the intersection.

If more than two LIS's need to be intersected, code similar to that given in Figure 5 can be generated.

COMPLEXITY

If the block size is greater than the coefficient of the normalized loop ($S_i \cdot c_i$), then at least one element is accessed per block. In this case the amount of work is proportional to the number of elements accessed, and the amount of work per element is constant.

If the block size is less than the coefficient of the normalized subscript then there may be blocks from which no element is accessed. In the worst case, no array elements accessed by an array reference may

reside on the processor. For example, consider a cyclic(1) distribution onto two processors. Processor $p = 0$ will get all of the odd elements, and processor $p = 1$ will get all of the even elements. If the normalized subscript is $2 \cdot i$, then no elements on processor 0 will be accessed, and b blocks will be visited.

On average,

$$\#b = \frac{U_a - L_a + 1}{bs \cdot \#p}$$

blocks reside on each processor. On average,

$$\#e = \frac{U_a - L_a + 1}{S_i \cdot c_i \cdot \#p}$$

elements will be accessed on each processor. Thus, the work per element accessed is $\#b/\#e$. Simplifying this fraction gives a work per element of $O((C_i \cdot S_i)/bs)$

The intersection takes, in the worst case, $O(bs^{r-1})$ time, where r is the number of references being intersected. In practice, the running time will be determined by the effectiveness of outer if statements in cycling through blocks without references. We hope to run experiments to gain some insight into the effectiveness of the if statements in improving the efficiency of the algorithm.

RELATED WORK

Gupta, et. al. [6] use a *virtual processor* approach to develop *virtual cyclic* and *virtual block* techniques. In the virtual cyclic approach, array elements are accessed in a different order than in the original loop. In the virtual block approach, processors potentially have to perform a scan on the order of the number of blocks in the entire system, or solve diophantine equations at run-time.

The technique of Ancourt, et. al. [1] have a technique that is not guaranteed to access the elements of the array in the same order that they were accessed in the original loop. It is, like the *virtual cyclic* approach mentioned above, therefore useful primarily for *INDEPENDENT* loops.

Chatterjee, et al. [4] developed a technique that uses a state machine to iterate over the accessed array elements owned by a processor. The construction of the state machine on each processor is a non-trivial task involving the solution of diophantine equations at run-time.

Stichnoth, et al. [11] provide a method for finding LIS's by treating the cyclic(bs) distribution as a collection of cyclic(bs) distributions. Unfortunately, the array elements accessed by the LIS are not enumerated in the same order they are visited in the original program. Thus the technique is limited to fully parallel loops.

Hiranandani, et al. [7] provide a method that is not valid for all programs. In particular, $c \mod \#p \cdot bs < bs$ must hold.

Kennedy, et al. [8] provide a method that visits members of the LIS in order. The method, however, requires solving Diophantine equations at run-time.

---

[5]If $S_i \cdot c_2 < 0$, an analogous result can be obtained using the function for U_{i_2}.

None of the general methods allow computed LIS's to be intersected. This means that references with coupled subscripts, e.g. when diagonals are accessed, cannot be handled. None of the general methods allow computed LIS's to be unioned. This means that scheduling loop iterations containing multiple cyclic(bs) distributed left-hand sides will be less precise. Our method allows both of these operations to be performed. The form of the loop bound is unimportant. Therefore our method works with trapezoidal loops, i.e. loops whose bounds are a function of outer loop induction variables, with no additional overhead. Finally, all of the other general methods require the solving of diophantine equations at run-time. Our method formulates the solution as closed form parametric equations, with the compiler emitting expressions resulting from substituting the parameters into this equation. Thus, diophantine equations do not need to be solved either at compile-time or run-time, and correct and efficient code can be generated even if the parameters to the problem are unknown at compile-time.

CONCLUSIONS

We have presented a method for computing local iteration sets of cyclic(bs) distributed arrays. The method is efficient – most of the overhead is subsumed by loop iteration – and unlike other methods does not require the solution of diophantine equations at run-time. The techniques presented extend the ability of current techniques by allowing intersections and unions of LIS's to be computed. This allows LIS's for diagonal references to be found, and allows efficient scheduling, using the owner computes rule, to be performed when multiple cyclic(bs) distributed left hand sides occur in a loop body. Because the form of the loop bound is irrelevant to the algorithm our technique works with both constant loop bounds, and trapezoidal loop bounds. Finally, because the LIS is represented symbolically at compile-time, input parameters to the problem – e.g. the number of processors, array bounds, block size and loop bounds – may be unknown at compile time and can even vary during the course of the program execution.

As well, the implementation of the algorithm is extremely simple. This allows simpler, more correct compilers to be built. This in turn increases the likelihood that compile-time handling of cyclic(bs) distributions will be implemented in a compiler.

ACKNOWLEDGEMENTS

I would like to thank Manish Gupta, Edith Schonberg and Peter Sweeney for their comments.

REFERENCES

[1] C. Ancourt, D. Coelho, F. Irigoin, and R. Keryell, *A linear algibra framework for static hpf code distribution*, in Proceedings of the Fourth Workshop on Compilers for Parallel Computers, Delft, The Netherlands, December 1993.

[2] U. Banerjee, *Dependence analysis for supercomputing*, Kluwer Academic Publishers, Boston, Mass., 1988.

[3] B. Chapman, P. Mehrotra, and H. Zima, *Vienna Fortran - a Fortran language extension for distributed memory multiprocessors*, tech. rep., Institute for Computer Applications in Science and Engineering, NASA Langley Research Center, 1991. Report 91-72.

[4] S. Chatterjee, J. R. Gilbert, F. J. E. Long, R. Schreiber, and S.-H. Teng, *Generating local addresses and communication sets for data-parallel programs*, in Proc. 4th annual ACM Symposium on Principles and Practice of Parallel Programming, San Diego, CA, May 1993.

[5] G. Fox, S. Hiranandani, K. Kennedy, C. Koelbel, U. Kremer, C. Tseng, and M. Wu, *Fortran D language specification*, tech. rep., Dept. of Comp. Sci, Rice University, 1990. Report COMP TR90-141.

[6] S. Gupta, S. Kaushik, C.-H. Huang, and P. Sadayappan, *On compiling array expressions for efficient execution on distributed memory machines*, tech. rep., Department of Computer and Information Sciences, The Ohio State University, April 1994. Technical Report OSE-CISRC-4/94-TR19.

[7] S. Hiranandani, K. Kennedy, J. Mellor-Crummey, and A. Sethi, *Compilation techniques for block-cyclic distributions*, in Proc. 1994 ACM International Conference on Supercomputing, Manchester, England, July 1994.

[8] K. Kennedy, N. Nedeljković, and A. Sethi, *A linear time algorithm for computing the memory access sequence in data-parallel programs*, tech. rep., Center for Research on Parallel Computation, Rice Univ., 1994. Tech Report CRPC-TR94485-S.

[9] C. Koelbel, D. Loveman, R. Schreiber, G. Steele Jr., and M. E. Zosel, *The High Performance FORTRAN Handbook*, The MIT Press, Cambridge, MA, 1994.

[10] C. Koelbel and P. Mehrotra, *Compiling global name-space parallel loops for distributed execution*, IEEE Trans. Parallel and Distributed Systems, 2 (1991), pp. 440–451.

[11] J. Stichnoth, D. O'Hallaron, and T. Gross, *Generating communication for array statements: design, implementation and evaluation*, in Proceedings of the Sixth Workshop on Languages and Compilers for Parallel Computing, August 1993.

[12] H. Zima and B. Chapman, *Compiling for distributed-memory systems*, Proceedings of the IEEE, 81-13 (1993), pp. 264–287.

Blocking Entry Points in Message-Driven Parallel Systems*

R. J. Richards   **B. Ramkumar**
Department of Electrical and Computer Engineering,
University Of Iowa,
Iowa City, Iowa 52242
{raze,ramkumar}@eng.uiowa.edu
Phone: 319-335-5957

Abstract – Message-driven parallel programming permits high processor utilization through small light weight tasks and latency tolerance. In this model, messages are processed to completion by tasks which execute in a non-premptive, non-blocking manner. As a result, it is not necessary to save the state of a task between context switches using run-time stacks as is usually done with multithreading.

Message-driven execution implies that messages may be processed in a nondeterministic order. If a predetermined order is desired, some awkward programming may be necessary to enforce such an order. An entry point is defined as a location in a task, (typically the name of an operation or method) to which a message may be sent to request remote processing. In this paper, we generalize entry points to efficiently support implicit blocking within the context of message-driven execution.

1 Introduction

Parallel programs on large scale distributed memory multiprocessors frequently suffer from poor processor utilization, even for problems exhibiting sufficient concurrency. In practice, considerable processing time remains unutilized during synchronization across processors. This is often due to cost of the inter-process communication needed to implement synchronization.

In an effort to improve processor utilization, a multithreaded approach to parallel programming has been proposed. Multithreading can be provided at different levels of granularity: the programmer can create coarse grained threads (e.g. Brown threads [12]), threads can be supported at the operating system level (e.g. Mach threads [20]), or fine grained threads can be supported with hardware support (e.g. Alewife [1], Tera [3]) The basic idea is to provide several threads of execution, each of which represents a lightweight process or task. Threads can be created and destroyed dynamically. Each thread is provided with its own run-time stack, and can be activated or interrupted by the scheduler. The state of an interrupted thread is preserved by freezing its run-time stack. Unlike UNIX processes, context switching between threads is fast and can be further improved by using additional hardware support like multiple register sets. Moreover, sharing of memory across threads is permitted. A potential weakness of this approach is that it relies on the programmer or the compiler creating sufficient threads to keep the processor utilization high. Moreover, this model is arguably more difficult to implement efficiently on nonshared memory systems.

In an alternative approach, several *message-driven* programming models have been proposed to improve processor efficiency. Message-driven execution is distinct from multithreading in two ways. Each message is sent to an *entry point* in a task which processes the message data using an indivisible non-interruptible piece of computation. Typically, an entry point is a function, procedure or method in the task. Moreover, a task is scheduled only when a message destined for one of its entry points is available. Typically, the task processes the message and relinquishes the processor. A task does not block awaiting a message, nor is it preempted or interrupted for scheduling purposes. Agha's actor model of concurrent computation [2] captures the essence of the message-driven approach. Like the multithreading approach, this model lends itself to very fast context switching. Moreover, it provides latency tolerance by ensuring that every processor is kept busy by continuously processing messages, thereby effectively overlapping communication with useful computation. A criticism of message-driven programming is that the dataflow nature of such programming models make them nondeterministic and less intuitive, and hence harder to program.

---
*This research was supported in part by the National Science Foundation under grant NSF CCR-9308108.

As with multithreading, message-driven execution can be supported at different levels of granularity. At the programming language level, the execution typically has much smaller granularity than threads. Typically, message-driven programs create thousands of tasks, even for relatively small execution times (e.g. Mentat [6], Charm [14]). This has the potential to lend itself to better processor utilization when compared with programmer specified threads. The *Active Messages* mechanism [21] supports finer grained message-driven execution on commercial architectures like the nCUBE/2 and the CM-5. The J-machine [11] and Monsoon [5] are examples of architectures that support fine grained message-driven execution.

Recently, hybrid systems have been proposed to explore the benefits of combining both approaches. The Threaded Abstract Machine [10] and the StarT project [4] are examples of such systems.

In this paper we focus on message-driven supported explicitly at the programming language level. Examples of such systems include ABCL/1 [22], Cantor [7] and Concurrent Aggregates [8], all which are designed to support fine-grained concurrent computing, as well as small to medium grained systems like Mentat [6] and Charm [15]. Other programming systems like Split-C [9] and Id90 compiled onto TAM [19] implicitly use message-driven support and are outside the scope of this research.

We present an extension to entry points that permits blocking in a message-driven framework *without* compromising the message-driven nature of the execution model or the high processor utilization it affords. We discuss how this is achieved and discuss applications of blocking entry points.

2 Intrepid

Intrepid is an experimental language and programming system which extends C with concurrent objects and restricted inheritance. The primary objectives of the Intrepid project is to investigate language, compiler and runtime support for debugging concurrent object-oriented programs.

A prototype compiler for Intrepid has been developed using the Charm programming language. The prototype supports Charm with Intrepid extensions.[1] Intrepid differs from Charm in that it supports both message-driven execution and SPMD execution (using blocking entry points) in the same framework. Another key difference is that Intrepid supports efficient task migration [13], both as a transparent system feature as well as a programming tool.

Intrepid supports two basic types of objects: messages and tasks. Programmers express parallel pro-

---

[1] A new compiler for Intrepid, independent of Charm, is currently under implementation.

```
chare Example{

    float temp, err;
    BOOLEAN recv_1, recv_2, recv_err, calc_done;
    TaskIDType parent;

    entry init : (message MESSAGE *msg){
        recv_1 = recv_2 = rec_err = calc_done = FALSE;
        parent = msg_i->pID;
    }/* end init */

    entry one : (message MSG1 *msg1){
        float result;
        if(! recv_2){
            recv_1 = TRUE;
            temp = msg1->data;
            return;
        }
        result = calc_err(msg1->data/temp);
        calc_done = TRUE;
        if(recv_err)
            send-msg(. . .,result, parent);
        else
            temp = result;
    }/* end one */

    entry two : (message MSG2 *msg2){
        float result;
        if(! recv_1){
            rev_2 = TRUE;
            temp = msg2->data;
            return;
        }
        result = calc_err(temp/msg2->data);
        calc_done = TRUE;
        if(recv_err)
            send-msg(. . .,result, parent);
        else
            temp = result;
    }/* end two */

    entry  three : (message MSG3 *msg3){
        err = msg3->data;
        recv_err = TRUE;
        if(calc_done){
            send-msg(. . .,calc_error(temp), parent);
        }
    }/* end three */

    private float calc_err(float num){
    float error;
        if(recv_err && calc_done)
            error = /* calculate error(msg3,num) */
            return(error);
        else
            return(num);
    }/* end private function */
}/* end chare */
```

Explanation:
    Three entry points need to acquire data.
    Boolean flags indicate what data is available.
    The flags recv_1 and recv_2 serve this purpose.
    The calc_done flag indicates if a partial result is available.

Figure 1: A simple *task* in Charm implemented without *blocking entry points*. Some liberties have been taken with the syntax for ease of exposition.

```
task EX{

    void init (MESSAGE *msg){
        float result;
        MSG1 *msg1;
        MSG2 *msg2;

        parent = msg->pID;
        entry one (MSG1 msg1);
        entry two (MSG2 msg2);
        result + = calc_err(msg1->data/msg2->data);
        send-msg(. . .,result, msg->parent_ID);
    }/* end init */

    private float calc_err(float num){
        MSG3 *msg3;
        float temp;

        entry three (MSG3 msg3);
        temp = /* calculate error(msg3, num) */
        return(temp);
    }/* end function */

}/* end task */
```

Figure 2: A simple *task* in Intrepid implemented with blocking entry points. This Intrepid task provides the same functionality as the chare shown in Figure 1.

grams as dynamically created tasks communicating with each other using messages. The mapping of tasks to processors is done automatically by Intrepid's run-time system, unless explicitly specified by the user. Multiple instances of these objects can be dynamically created and destroyed at run time.

3 Blocking Entry Points

Message-driven programming does not guarantee the order of message processing. Often, this is unimportant. For example, when a task needs to receive n ($n > 1$) messages and the messages can be processed in any order, message-driven execution handles such situations very efficiently. However, when message ordering is important, the programmer is forced to devise solutions within the program to enforce the desired ordering.

We compare blocking entry points in Intrepid with pure message-driven execution in Charm. Charm permits the dynamic creation of objects called *chares*. Figure 1 describes the basic syntax of a *chare* in Charm. In the figure, a chare awaits the arrival of three messages. Although the messages can arrive in any order, processing of the messages follow a strict ordering. To accomplish this, four flags, *recv_1*, *recv_2*, *recv_err* and *calc_done* are used keeping track of which messages have arrived, and what phase of the calculation has been completed. A *temp* variable is used to buffer the message arriving early.

This problem can easily be addressed if it were possible for the programmer to enforce an ordering on the processing of messages. We show how this is done in Intrepid with blocking entry points in Figure 2.

Syntax and Semantics

An entry point in Charm has a block of C-code associated with it (see Figure 1). This block is sequentially executable and non-interruptible. Nesting of entry points is not permitted in Charm.

In Intrepid, entry points are defined differently: task methods are *implicit* entry points. They may be called directly or be invoked via messages. An entry point may also be inserted in a task method's C-code block. Blocking entry points differ from implicit entry points in two ways.

1. Unlike methods, blocking entry points do not have code blocks associated with them. The syntax of a blocking entry point is as follows:
   entry *entry_name* (message *MsgType *msg*);
   Messages can be sent to blocking entry points in exactly the same manner as to methods. In fact, it is not necessary for a sender to know whether the destination of a message is a method or a blocking entry point.

2. They have blocking semantics. Consider a task C with a method E which has blocking entry points $E_i, i = 1, 2, \ldots$ within its C-code block. Since E's C-code block is executed sequentially, a total ordering is imposed on all $E_i, i = 1, 2, \ldots$ that appear in it. Thus, if one of these entry points awaits the receipt of a message that is as yet unavailable, execution cannot continue until the arrival of that message.

In order to meet the message-driven requirements of Intrepid's programming model, control is relinquished as soon as a blocking entry point is encountered. A blocking entry point may thus partition E's C-code block into smaller non-blocking, non-interruptible blocks of sequentially executable code.

These semantics have been implemented in Intrepid with some simple optimizations as follows.

- If a message for a given blocking entry point E_b arrives before E_b is encountered in execution, the message is buffered by the task until needed.
- Upon encountering a blocking entry point E_b, a check is performed to see if a message for E_b has already arrived and been buffered. If so, the message is picked up and execution continues without blocking.
- If a blocking entry point E_b is encountered before a message arrives for it, the processor is relinquished. When the awaited message is picked up for processing, execution resumes at E_b.

4 Performance

Three example programs have been written to evaluate blocking entry points. These programs are synthetic benchmarks that permit controlled experiments to isolate the overheads incurred.

The first program computes the number of prime numbers between 2 and a user-specified upper limit. It uses a divide and conquer algorithm that recursively partitions the interval, creating a binary tree to represent the problem space. It can be viewed as representative of all such programs. This program executes very efficiently using non-blocking entry points in Charm, yielding excellent speedups over an efficient uniprocessor implementation. A blocking entry point version of this benchmark will show any degradation in performance that may be evident due to the support and overhead associated with blocking.

The second program creates exactly one *task* on each processor. Each *task* then sends a message to the next g modulo #pes and blocks until it receives g messages. This process is repeated until a total of 100,000 messages are sent. Both g and #pes are varied to see their effect on performance. Note that since there is only one task per processor, each processor will have to idle when a blocking entry point awaiting a message is encountered. A blocking entry point version of this benchmark will expose the raw cost of supporting blocking entry points over a non-blocking version of the same program.

The third program is a modified version of the prime number counter, where combining the results at nodes internal to the tree are performed by a recursive function call. Specifically the entry point calls a recursive private function, which calls itself several times before calling yet another private function, which performs the combination. The version that uses non-blocking entry points passes the pointer to the incoming message along the path of function calls, while the version using blocking entry points simply has an entry point at the deepest level of recursion. The purpose of this experiment is to demonstrate that the state of an arbitrary path of function calls can be efficiently saved and restored.

Two additional experiments were conducted with the modified prime number counter benchmark. In the first experiment, this was used to measure the overhead of saving and restoring varying amounts of state with the use of blocking vs non-blocking entry points. This was conducted on one processor by varying the depth of recursion before a blocking entry point was encountered. In the second experiment, for a fixed recursive depth of 1000 calls, the number of processors was varied to observe its effect on the overhead.

For each experiment, the respective benchmark pro-

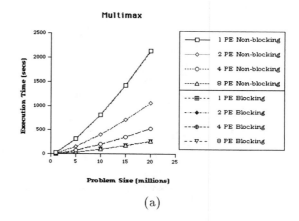

(a)

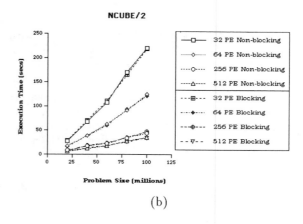

(b)

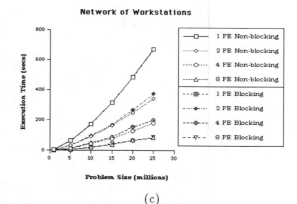

(c)

Figure 3: The performance of the prime counter benchmark for each of the three machines as the problem size is varied. The x-axis indicates the upper limit to which the search for primes is made. The number of processors is fixed. On the nCUBE/2 the sequential calculation is performed on partitions of $\leq 26,250$ numbers, the other two architectures use partitions of $\leq 42,000$ numbers.

PEs		1 msg	50 msgs	500 msgs
1	Blocking	101.062	100.462	100.762
	Non-blocking	69.070	64.751	66.969
2	Blocking	125.780	51.987	53.251
	Non-blocking	94.411	49.778	52.012
4	Blocking	187.432	34.546	40.654
	Non-blocking	150.621	32.971	35.628
6	Blocking	365.943	29.765	34.567
	Non-blocking	212.114	29.818	32.842

(a) Encore Multimax.

PEs		1 msg	50 msgs	500 msgs
1	Blocking	23.143	22.831	22.973
	Non-blocking	18.08	18.107	18.144
2	Blocking	64.395	23.419	23.580
	Non-blocking	61.851	21.018	21.077
4	Blocking	68.905	14.394	14.458
	Non-blocking	66.648	13.182	13.131
8	Blocking	73.485	7.864	8.008
	Non-blocking	69.588	7.256	7.285
16	Blocking	65.534	4.179	4.290
	Non-blocking	61.630	3.880	3.962
32	Blocking	64.636	2.424	2.529
	Non-blocking	64.778	2.281	2.432
64	Blocking	70.663	2.471	2.370
	Non-blocking	68.526	2.397	2.308
128	Blocking	65.212	5.801	5.151
	Non-blocking	63.770	5.729	5.374
256	Blocking	84.695	19.043	18.391
	Non-blocking	85.810	19.098	18.321
512	Blocking	133.856	72.107	71.626
	Non-blocking	134.948	71.983	71.129

(b) nCUBE/2.

PEs		1 msg	50 msgs	500 msgs
1	Blocking	2.654	2.525	2.538
	Non-blocking	2.378	2.306	2.340
2	Blocking	14.680	5.276	4.526
	Non-blocking	29.567	5.613	5.375
4	Blocking	7.62	4.12	3.65
	Non-blocking	51.505	4.980	5.093

(c) Network of Workstations.

Table 1: Round Robin Message transmission with 1 task/processor. Execution time is in seconds.

gram was run *without change* on three machines: a shared memory 8-processor Encore Multimax, a 1024 processor nonshared memory nCUBE/2 (non-shared memory) hypercube, and a network of 8 Sun Sparc 2 workstations.

Prime Number Counter

The problem is broken in half at each internal node of the tree, until the pieces of the problem are less then some user-specified limit. The number of primes at each leaf node of the problem decomposition tree is calculated using a sequential sieve algorithm.

Results. The graph in Figure 3(a) shows the overhead on the Encore Multimax with 8 processors as both the problem size and the number of processors are varied. As can be seen, the overhead is insignificantly small. In all cases the line representing the use of blocking entry points is indistinguishable from the non-blocking entry point case.

On the nCUBE/2, a larger instance of the prob-

lem was run (upper limit of 100,000,000 instead of 20,000,000) as more processors were available. This was done primarily to avoid the effect of processor starvation on the measured data. Recall that this benchmark was designed to expose any overheads brought about by the use of blocking entry points.

Due to limited memory, it was not possible to run the larger instance of the benchmark for less than 32 processors of the nCUBE/2 (Figure 3(b)). As is clear from the graphs, the overhead, once again, is insignificantly small.

On a network of Unix workstations, the effects of time-slicing Unix processes and background network traffic can cause significant variation in the observed performance. However, given the observed behavior of the benchmark on the other two types of architectures, and the low overheads evident from Figure 3(c) serve to corroborate the observations made on the other two machines. We present this data more as a proof of concept; it can not be used independently to draw any major conclusions.

Round-Robin Message Transmission

Since this benchmark creates only one task per processor, a blocking entry point version of this benchmark is expected to perform worse than its non-blocking equivalent. The two will perform comparably only when the state saving necessary during a context switch in the blocking case is minimal. This will only happen when most of the messages arrive before the task they are destined for is blocked. This benchmark is thus designed to determine the effectiveness of the underlying message-driven execution in eliminating the need to save state in the blocking case.

We vary the number of messages that each task must send as a group before it is forced to block. The task must then receive the same number of messages before it can send the next group of messages. If n is the number of messages a task is required to send in a group, it sends one to each of the tasks on the next n processors in round-robin fashion.

Results. On shared memory (Table 1(a)), with one message being transmitted at a time, the overhead of using a blocking entry point is high due to the amount of state information that needs to be frequently saved. However, this cost becomes insignificant as the number of messages is increased.

The one processor results on the nCUBE/2 (Table 1(b)) show that no matter how the messages are grouped, a small overhead of blocking over non-blocking is evident. However, increasing the number of processors makes the two cases indistinguishable. As the number of processors is increased the time it takes to send one message $100,000$ times increases due to communication costs.

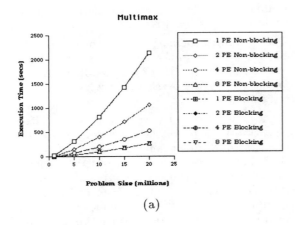

(a)

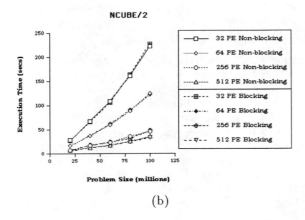

(b)

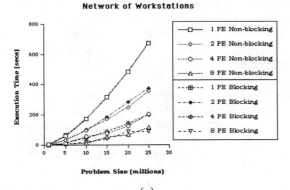

(c)

Figure 4: Modified Prime Number Counter: The experiment parameters are the same as those for the Prime Number Counter, Figure 3. The program contains a blocking entry point in a function that is six levels deep in a recursion.

As the number of messages sent is increased, on multiple processors, the amount of idle time per processor decreases, until the number of messages sent per group is the same as the number of processors used. Then the processors reach a state of always being active(have a message to process). At this point the overhead of blocking becomes insignificant.

The results on a network of workstations are interesting especially for the case when only one message per group can be transmitted. (Table 1(c)). The non-blocking version of this program takes significantly longer to execute than the blocking version. We believe this is due to the increased amount of contention for the Ethernet due to background traffic. However, this effect disappears as the number of messages per group is increased.

Blocking Entry Points in Recursive Functions

For the third experiment, the prime number counter was modified to demonstrate the use of blocking entry points inside a function. The primes computation counts the number of primes in the user-specified interval by dividing the problem down a binary tree. When the results are being collected back up the tree, the combining computation is done at the end of a chain of function calls. This chain of function calls is generated by a function calling itself recursively, then finally calling a different function, which performs the computation. This benchmark attempts to measure the overhead saving and restoring state of recursive functions calls.

Results. The extra effort required in saving and restoring the state of recursive function calls does not translate into longer execution times. Figure 4 shows no degradation in performance on the Encore Multimax and nCUBE/2, and very little degradation in performance on the network of workstations.

State Saving Overheads

In this experiment, the number of times a recursive function was allowed to call itself was varied. A blocking entry point was inserted at the deepest level of recursion thereby necessitating state saving at every level of the recursion. This shows how the overhead increases as the amount of state to save and restore increases. This experiment was run on a single processor, so the entire overhead could be observed.

The second experiment varies the number of processors for a fixed number (1000) of recursive calls. For both parts of this experiment, an upper limit of 20,000,000 was used. Both experiments were conducted on the nCUBE/2.

Results. As expected, the overhead of saving and restoring state increases linearly as the amount of state increases, see Figure 5(a). With a large amount of state to be saved and restored, increasing the num-

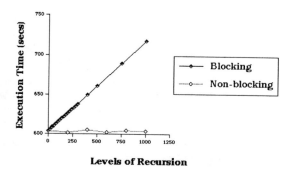

(a) On one processor of the nCUBE/2, the overhead of saving and restoring state increases as the depth of recursion is increased.

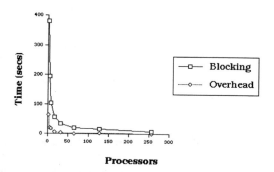

(b) For a fixed depth of recursion (1,000), the overhead of saving and restoring state as number of processors is increased on the nCUBE/2 from 2 to 256. The upper curve indicates the total execution time of the program as the number of processors is varied, and the lower curve (dotted) indicates its overhead component.

Figure 5: Recursion and blocking entry points.

ber of processors causes a sharp drop in the absolute overhead. In Figure 5(b) we plot the total execution time and its overhead component as the number of processors is varied from 2 to 256 on the nCUBE/2. This shows that the overhead decreases dramatically as the number of processors is increased.

5 Applications

We have focused on the Intrepid environment as the test-bed for this research. We believe that the applications presented below, although expressed in the context of Intrepid, have wider applicability, albeit with modified syntax and semantics.

Information Sharing Abstractions

Kale *et al* have proposed several information sharing abstractions to ease parallel programmability [16]. These abstractions include *read-only* variables, *write-*

once variables, *accumulator* variables, *monotonic* variables and *distributed* tables.

We have shown elsewhere [18] that blocking versions of these abstractions can be implemented more efficiently using Intrepid, than their non-blocking counterparts in Charm. Moreover, they are simpler to use and scale better. For example, for write-once variables, our experiments show more than a 100% improvement in the cost of installing write-once variables on 2 processors of a Encore Multimax and 8 processors of an nCUBE/2. This improvement increases to 140% on 8 processors of the Encore and over 300% on 128 processors of the nCUBE/2. The blocking implementations were found to perform better to varying levels for all the abstractions.

SPMD Programming

The SPMD programming model is a widely used paradigm for nonshared memory programming. However, porting such programs to shared or distributed shared memory machines requires major changes to the source code to retain efficiency.

Blocking entry points are analogous to blocking receives in SPMD programming. Since Intrepid code is largely C-based, the differences in syntax are minor. Thus, converting a SPMD program written in C to Intrepid is relatively easy. Once accomplished, this provides several additional advantages. Porting code across both shared and nonshared memory machines is now automatic. Also, most SPMD systems do not provide the information sharing abstractions discussed earlier for ease of programming. Blocking entry points in Intrepid can offer this feature to SPMD programmers. Sophisticated compiler support for debugging and performance monitoring are no longer specific to any one target architecture.

Intrepid also permits experimentation with multiple SPMD processes per processor, even if this is not supported by the underlying operating system (e.g. on Intel i860 and nCUBE/2). It also provides the SPMD programmer an option to experiment with message-driven programming to improve the performance of their programs.

6 Conclusions and Future Work

We have successfully implemented blocking entry points in a message-driven framework, thereby permitting a programmer to impose an ordering on message arrival if desired. We have described the syntax and semantics of blocking entry points, and discussed how they were implemented, and how they differ from methods. We have also demonstrated that it is possible to efficiently support this feature using a variety of benchmark programs designed to expose any weaknesses of blocking entry points. This efficiency has been accomplished by exploiting the message-driven execution model to provide high processor utilization. The results demonstrate that it is possible to exploit the convenience of blocking entry points whenever applicable without sacrificing performance.

We have presented two interesting applications of blocking entry points and have shown how it can be used to enhance expressibility in parallel programs. We have validated these results further by comparing implementations of several information sharing abstractions using blocking entry points in methods [18]. We have also implemented several sections of the Intrepid run-time system using blocking entry points. We have also implemented a blocking entry point version of parallel circuit extraction, previously implemented using non-blocking entry points in Charm [17].

Acknowledgments

We would like to thank the Weeg Supercomputing Center at the University of Iowa for access to their Encore Multimax. Our thanks also to Sandia National laboratories where the nCUBE/2 experiments were conducted.

References

[1] Agarwal A., Chaiken D. *et al*. The MIT Alewife Machine: A Large Scale Distributed-Memory Multiprocessor. In *Proceedings of the Workshop on Scalable Shared-Memory Multiprocessors*, 1991.

[2] Agha, G.A. *Actors: A Model of Concurrent Computation in Distributed Systems*. MIT press, 1986.

[3] Alverson R., Callahan D., Cummings D., Koblenz B., Porterfield A., Smith B. The Tera Computer System. In *Proceedings of the International Conference on Supercomputing*, pages 1–6, June 1990.

[4] Ang B.S., Arvind, Chiou D. StarT the Next Generation: Integrating Global Caches and Dataflow Architecture. Tech. Rep. CSG Memo 354, Laboratory for CS, Mass. Inst. of Tech., Feb. 1994.

[5] Arvind, Culler D.E., Maa G.K. Assessing the Benefits of Fine-Grained Parallelism on Dataflow Computers. *International Journal of Supercomputer Applications*, vol. 2, no. 3, Nov. 1988.

[6] G. A.S. *Mentat: An Object-Oriented Macro Dataflow System*. PhD thesis, Dept. of Computer Science, University of Illinois, Urbana-Champaign, June 1988.

[7] Boden, N. A Study of Fine-Grained Programming using Cantor. Master's thesis, Dept. of Computer Science, California Institute of Technology, Nov. 1988.

[8] Chien, A. *Concurrent Aggregates: Supporting Modularity in Massively Parallel Programs*. MIT press, 1993.

[9] Culler D.E., Dusseau A., Goldstein S.C., Krishnamurthy A., Lumetta S., vonEicken T., Yelick K. Parallel programming in split-c. In *Proceedings of Supercomputing*, Nov. 1993.

[10] Culler D.E., Goldstein S.C., Schauser K.E., von Eiken T. TAM – A Compiler Controlled Threaded Abstract Machine. *JPDC*, 18 no. 3:347 – 370, 1993.

[11] Dally, W.J. Fine-Grain Message-Passing Concurrent Computers. In *The Third Conference on Hypercube Concurrent Computers and Applications*, Pasadena, California, Jan. 1988.

[12] Deoppner, T.W. Jr. A Threads Tutorial. Technical report, Brown University, Providence, Rhode Island, 1987.

[13] Doulas, N., Ramkumar B. Efficient Task Migration for Message-Driven Parallel Execution on Nonshared Memory Architectures. In *International Conference on Parallel Processing*, Aug. 1994.

[14] Fenton, W., Ramkumar, B., Saletore, V.A., Sinha A.B., Kalé, L.V. Supporting Machine Independent Programming on Diverse Parallel Architectures. In *ICPP*, pages II:193–201, Aug. 1991.

[15] Kale L.V., Ramkumar B., Sinha A.B.,Gursoy A. The CHARM Parallel Programming Language and System: Part I – Description of Language Features. *IEEE TPDS*, 1994.

[16] Kale L.V., Sinha A.B. Information Sharing Mechanisms in Parallel Programs. In *Proceedings of International Parallel Processing Symposium*, pages 461–468, Apr. 1994.

[17] Ramkumar, B., Banerjee, P. A Portable Parallel Algorithm for VLSI Circuit Extraction. In *Proceedings of the 7th IPPS*, Apr. 1993.

[18] Richards, R.J., Ramkumar B. Blocking Data Sharing Abstractions for Parallel MIMD Programming. Tech. rep., Univ. of Iowa, May 1995.

[19] Schauser K.E., Culler D.E., von Eiken T. Compiler-Controlled Multithreading for Lenient Parallel Languages. In *Proceedings of the Conference on Functional Programming languages and Computer Architecture*, Aug. 1991.

[20] Tevanian A., Rashid R.F., Golub D.B., Black D.L., Cooper E., Young M.W. Mach Threads and the Unix Kernel: The Battle for Control. In *USENIX Association Summer Conference Proceedings*, June 1987.

[21] vonEicken T., Culler D., Goldstein S.C., Schauser K.E. Active messages: a mechanism for integrated communication and computation. In *Proceedings of the 19th ISCA*, pages 256–266, May 1992.

[22] Yonezawa A. *et al*. Modeling and Programming in an Object-Oriented Concurrent Language ABCL/1. In *Object-Oriented Concurrent Programming*. MIT Press, Cambridge, Mass., 1987.

Compiling Portable Message-Driven Programs*

B. Ramkumar[+]   **C. P. Forbes**[+]   **L. V. Kalé**[++]

[+]Dept. of Electrical and Computer Eng.
University of Iowa
Iowa City, IA 52242
E-mail: ramkumar@eng.uiowa.edu

[++]Dept. of Computer Science
University of Illinois
Urbana, IL 61801
Email: kale@cs.uiuc.edu

Abstract

We describe the design of a compiler for the portable execution of concurrent object-oriented programs on MIMD architectures. The compiler has been implemented for the Intrepid parallel programming system under development at the University of Iowa. The Intrepid project is investigating language, runtime support and programmability requirements needed to make concurrent object-oriented programs portable and debuggable. The compiler has been implemented using multiple phases to permit the modular addition of new features for analysis and debugging support. Salient features of the translation and how efficient portability is achieved are described.

1 Introduction

The increasing availability and affordability of parallel machines has given rise to expectations of significant improvements in performance at relatively low expense. This expectation has remained largely unrealized, primarily due to the high cost of parallel program development. This cost can be attributed to three significant causes. First, parallel algorithms tend to be considerably more complex than comparable sequential algorithms. Second, parallel architectures exhibit widely differing characteristics. As a result, parallel programs tend to perform poorly on architectures other than the one they are designed for. The fast evolution of parallel architectures has only served to compound this problem. Third, the presence of multiple threads of control in parallel programs make them difficult to debug, thereby contributing to a longer development time.

This paper focuses on the second of these problems: the dependency between parallel programs and the target architecture. In an effort to address this problem, several researchers have been studying the problem of providing portability of parallel program across MIMD architectures. The primary objective common to all work in this area has been to achieve *effective portability*. This term was first defined by Alverson and Notkin [4] and stipulates that a portable parallel program must run reasonably efficiently *without change* on all target architectures.

Several approaches, differing significantly in their assumptions and objectives, have been proposed to address this problem. This effort has ranged from providing more sophisticated parallelizing compilers for existing sequential languages, e.g. [3, 6, 16], sequential languages supplemented with machine independent message passing primitives, e.g. [13, 15, 23], to the design of explicitly parallel languages and programming environments, e.g. [1, 10]. Some systems restrict themselves to shared memory systems, e.g. [2, 8, 18], others subscribe to a shared memory computational model, e.g. [4, 5, 7], while yet others have proposed systems based on a message passing communication model, e.g. [10, 13, 15, 17, 23]. Some systems are object-oriented or object-based, e.g. [4, 6, 9], while others are procedural, e.g. [3, 7, 15, 23]. Finally, some systems are based on a SPMD execution model, e.g. [3, 15, 23], others are multi-process or multithreaded, e.g. [7, 4, 10] while yet others are message-driven or coarse-grained dataflow in nature, e.g. [1, 6].

The *Intrepid* project at the University of Iowa is investigating techniques to make concurrent object-oriented programs portable and debuggable. In this paper, we focus on the language and compilation aspects of our research. We describe the design of a multi-phase translator for the efficient portable execution of the Intrepid language on shared and non-shared memory MIMD architectures. The design also allows for the modular expansion of the compiler to provide new functionality and static analysis capabilities.

2 Intrepid

Intrepid is an integrated environment for performance-prediction, instrumentation and debugging of concurrent object-oriented programs. It

---

*This research is supported in part by the National Science Foundation grant CCR-9308108.

shares some features with the Charm programming environment [17, 20], notably its selective packing for messages. However, it primarily differs from Charm in that it combines the message-driven and SPMD styles of programming [21] while retaining efficient portability. Another key difference is that Intrepid supports efficient task migration [14], both as a transparent system feature as well as a programming tool.

Intrepid's language design has been driven by three important principles: simplicity, portability and debuggability. Rather than support full object-orientation, we have extended the C language with restricted inheritance and concurrent objects. Intrepid's object types are all derived from a single base type called a *portable structure* (for details, see [19]).

A prototype compiler for Intrepid providing extensive debugging support was developed using the Charm parallel programming environment[1][20]. The compiler translates Intrepid programs into C. The translated C program is then compiled and linked with Intrepid's run-time library to generate the executable code. Language extensions to improve programmability and debuggability were identified and implemented [14, 21, 22]. A new environment for performance debugging of portable programs was implemented and was shown to be able to predict performance to within 5% of actual program behavior on realistic applications both shared and message passing architectures [11, 12]. A graphical framework for interactive debugging of concurrent object-oriented programs is currently under development.

Intrepid supports two basic types of objects: messages and tasks. Programmers express parallel programs as dynamically created task instances communicating with each other using messages. The mapping of tasks to processors is done automatically by Intrepid's run-time system, unless explicitly specified by the user. Multiple instances of these objects can be dynamically created and destroyed at run time.

A large program may comprise several task and message definitions. Task and message definitions may be logically grouped into distinct *modules*, which provide distinct name spaces. Each module definition must export an interface which identifies the names and prototypes of messages and task methods that are visible to other modules. Entities not exported remain private to the module.

2.1 Intrepid's Runtime Behavior

The Intrepid runtime system maintains a pool of messages that need to be processed. When a message is created for delivery to a named method in a named task instance, it is placed in this pool. The Intrepid kernel continually runs a loop on each processor which picks a message and delivers it to the correct instance of the task at the appropriate method. Processing a message implies executing its destination method without interruption with the message as its argument. The execution of the method may result in the creation of more concurrent tasks and/or messages. When execution of the method is complete, the runtime system picks the next message from the pool and repeats this process.

3 The Intrepid Compiler

The Intrepid compiler consists of several phases. Such a multi-phase implementation permits new phases that can provide new features to be added with ease. It also allows optional use of phases, a necessary feature for debugging support.

3.1 Multiphase Compilation

Each phase of the Intrepid compiler performs a separate part of the translation process. The first phase performs preprocessing of an Intrepid program. This phase performs three tasks. The first is to the C preprocessor to handle C directives like "#include" and "#define". The second is to provide support for functions with optional parameters with default values. The third task is provide name space support for modules in Intrepid. Each module has its own name space and its components can be accessed from outside the module by using a "module extension" in front of the component name. The Intrepid preprocessor mangles this "extended name" into a flat C name.

The second phase uses the postprocessed file and creates a symbol table for the Intrepid program. Two additional phases provide support for data sharing abstractions supported in Intrepid [22] and blocking primitives for SPMD programming [21]. The final phase uses the symbol table and the translated Intrepid program generated by the previous phases to generate a C program that that is equivalent to the original Intrepid program. This C program is subsequently compiled using an ANSI compliant compiler and then linked to the Intrepid run-time system on the desired target machine.

3.2 Program Translation

The translation of an Intrepid program requires transforming Intrepid constructs into analogous C constructs. As mentioned earlier, Intrepid extends C with concurrent objects called tasks. Each task has its own persistent data, public methods that can be invoked by other tasks directly (via a function call) or via a message, and private methods that can only be called from within the task scope.

A task's persistent data is translated into a C-struct. This data is global to all the methods in the task scope and is accessed directly. Task methods are translated into C functions with an extra parameter:

---

[1] The prototype supported Charm with Intrepid extensions. A new compiler for Intrepid, independent of Charm, is under development.

a pointer to the persistent data of the task. All references to the persistent data in the task scope are then translated into a pointer dereference.

3.2.1 The Translator Table

At runtime the Intrepid kernel needs information about tasks and messages that may be created and used during execution. The Intrepid compiler generates C-functions that are used to fill a *translator table* with this information during run-time initialization. The information in the table can be divided into two categories: message information and task information.

For each message, information about the base type (a portable structure) it is derived from is maintained. For each portable structure [19], its size (for initial allocation), and function pointers to its *pack*, *unpack* and *current-size* functions are maintained. The pack function packs a message with possibly dynamic data fields into a contiguous buffer whose size is computed by the current-size function. The unpack function restores a packed message into its original form. These functions are called by the Intrepid run-time system only when needed.

Like messages, tasks are also derived from portable structures: a task's persistent data is an instance of such a structure. Thus, tasks can be easily migrated across processor boundaries transparently. This feature is essential for checkpointing and replay during debugging.

For each task, a function pointer table is created on each processor to store references to its public methods. C-functions to fill in this table are generated by the compiler and called during system initialization. All references to task and method names in Intrepid programs are translated to integer variables that store the corresponding indices in this function pointer table. During system initialization, these variables are assigned the appropriate indices into the table. This approach is also used to support separate compilation of modules.

At run time, when a reference to a task method is made in an Intrepid program, messages destined for a method carry the corresponding index into this table. This index is used by the the runtime system to locate and invoke the appropriate method to "process" the message.

3.3 Debugging Support

An important objective in the design of the Intrepid compiler is support for user-level debugging. Some of this support falls into the category of type consistency checking. In addition, support is also provided for certain checks that can only be performed at run-time. Finally, compiler generated functions also make interactive debugging possible.

The Intrepid translator detects declaration errors during the symbol table generation phase. During this phase the Intrepid translator compares the prototypes of tasks, including all public and private methods within the task. The translator identifies and flags any discrepancies that may exist.

When creating a new task or sending a message to a task method, an invalid method name could be specified, or the type of the message being sent might not match the type expected at the method. The Intrepid compiler flags these as errors at compile-time by matching module interface information with the type at the call site. When sending a message to an existing task it is also possible for the user to specify either a nonexistent or dead task instance. This check is performed at run-time using information generated at compile-time.

In Intrepid, task method names are first-class objects. It is therefore possible to declare method variables that store method names, and send method names as message fields from one task to another. (Recall that a method name is translated into a index into a function pointer table (see Section 4.) Consequently, it is possible for a task to use a method variable as a destination for a message. The Intrepid compiler generates information stored in the translator table to detect validity of destination method at run-time even for such cases.

One of the basic operations of an interactive debugger is the ability to show the user the contents of variables in the program. Requests for variable information will be in terms of the name of the variable as it appears in the Intrepid program being debugged. As described in Section 3.1, a given identifier may undergo mangling during the process of translation.

In order for the debugger to determine the translated name from an original name in the Intrepid program, the debugger must have access to the symbol table and the current context. The debugger can unmangle a translated name or mangle an Intrepid identifier using compiler-generated functions and symbol table interface functions This is used by the debugger to read user commands and display results.

4 Conclusions and Future Work

Any compiler is necessarily language specific. However, this research makes several contributions of interest to researchers in parallel programming languages, environments and debuggers.

- This research demonstrates that a source-to-source translation as manifested in the Intrepid translator is an efficient technique for providing portability across MIMD architectures for parallel programs.

- The approach is practical and cost-effective. The alternative of writing and supporting a compiler for each target machine is prohibitively expensive in both time and cost.

- Its support for the easy addition of new phases provides a pragmatic way to experiment with new parallel extensions to existing languages.

- It permits a research quality compiler to directly benefit from commercial architecture-specific compilation utilities with minimal effort.

- The mechanism proposed to support distinct name spaces via modules can be used for other sequential and parallel languages. The use of partitioned index structures is both efficient and is very useful in providing debugging support.

- This approach separates the user from having to contend with differences between compilers and programming environments on different MIMD systems. One compiler can be used on *all* target platforms.

We have shown how general Intrepid objects and messages are supported on top of C with translator and runtime support. We have also shown how translator support for the message-driven execution was implemented. This approach can be applied to other languages like Fortran and C++ with relative ease.

References

[1] Agha, G.A. *Actors: A Model of Concurrent Computation in Distributed Systems.* MIT press, 1986.

[2] Alaghband G., Benten M.S., Jakob R., Jordan H.F., Ramanan A.V. Language Portability across Shared Memory Processors. *IEEE TPDS*, 4 no. 9:1064–1072, September 1993.

[3] Allen F.E., Kennedy K. Automatic Translation of Fortran Programs to Vector Form. *ACM TOPLAS*, 9:4, October 1987.

[4] Alverson G.A., Notkin D. Program Structuring for Effective Parallel Portability. *IEEE TPDS*, 4 no. 9:1041–1059, September 1993.

[5] Andrews G.R., Olsson R.A. *et al.* An Overview of the SR Language and Implementation. *ACM TOPLAS*, 10 no. 1, January 1988.

[6] Grimshaw A.S. *Mentat: An Object-Oriented Macro Dataflow System.* PhD thesis, Dept. of Computer Science, University of Illinois, Urbana-Champaign, June 1988.

[7] Bal H.E., Kaashoek M.F., Tanenbaum A.S. Orca: A language for parallel programming of distributed systems. *IEEE TSE*, 18 no. 3:190–205, March 1992.

[8] Beck B. Shared-Memory Parallel Programming in C++. *IEEE Software*, July 1990.

[9] Bershad B., Lazawska E., Levy H. Presto: A System for Object-oriented Parallel Programming. *Software - Practice and Experience*, August 1988.

[10] Carriero, N., Gelernter, D. How to Write Parallel Programs: A Guide to the Perplexed. *ACM Computing Surveys*, 21 no. 2:323–357, September 1989.

[11] Chillariga, G., Ramkumar B. Debugging the Performance of Portable Parallel Programs. Tech. rep., Univ. of Iowa, Dept. of Elec. and Comp. Eng., March 1995.

[12] Chillariga, G., Ramkumar B. Performance Prediction for Portable Parallel Execution on MIMD Architectures. In *Proceedings of IPPS*, April 1995.

[13] Clark L., Glendinning I., Hempel R. The MPI Message Passing Interface Standard. Tech. rep., University of Southampton, U.K., March 1994.

[14] Doulas, N., Ramkumar B. Efficient Task Migration for Message-Driven Parallel Execution on Nonshared Memory Architectures. In *Proceedings of ICPP*, August 1994.

[15] Flower, J., Kolawa, A., Bharadwaj S. The Express Way to Distributed Processing. In *Supercomputing Review*, pages 54–55, May 1991.

[16] M. Gupta and P. Banerjee. Automated Data Partitioning on Distributed Memory Multiprocessors. *Proc. 6th Distributed Memory Multicomputers Conference (DMMC6)*, May 1991.

[17] Kale L.V., Ramkumar B., Sinha A.B.,Gursoy A. The CHARM Parallel Programming Language and System: Part I – Description of Language Features. *IEEE TPDS*, 1994. (Submitted.)

[18] Lusk E.L. *Portable Programs for Parallel Processors.* New York: Host, Rinehart and Winston, 1987.

[19] Ramkumar B. *Portable Structures:* A Distributed Data Type for Efficient Portable Parallel Programming. Tech. rep., Univ. of Iowa, Dept. of Elec. and Comp. Eng., December 1994.

[20] Ramkumar B., Sinha A.B., Kale L.V., Saletore V.A. The CHARM Parallel Programming Language and System: Part II – Implementation and Performance. *IEEE TPDS*, 1994. (Submitted.)

[21] Richards, R.J., Ramkumar B. Blocking Entry Points in Message-Driven Parallel Systems. In *Proceedings of ICPP*, August 1995. (to appear).

[22] Richards, R.J., Ramkumar B. Efficient Data Sharing Abstractions for Portable Parallel MIMD Programming. Tech. rep., Univ. of Iowa, Dept. of Elec. and Comp. Eng., March 1995.

[23] Sundaram V. PVM: A Framework for Parallel Distributed Computing. *Concurrency: Practice and Experience*, vol. 2, no. 4, December 1990.

Architecture-Dependent Loop Scheduling via Communication-Sensitive Remapping *

Sissades Tongsima   Nelson L. Passos   Edwin H-M. Sha

Dept. of Computer Science & Engineering
University of Notre Dame
Notre Dame, IN 46556
E-mail: nung@cse.nd.edu, npassos@darwin.cc.nd.edu, hms@cad.cse.nd.edu

Abstract   *In this paper, we propose a novel efficient technique called cyclo-compaction scheduling, taking into account the data transmission delays and loop carried dependency associated with specific target architectures. This technique uses the retiming technique (loop pipelining), implicitly applied, and a task remapping to appropriate processors in order to compact the schedule length and improve the parallelism iteratively while handling the underlying imposed communication environment and resource constraints. Algorithms and the corresponding theorems are presented. Experimental results for different architectures show the effectiveness of our algorithm.*

INTRODUCTION

The achievement of high performance via parallel computing requires an efficient scheduling which considers both architectural and communication aspects of the system. The appropriate processor assignment is part of the solution that can enhance the performance of computation-intensive applications running in a parallel computer. Most applications requiring the usage of parallel systems are usually iterative or recursive. This recursivity can be represented by *cyclic* data-flow graphs with loop carried dependencies. Loop pipelining is a common optimization technique inherent to cyclic applications. This technique must be explored in order to improve the parallelism across iterations [1]. In this paper, we present a novel loop scheduling algorithm, called *cyclo-compaction scheduling*, that considers the communication overhead and loop-carried constraint while performing loop pipelining to compact a schedule repeatedly.

Many heuristic methods have been developed for scheduling acyclic data-flow graphs [2–4] such as critical path heuristics, list scheduling heuristics and graph decomposition heuristics. All of these methods do not

*This work was supported in part by Royal Thai Scholarship, Mensch Graduate Fellowship, and an ORAU Faculty Enhancement Award.

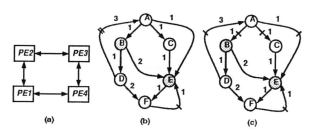

Fig. 1: Target architecture, CSDFG and its retimed graph

consider the parallelism and pipelining across iterations. For scheduling cyclic data-flow graphs, most of the techniques developed for parallel compilers do not consider the communication cost. For example, the technique of software pipelining is used to find a repeating pattern as a static schedule [1, 5]. Likewise, the polynomial time rate-optimal scheduling without resource constraint for cyclic data-flow graphs proposed by Chao and Sha also does not consider the communication cost [6].

For scheduling tasks on parallel systems, Shukla and Little proposed an algorithm for scheduling task data-flow graphs not considering the imposed communication cost [7]. Munshi and Simons proposed a polynomial time scheduling algorithm for cyclic DFGs without considering communication costs with respect to the processor architecture [8]. Hoang and Rabaey presented a scheduling algorithm which addresses the inter processor communication delays yet the output schedule may be an infeasible schedule if there are smaller number of available processors than its requirement [9].

Chao, LaPaugh and Sha introduced an efficient methodology called the *rotation* scheduling [10]. This technique optimizes the schedule length by using loop pipelining, but does not consider the communication between proces-

Start-up schedule

cs	pe1	pe2	pe3	pe4
1	A			
2	B			
3	B	C		
4	D			
5	E			
6	E			
7	F			
8	A			
9	B			
10	B	C		
11	D			
12	E			
13	E			
14	F			
⋮				

(a)

First iteration

cs	pe1	pe2	pe3	pe4
X	A			
1	B	A		
2	B	C		
3	D			
4	E			
5	E			
6	F			
7	B	A		
8	B	C		
9	D			
10	E			
11	E			
12	F			
⋮				

(b)

Fig. 2: Transformation of the start-up schedule table

Second iteration

cs	pe1	pe2	pe3	pe4
X	A			
X	B	A		
1		C		
2	D	B		
3	E	B		
4	E			
5	F			
6		A		
⋮				

(a)

Third iteration

cs	pe1	pe2	pe3	pe4
X	A			
X	B	A		
X	C			
1	D	B		
2	E	B	C	
3	E			
4	F			
5		A		
⋮				

(b)

Fig. 3: Second and third iteration of the cyclo-compaction

sors. A modified *rotation* scheduling strategy dealing with the communication cost was previously proposed for *unit-time* data-flow graphs and completely connected architectures [11]. In this paper, we propose a new and improved scheduling algorithm called cyclo-compaction scheduling which optimizes the schedule length of *general-time* DFGs with loop carried dependency considering a target architecture and the communication delays between processors. This method focuses on loop pipelining while considering the processor availability and the communication overhead regarding to each particular architecture. The results obtained through the use of this technique are directly applicable to tightly coupled parallel systems, as well as high level synthesis of multi-chip systems. The cyclo-compaction scheduling algorithm schedules nodes in a DFG without violating dependencies and communication constraints.

The communication cost is measured by the number of time units required for transmission between processors and the distance between one processor element (sender) and a target processor (receiver).

A general-time cyclic DFG shown in Figure 1(b) comprising of six tasks (nodes) needs to be scheduled in a *2-D Mesh* architecture. The target system is shown in Figure 1(a) with four processor elements. The computation time of nodes A, C, D, F are assumed to be 1 time unit, nodes B and E are 2 time units. A modified list scheduling process gets the initial schedule shown in Figure 2(a). This algorithm schedules the nodes by taking into account communication costs with respect to the architecture. Notice that in a system not considering the communication overhead, node C could be parallel to the first cycle of node B. However, the communication cost with respect to the

2-D Mesh architecture and the dependency between A and C require node C to be scheduled at control step 3 either under processor 2 (*pe2*) or 4 (*pe4*) or at control step 4 under processor 3. The algorithm then selected *pe2*. This initial schedule is then submitted to the novel technique called *cyclo-compaction scheduling* algorithm. This algorithm first reschedules node A from control step 1 while implicitly moving one *delay* from its incoming edges to all of its outgoing edges (see Figure 1(c)). Node A is then assigned to control step 2 (or control step 1 after renumbering the table, Figure 3(a)) under *pe2*. After the third iteration, the schedule length is 2 control steps shorter than the initial schedule (see Figure 3(b)). Figure 2 and 3 depict the transformation of the initial schedule table from seven control steps down to five. The remaining of this paper is organized as follows: basic concepts and specifications of target architectures are presented in Section . Section presents mathematical models and basic functions for the initial scheduling process. Section deals with the optimization of the initial schedule by the technique called *cyclo-compaction scheduling*. Section provides examples of application to different architectures. Finally, a conclusion summarizes the concepts presented in this paper.

PRELIMINARIES

This section introduces the definition of *communication sensitive data-flow graph* (CSDFG) and other basic terminologies used in this paper.

Definition 1 *A CSDFG* $\mathcal{G} = (\mathcal{V}, \mathcal{E}, d, t, c)$ *is a node-weighted and edge-weighted directed graph, where* $\mathcal{V}$ *is the set of nodes,* $\mathcal{E} \subseteq \mathcal{V} \times \mathcal{V}$ *is the set of dependence edges, d is the delay between two nodes, t is a function from* $\mathcal{V}$ *to the positive integers representing the computation time of each node, and c is a function from* $\mathcal{E}$ *to the positive integers representing the data volume transferred between two* *nodes when they are assigned to different processors.*

Figure 1(b) depicts an example of CSDFG, where $\mathcal{V} = \{A, B, C, D, E, F\}$ and $\mathcal{E} = \{e1 : (A, B), e2 : (A, C), e3 : (A, E), e4 : (B, D), e5 : (B, E), e6 : (C, E), e7 : (D, A), e8 : (D, F), e9 : (E, F), e10(F, E)\}$. The delay functions are $d(e1) = d(e2) = d(e3) = d(e4) = d(e5) = d(e6) = d(e8) = d(e9) = 0, d(e7) = 3$ and $d(e10) = 1$, the execution time are $t(A) = t(C) = t(D) = t(F) = 1$ and $t(B) = t(E) = 2$, and the communication cost for assignment to different processors is $c(e1) = c(e2) = c(e3) = c(e4) = c(e6) = c(e9) = 1$, $c(e5) = c(e8) = 2$ and $c(e7) = 3$. The notation $u \xrightarrow{e} v$ conveys that e is an edge from node u to node v. The notation $u_{p_i} \xrightarrow{m} v_{p_j}$ conveys that there exists a transferring data volume cost m whenever u, executed by processor p_i, and v is executed by processor p_j.

An *iteration* is the execution of each node in $\mathcal{V}$ exactly once. Iterations are identified by an index i starting from 0. Inter-iteration dependencies are represented by the weighted edges. An iteration is associated with a static schedule. A static schedule must obey the precedence relations defined by the DFG. For any iteration j, an edge e from u to v with delay $d(e)$ conveys that the computation of node v at iteration j depends on the execution of node u at iteration $j - d(e)$. An edge with no delay represents a data dependency within the same iteration. A legal DFG must have strictly positive delay cycles, i.e., the summation of the delay functions along any cycle cannot be less than or equal to zero.

The clock cycle is the synchronization time interval in the multiprocessor system. A clock cycle is equivalent to one control step in the static schedule. A task that is longer than one clock cycle requires the allocation of resources for multiple control steps. A processing element that has a pipeline design can execute a new task before the completion of the previous one.

The *retiming* technique is a commonly used tool for optimizing synchronous systems [12]. A retiming r is a function from $\mathcal{V}$ to integers. The value of this function is the number of delays taken from all incoming edges of node v and moved to each of its outgoing edges and vice versa. An illegal retiming function occurs when one of the retimed edge delays becomes negative. This situation implies a reference to a non-available data from the future iteration. An example of retiming is shown in Figure 1(b) and 1(c). One delay is drawn from the incoming edge of node A and pushed to all of its outgoing edges, see Figure 1(c). Delays are represented as bar lines over the graph edges.

A *prologue* is the set of instructions that must be executed to provide the necessary data for the iterative process after it has been successfully retimed. In our example, the instruction A becomes the prologue. An *epilogue* is the

other extreme, where a complementary set of instructions will need to be executed to complete the process. We may assume that the time required to run the prologue and epilogue are negligible, when compared to the total computation time of the problem.

START-UP SCHEDULING

In this section, we modify a traditional *list-based scheduling* algorithm to be able to deal with both data dependencies and communication delays with respect to a target architecture. Such constraints are parameters in the functions described in this section. Some mathematical models which are required in the algorithm are also presented.

We begin by introducing two basic functions which provide information on the execution time of a scheduled node. Since the CSDFG represents a general-time type of dataflow graph, the beginning and the end of the execution of each node may be associated with different control steps respectively designated by the functions CB and CE. We define these functions as following:

Definition 2 *Given a CSDFG $\mathcal{G} = (\mathcal{V}, \mathcal{E}, d, t, c)$ and a node $u \in \mathcal{V}$, the function $CB(u)$, from $\mathcal{V}$ to the positive integers, indicates the control step to which u is assigned by the scheduling process, relative to the starting control step of the current iteration.*

Definition 3 *Given a CSDFG $\mathcal{G} = (\mathcal{V}, \mathcal{E}, d, t, c)$ and a node $u \in \mathcal{V}$, the function $CE(u)$, from $\mathcal{V}$ to the positive integers, indicates the control step at which u finishes its execution, relative to the starting control step of the current iteration, i.e., $CE(u) = CB(u) + t(u) - 1$.*

In the example shown in Figure 2(a), $CB(B)$ is 2 and $CE(B)$ is 3. Since the architecture is the other factor needed to take into account, it is necessary to provide a function that returns the assigned processor number for a specific task. We define such a function namely PE as following:

Definition 4 *Given a CSDFG $\mathcal{G} = (\mathcal{V}, \mathcal{E}, d, t, c)$, the function $PE(u)$, from $\mathcal{V}$ to the set of processors, defines the processor assigned to execute the task represented by node u. If no previous assignment exists, the value of $PE(u)$ is undefined.*

In example shown in Figure 2(a), $PE(B) = 1$. In order to obtain an initial schedule considering the communication overhead, a list-based scheduling technique is modified. Since the traditional list-based scheduling algorithm does not cover the underlying processor communication constraints, the *priority function* for this environment needs to be tailored. The following definitions introduce the functions used to determine the scheduling priority.

Definition 5 *The mobility of a node v, $MB(v)$, is defined as the difference between the current control step and the as-late-as-possible control step, a control step in which a node is able to be scheduled as late as possible, that node v can be scheduled without increasing the schedule length.*

In order to make the underlying communication cost play a significant role, we assume that the communication cost is proportional to the transmitted data volume and the number of links in which transmitted data traverse through. Also the communication channels are multiple so that there is no congestion in transmitting data. The following is the definition for the communication overhead associated with an architecture.

Definition 6 *For a given dependency $u \xrightarrow{m} v$ in a CS-DFG $G = (V, \mathcal{E}, d, t, c)$, the communication function, $M(p_i(u), p_j(v))$, is the product of the number of links that some data transmitted from processor p_i to processor p_j traverses through by the transferring data volume m.*

As an example, Figure 1(b), if node B was scheduled on $pe1$ and node E was scheduled on $pe3$ in the 2-D Mesh environment, the communication function would result $2 \times 3 = 6$. From aforementioned observations, it is clear that the topology of the architecture plays an important role in the computation of such functions. For instance, in a linear array of N processors, in order to route from one terminal node to the other end node, the data needs to traverse through $N - 1$ links.

Consequently, we can express the priority function for the initial scheduling algorithm which is modified in order that the function preserves both mobility and communication factors. Note that the communication cost in this function is not obtained from the communication function M but m since there is no such an information telling which processor a node is assigned to at this stage. Below we formulate the relationship of all factors as components of the *priority function PF*.

Definition 7 *Given a CSDFG $G = (V, \mathcal{E}, d, t, c)$, a node $v \in V$, a set of nodes $u_i \in V$ where u_i represents each of the predecessors of v, a processor p_j where v can be scheduled, and the data volume m_i between u_i and v, the priority function $PF(v)$ is defined as following:*

$$PF(v) = \max_i \left\{ m_i - (cs_{cur} - (CE(u_i) + 1)) - MB(v) \right\}$$

where cs_{cur} is the current control step being scheduled.

The PF function considers all relevant factors that may affect the final schedule. The $cs_{cur} - (CE(u_i) + 1)$ factor implies the number of control steps between the predecessors of u_i and the current control step, i.e., how long task v

has been delayed after its predecessors have been executed. The the data volume cost m is reduced its effect proportionally to the deferred time. The value of $MB(v)$ function reduces the priority proportionally to how long node v can be delayed without affecting the total execution time. A node with higher mobility (ability of being scheduled later) implies a lower PF. The higher returned value from PF regards as the higher priority of the node to be scheduled.

Algorithm

At this point, we now present the *start-up scheduling* algorithm based on PF. The input of this algorithm is the CSDFG describing the problem, with no feedback edges. Figure 1(b) shows the input graph with respect to our example. This algorithm computes an *initial schedule* considering communication costs and the target architecture. All nodes which are ready to execute are inserted into a *ready list* and rearranged according to the priority given from PF function. The algorithm for this initial schedule is shown below.

Algorithm START-UP-SCHEDULING(G)
Input : CSDFG $G = (V, \mathcal{E}, d, t, c)$ no feedback edges.
Output : An initial schedule for G.
begin
    $cs \leftarrow 0$; $list \leftarrow \phi$;
    while $V \neq \phi$ **or** $list \neq \phi$
    do $dlist \leftarrow \phi$
        if there exists $u \in V \rightarrow$ no PRED(u)
        or all PRED(u) have been scheduled.
            then INSERTLIST($u, list$)
        ARRANGE($list$);
        while $list \neq \phi$ **and** Processors-Available
        do $node \leftarrow$ EXTRACT($list$);
            $cm \leftarrow \min_j\{\max_i\{CE(\text{PRED}_i(node)$
            $+M(PE(\text{PRED}_i(node)), p_j\}\}$;
            if $cm < cs$
                then SCHEDULE($node, cs, p_j$);
                else $dlist \leftarrow dlist \cup \{node\}$;
            $list \leftarrow list - \{node\}$;
        endwhile
        $list \leftarrow dlist \cup list$; cs^{++};
    endwhile
end

On the above algorithm, the PF function is applied inside the routine ARRANGE. Therefore, the ready node that has the highest priority is scheduled first. The algorithm attempts to schedule nodes from the *list*, one by one, according to the priority. It also checks the validity of scheduling the node to an available control step under an available processor by calculating the minimum possible scheduling control step for that specific position. In the pseudo code above, the algorithm computes the possible control step for the tentative scheduling node by adding the last control step that the parent of the node resided with the communication overhead with respect to a specific architecture (M). Since it is possible that the node can have multiple number of predecessors, the max function are useful to choose the largest cost among parents. Unlike the cost from multiple predecessors, the min function is applied to determine which processor the node should be scheduled to. The node that cannot be scheduled in the algorithm iteration will be postponed to the next iteration. Finally, the algorithm stops when all nodes have been scheduled.

For instance, in the example of Figure 1(b), the algorithm considers the input graph as an *acyclic* CSDFG. Node A is assigned to the first available position which is control step 1 under $pe1$, since node A is the only *root* of the graph. Thereafter node B and C are ready to be inserted into the *list*. Both node B and C have equal opportunity to be scheduled. Node B is selected to be scheduled first at control step 2 under $pe1$. Nevertheless, because of the communication cost from node A to C, node C cannot be scheduled at the same time as node B; therefore node C is deferred to the next iteration. After node B has been scheduled, node D, the successor of node B becomes ready and it is inserted into the ready list. However the control step 2 is no longer valid for any node inside the list. The algorithm carries out this problem by looking for the next available control step. The algorithm stops when all nodes from the CSDFG have been scheduled. Since the distance between source and destination in *2-D Mesh* architecture is not regular, assigning a node to different processors causes different results. In Figure 2(a), if node C had been spawned under processor $pe3$, it would need to be scheduled at least at control step 4. Therefore, the algorithm selected to assign node C to control step 3 under $pe2$. The initial schedule is consequently issued to the cyclo-compaction phase introduced in the next section.

CYCLO-COMPACTION SCHEDULING

This section introduces our optimization algorithm called *cyclo-compaction scheduling* algorithm. The initial schedule from the previous section will be compacted by this systematic algorithm. The algorithm determines the underlying architecture as part of the rescheduling process. The optimizing system consists of two phases: a *rotation* phase which implicitly applies the *retiming* operation [6] and the *re-mapping* phase which re-assigns retimed nodes from the rotation stage back to a schedule table.

The cyclo-compaction scheduling deallocates nodes from the first row of a schedule table and partially reschedules these nodes to new positions. The rotation phase infers to the movement of delays in the graph. Such a de-

lay movement yields the better pipelining capability of the graph known as retiming. The re-mapping is essentially a partial scheduling process.

The following are the definition for the rotation stage which extracts nodes from the schedule table while moving delays associated with the nodes in the input graph systematically.

Definition 8 *Given a CSDFG $\mathcal{G} = (\mathcal{V}, \mathcal{E}, d, t, c)$ and $\mathcal{J}$ a subset of $\mathcal{V}$, the* rotation *of set $\mathcal{J}$ draws one delay from every incoming edge and pushes it to every outgoing edge of $\mathcal{J}$. The CSDFG is transformed into a new CSDFG, $\mathcal{G}_{\mathcal{J}}$, in a schedule table of length L. This is equivalent to moving the row number 1 to the position $L + 1$.*

The rotation holds every property of the retiming operation. Therefore, *all* of incoming edges of the rescheduled node must have delays greater than zero; otherwise the rotation is regarded as an illegal operation. Note that the rotation operation here is scheduling computing environment in which we are able to consider an intermediate result as an output format. Hence it facilitates the means of solving this problem. The following lemma presents that the rotation itself does not change the schedule length.

Lemma 1 *Given a CSDFG $\mathcal{G}$, scheduled to be executed according to a schedule table H with length equal to L control steps, the rotation operation retimes $\mathcal{G}$ creating a new CSDFG $\mathcal{G}_r$ which has the same schedule length L.*

As an example, Figure 2(a) and (b) represents the rotation phase of node A, by moving node A from $cs1$ down to $cs8$. The re-mapping operation re-schedules it back to $cs2$ which will be $cs1$ of the next iteration after renumbering the control steps.

There are two approaches of dealing with the re-mapping, the re-mapping with relaxation and without relaxation. The former is to allow the schedule length growing larger than the initial one in the intermediate state. The latter is to guarantee that the schedule length will not be longer than the previous one. Both of these methodologies, in fact, produce the final more compact schedule length by storing the shortest schedule table from every iteration of intermediate steps.

The basic notion of two re-mappings are the same. The following definition represents the concepts of two re-mapping operations.

Definition 9 *Given a CSDFG $\mathcal{G} = (\mathcal{V}, \mathcal{E}, d, t, c)$, $\mathcal{J}$ a subset of $\mathcal{V}$ and a current schedule length L_i, the* re-mapping without relaxation *operation is defined as assigning nodes in $\mathcal{J}$ to the schedule table yielding the shorter or at least the same schedule length L_i. Unlike without relaxation, the* re-mapping with relaxation *allows to have a longer schedule length $L_n > L_i$.*

Intuitively, when nodes get re-scheduled with changing delay assignments (retiming) the dependencies within an iteration and between an iteration tend to be modified accordingly. The underlying communication cost may imposes an extra length to the new schedule table. Hence, without the relaxation method, if the nodes from $\mathcal{J}$ are not able to be scheduled to a table in such a way that the schedule length L will be shorter than the former one, the relaxation method will abandon itself. This results at least same schedule length as before according to the lemma 1. In other words, the re-mapping phase does not occur in this case. On the other hand, the re-mapping with relaxation, is always taken place even if the schedule length becomes longer.

Since the communication overheads cause difficulties of parallelizing data dependent nodes, every rotated node needs to conjecture the position in the schedule table according to the dependencies as well as the communication overheads. The *delays* of an edge $u \rightarrow v$ imply a dependency between an iteration and if there are no delay, it implies to a direct dependency in the same iteration. For example, in Figure 1(b), the edge $D \rightarrow A$ has 2 delays which means that node A receives the data from a previous execution of node D after two iterations. The edge $A \rightarrow B$ in Figure 1(b) has no delay between them; therefore, both of them must be executed in the same iteration. We formulate the strategy to guess the above constraints as an *anticipation* function AN described below:

Definition 10 *Given a CSDFG $\mathcal{G} = (\mathcal{V}, \mathcal{E}, d, t, c)$, and k a number of delays in the edge $u \xrightarrow{e} v \in \mathcal{G}$, the anticipation function of u, $AN(u)$, results in the first possible control step under a specific processor such that if node u is assigned to that position, the dependencies and communication constraints between the current iteration and the kth iteration will be held while the resulting schedule length is decreased by one control step.*

The following lemma provides a concrete way of calculating a safe schedule position, i.e., which processor element and at which control step.

Lemma 2 *Given a CSDFG $\mathcal{G} = (\mathcal{V}, \mathcal{E}, d, t, c)$, a rotated node $v \in V$, the set of incoming edges of v, $u_i \xrightarrow{e_i} v \in \mathcal{G}$, the delays of the edges e_i after rotation, $d_r(e_i) > 0$, and the schedule length L of the current iteration, the first valid control step for assigning v under a specific processor p_j, such that $CB(v)$ is inside a schedule length $L - 1$ with all of the constraints preserved is given by function AN where AN is*

$$\max_i \left\{ M(PE(u_i), p_j) - (d_r(e_i)(L-1) - CE(u_i)) + 1 \right\}$$

The anticipation function computes the expected control step that complies with a schedule length shorter by one.

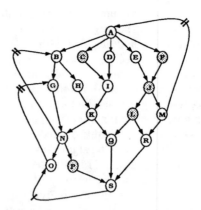

Fig. 4: An example of 19-node CSDFG

Both of the approaches attempt to partially schedule nodes from $\mathcal{J}$ according to the minimum value returned from the anticipation function. Nonetheless, if the target processor is not available at that returned value, the re-mapping operations will search for the *next-minimum-available* processor which can be the next minimum returned value from others processors or the next control step under the same resource. For example, in Figure 3(a), node C is rotated and ready to be re-scheduled. The re-mapping phase computes the possible $AN(C)$ for every processor. The $AN(C)_{pe1} = 1 - ((6-1)-6)+1 = 3$, $AN(C)_{pe2} = 0-((6-1)-6)+1 = 0$, $AN(C)_{pe3} = 1 - ((6-1)-6)+1 = 3$ and $AN(C)_{pe4} = 2 - ((6-1)-6)+1 = 4$. The minimum returned value of this example would be zero under the same processor as node C's predecessor ($pe2$); however, the first available control step of this resource is 4 which is larger than $AN(C)_{pe3} = 3$. Consequently, node C is re-assigned to $cs3$ under $pe3$. The partial result of this re-mapping method after the third iteration is shown in Figure 3(b).

Nevertheless, for every iteration of the *cyclo-compaction* algorithm , the deallocation of the first row from the schedule table may destroy the dependencies between two iterations due to the rotation phase. Therefore, the function called *projected schedule length* for general-time CSDFGs, PSL_G, is applied to compute the required minimum schedule length for each iteration of the cyclo-compaction. Whenever the PSL_G value is greater than the re-mapping-phase schedule length, at most one empty control step is added to the end of the schedule table. In other words, the algorithm will assign empty control step to *compensate* the communication requirements.

It becomes clearer that by the end of the cyclo-compaction iteration, if the re-mapping without relaxation method is applied and the PSL_G returns the longer schedule length, the nodes will not be re-mapped resulting the same length as the previous iteration. On the contrary, the re-mapping with relaxation technique would permit to have that PSL_G length. The PSL_G function then is defined as following:

Definition 11 *Given a CSDFG $\mathcal{G} = (\mathcal{V}, \mathcal{E}, d, t, c)$ and a node $u \in \mathcal{V}$, the projected schedule length with respect to u, $PSL_G(u)$, from $\mathcal{V}$ to the positive integers, is the minimum schedule length required to satisfy the data dependency and communication constraints.*

We formulated the $PSL_G(u)$ as the following lemma.

Lemma 3 *Given a CSDFG $\mathcal{G} = (\mathcal{V}, \mathcal{E}, d, t, c)$, an edge $u \xrightarrow{e} v \in \mathcal{G}$, $PE(u) \neq PE(v)$, and $d(e) = k$ for $k > 0$, for any iteration i, a legitimate schedule length for G must be greater than or equal to $PSL_G(u)$, where $PSL(u)$ is*

$$\left\lceil \frac{M(PE(u), PE(v)) + CE^{(i)}(u) - CB^{(i+k)}(v)}{k} \right\rceil$$

From the observation of the re-mapping without relaxation methodology, the cyclo-compaction algorithm will not produce such a longer schedule length for the next optimizing iteration. Hence, we state this property as the following theorem:

Theorem 4 *Given a CSDFG $\mathcal{G} = (\mathcal{V}, \mathcal{E}, d, t, c)$, for every pass of the cyclo-compaction scheduling algorithm via the re-mapping without relaxation method, the resulting schedule length is always less than or equal to the one computed in the previous iteration.*

At this point, we can summarize the cyclo-compaction scheduling algorithm as following pseudo-code. Basically, both re-mapping approaches have the same major scheduling concept except the validity checking of the schedule length condition.

Algorithm CYCLO-COMPACT$(\mathcal{G}, z)$
Input : CSDFG $\mathcal{G} = (\mathcal{V}, \mathcal{E}, d, t, c)$.
Output : A more compact schedule S.
begin
    Compute the latest starting time for every $u \in \mathcal{V}$;
    $S \leftarrow$ START-UP-SCHEDULING$(\mathcal{G})$;
    $Q \leftarrow S$;
    for $(n = 1$ to $z)$
        $(\mathcal{G}, S) \leftarrow$ ROTATE-REMAP$(\mathcal{G}, S)$;
        if (LENGTH$(S) <$ LENGTH(Q)) **then** $Q \leftarrow S$;
    end
    Return(Q)
end
Procedure ROTATE-REMAP$(\mathcal{G}, S)$
begin
    $\mathcal{J} \leftarrow \{$Deallocated nodes from the schedule $S\}$;
    $\mathcal{G}_{\mathcal{J}} \leftarrow$ RETIME$(\mathcal{G}, \mathcal{J})$;

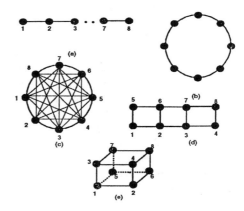

Fig. 5: Five target architectures

    $S \leftarrow$ REMAPPING$(\mathcal{G}_{\mathcal{J}}, \mathcal{J}, S)$;
    Return$(\mathcal{G}_{\mathcal{J}}, S)$
end
Procedure REMAPPING$(\mathcal{G}, \mathcal{J}, S)$
begin
    Compute $AN(u) \ \forall u \in \mathcal{J}$ and $\forall$ processors
    Re-schedule $u \in J$ such that $CB(u) \geq AN(u)$
    and $CE(u) <$ LENGTH(S)
    (optional) **and** $\forall v \ PSL_G(v) \leq$ LENGTH(S)
end

EXPERIMENTS

In this section, we present the examples tested on specific architectures which are *linear array, 2-D Mesh, ring, completely connected* and *cube* [13] as shown in Figure 5. We first begin with the example shown in Figure 4. A more complex CSDFG comprising of nineteen *general-time* nodes to be scheduled to 8 processors connected as the topologies presented in Figure 5. In this example, $t(C) = t(F) = t(J) = t(L) = t(P) = 2$ time unit and the other nodes have execution time equal to 1 time unit shown in Figure 4. Data volume associated with every edge is one.

The following table presents the results obtaining from applying the cyclo-compaction algorithm. The two re-mapping approaches are also compared.

Table 1: Results from the 19-node example

Comparing two re-mapping methods										
Relax.	complete		linear		ring		2-D Mesh		cube	
	$\mathcal{S}^a$	$\mathcal{R}^b$	$\mathcal{S}$	$\mathcal{R}$	$\mathcal{S}$	$\mathcal{R}$	$\mathcal{S}$	$\mathcal{R}$	$\mathcal{S}$	$\mathcal{R}$
with	12	5	13	7	15	7	13	6	13	6
without	12	9	13	11	15	11	13	11	13	10

[a] Start-up schedule
[b] After applying cyclo-compaction for 200 passes

The examples above have demonstrated that the performance of the system would be better in the completely connected architecture than the other architectures because of the uniformity of communication cost among processors. We modified the input DFG of two benchmarks, the 5th elliptic and *lattice* filter, to be a CSDFG. These benchmark graphs are tailored slightly by enlarging the size of those input cyclic DFG by using the *unfolding*[1] technique and by grouping the unfolded nodes to create different data volume costs. Table 2 presents the experimental results on these benchmarks, with a *slow down* factor of three [12]. Two re-mapping strategies with respect to the four architectures, *completely connected, ring, 2-D Mesh,* and *cube* are compared. The $\Im$ column represents the start-up schedule length for each application and the $\Re$ column shows the resulting schedule length.

Table 2: Applying cyclo-compaction on two benchmarks

Testing on the modified fifth elliptic and lattice filter								
Applications	Complete		Ring		2-D Mesh		Cube	
	$\Im^a$	$\Re^b$	$\Im$	$\Re$	$\Im$	$\Re$	$\Im$	$\Re$
Lattice (w/)	105	35	105	71	105	63	105	37
Lattice (w/o)	105	100	105	103	105	103	105	100
Elliptic (w/)	126	57	126	102	126	95	126	71
Elliptic (w/o)	126	126	126	126	126	126	126	126

[a] Start-up schedule
[b] After applying cyclo-compaction for 400 passes

From the information in Table 2, the completely connected architecture seems to give out the shortest schedule table of all. The re-mapping scheme *with* relaxation yields the better result even though the intermediate scheduling process may give the longer schedule table according to the PSL_G.

CONCLUSION

Previous scheduling mechanisms are restricted to noncyclic data flow graphs or do not consider communication overhead that exists when two tasks, i.e., nodes in the data flow graphs are spawned to different processors. This paper presented a new algorithm called cyclo-compaction scheduling to schedule cyclic data flow graphs with loop carried dependencies, taking into account communication costs for a specific architecture. Our algorithm compacts a given initial schedule iteratively by rotation and remapping. The retiming (or loop pipelining) is implied in each of our rotation steps, and remapping is performed to map rotated nodes to the best processor while communcation overhead and dependency legality are considered. According to our experiments, this algorithm gives a much shorter legal schedule for all kinds of parallel architectures.

---

[1] Unfolding is also called *unrolling* or *unwinding* in compiler design where a **for** or **while** loop is considered.

REFERENCES

[1] M. Lam, "Software pipelining," in *Proc. of the ACM SIGPLAN Conference on Progaming Language Design*, pp. 318–328, 1988.

[2] A. A. Khan, C. L. McCreary, and M. S. Jones, "A comparison of multiprocessor scheduling heuristics," in *1994 Int. Conference on Parallel Processing*, vol. II, pp. 243–250, 1994.

[3] B. Shirazi, K. Kavi, A. R. Hurson, and P. Biswas, "Parsa: A parallel program scheduling and assessment environment," in *1993 Int. Conference on Parallel Processing*, January 1993.

[4] R. A. Kamin, G. B. Adams, and P. K. Dubey, "Dynamic list-scheduling with finite resources," in *Int. Conf. on Computer Design*, pp. 140–144, Oct. 1994.

[5] A. Aiken and A. Nicolau, "Optimal loop parallelization," in *Proc. of the ACM SIGPLAN Conf. on Programming Languages Design and Implementation*, pp. 308–317, June 1988.

[6] L. Chao and E. Sha, "Unified static scheduling on various models," in *1993 Int. Conference on Parallel Processing*, pp. 231–235, August 1993.

[7] S. Shukla, B. Little, and A. Zaky, "A compile-time technique for controlling real-time execution of task-level data-flow graphs," in *1992 Int. Conference on Parallel Processing*, vol. II, pp. 49–56, 1992.

[8] A. A. Munshi and B. Simons, "Scheduling sequential," *SIAM Journal of Computing*, vol. 19, pp. 728–741, August 1990.

[9] P. D. Hoang and J. M. Rabaey, "Scheduling of dsp programs onto multiprocessors for maximum throughput," *IEEE Transactions on Signal Processing*, vol. 41, June 1993.

[10] L. Chao, A. LaPaugh, and E. Sha, "Rotation scheduling: A loop pipelining algorithm," in *Proc. of ACM/IEEE Design Automation Conference*, (Dallas, TX), pp. 566–572, June 1993.

[11] S. Tongsima, N. L. Passos, and E. Sha, "Communication sensitive rotation scheduling," in *Int. Conference on Computer Design*, pp. 150–153, October 1994.

[12] C. E. Leiserson and J. B. Saxe, "Retiming synchronous circuitry," *Algorithmica*, pp. 5–35, 1991.

[13] K. Hwang, *Advanced Computer Architecture: Parallelism, Scalability, Programmability*. New York, NY: McGraw-Hill Series in Computer Science, 1993.

Processor Management Techniques for Mesh-Connected Multiprocessors*

Byung S. Yoo[†], Chita R. Das[†] and Chansu Yu[‡]

[†]Dept. Computer Science and Engineering
The Pennsylvania State University
University Park, PA 16802
E-mail: {yoo | das}@cse.psu.edu

[‡]Information Technology R&D Laboratory
GoldStar Company
Seoul, Korea 137-140
E-mail: cyu@pc.isl.goldstar.co.kr

Abstract

This paper investigates various processor management techniques for improving the performance of mesh-connected multiprocessors. Three different techniques are analyzed. First, we use the smallest job first (SJF) policy to improve the spatial parallelism in a mesh. Next, a policy called multitasking and multiprogramming (M^2) is introduced. The M^2 scheme allows multiprogramming of jobs on various submeshes. Finally, a novel approach, called "limit allocation" is used for job allocation. With this policy, a job (submesh) size is reduced if the job cannot be allocated. While all of the three approaches are viable alternatives, which in conjunction with any allocation algorithm can improve system performance beyond what is achievable with a complex allocation scheme and the usually assumed FCFS scheduling, the M^2 and limit allocation techniques are especially attractive for providing some additional features. The M^2 policy brings in the concept of time-sharing execution for better efficiency and the limit allocation shows how job size restriction can be beneficial for performance and fault-tolerance in a mesh topology. Moreover, the limit allocation scheme can outperform any other approach even using the simplest allocation policy.

1 Introduction

A critical issue in designing a multiprocessor operating system (OS) is the underlying resource management policy to support multiple users. In a dynamic environment, where different jobs could have different computational requirements in terms of number of processors and execution time, efficient assignment of system resources to various jobs is admittedly quite complex. On the other hand, an efficient resource management policy is crucial to improve system performance. Resource management in multiprocessors consists of two steps - job or process scheduling and processor allocation. Scheduling is concerned with selecting the next job or process for execution. Processor allocation is concerned with choosing the required

number of processors for the job or process. These two components complement each other in utilizing system resources. Ideally, one would like to design a resource manager that can provide high throughput and high resource utilization while the management policy should be simple (low time and space complexity), fair, and starvation free. High efficiency and low complexity are conflicting requirements. Therefore, various allocation and scheduling schemes proposed in the literature exhibit a compromise between efficiency and complexity.

The focus of this paper is on developing processor management policies for mesh-connected multiprocessors. Almost all prior work on resource management in mesh systems has focused on processor allocation [1, 2, 6, 7, 11]. These policies attempt to allocate a required submesh to an incoming job. These space-division multiplexing techniques differ in their submesh recognition ability and use the FCFS scheduling for assigning the incoming jobs. While a better allocation policy can improve system performance, an allocation policy alone cannot boost performance significantly [4]. Response time behavior of all prior allocation policies support this fact. The FCFS scheduling policy has what is known as the 'blocking' property. A request for a large submesh that cannot be allocated may block subsequent smaller submesh requests, which are serviceable. This results in higher fragmentation and hence low resource utilization. It is therefore essential to investigate various job scheduling strategies, which along with a simple allocation scheme, can improve system performance beyond what is achievable with a complex allocation scheme and the FCFS scheduling.

The only reported scheduling policy for mesh-connected system is based on a combination of priority and reservation techniques [8]. It is shown that using this policy the average waiting time of a job is reduced considerably from the simple FCFS policy. However, this policy is quite complex and violates the fairness rule. The reservation policy that searches for all possible allocated submeshes alone does not improve performance, while it increases the search time. Priority with reservation provides better performance for

*This research was supported in part by the National Science Foundation under Grant No. MIP-9406984.

moderate traffic while only the priority policy yields the best results for high traffic. The scheduling policy needs to be changed with the traffic.

In this paper we discuss three types of resource management schemes for mesh-connected multiprocessors. First, we show that under moderate to heavy load, average job response time can be reduced by almost half by changing the scheduling strategy from FCFS to *smallest job first* (SJF). Next, we propose a generic processor management scheme called M^2 (multitasking and multiprogramming). Multitasking here refers to allocation and multiprogramming refers to temporal scheduling. This scheme uses both space-division and time-division multiplexing to manage jobs. The motivation for combining the two schemes is to take advantage of both policies. Multitasking (spatial allocation) can reduce interconnection latency while multiprogramming can increase throughput and minimize I/O latency.

Three parameters are used to explicitly define a processor management scheme using the M^2 framework. These are *scheduling unit* (processor job-based), *multiprogramming width* (processor-, subsystem-, or system-wide), and *partitioning mechanism* (static- or dynamic-partitioning). A three-tuple notation ($M^2_{unit,width,partition}$) is used to describe the M^2 approach. The M^2 policy we describe here is denoted as $M^2_{job,pr,dyn}$ and is called *virtual mesh*. An incoming job is allocated to one of the virtual meshes based on its form. The virtual meshes are scheduled in a round-robin fashion. When a virtual mesh is scheduled for a time quantum, jobs allocated to the virtual mesh are executed simultaneously during that time quantum.

The third approach that can reduce fragmentation and hence improve system performance is to reduce the job size (submesh size). Reduced job size increases the probability of finding a smaller submesh for a job and hence minimizes the job waiting time. We call it *limit allocation* since it limits the job size if a required submesh cannot be allocated. Another advantage of the limit allocation scheme is that it is less susceptible to processor failure than any other scheme since the fragmentation caused by a few faulty processor(s) hardly affects its performance. Such robustness makes the limit allocation highly attractive in fault-prone environment.

We simulated a 32 × 16 mesh to study the proposed schemes. The output parameters obtained were average response time and average processor utilization. All the policies use the first fit (FF) allocation rule [11]. The fault-tolerant ability of each scheme is also examined through simulation. Job rejection ratio as well as processor utilization of proposed schemes were obtained for various number of faulty processors. The simulation results indicate that all the three resource management schemes are viable approaches for efficient resource management. Using these three schemes, job response time can be reduced by 50% to 90% compared to the FCFS policy. The attractive feature of the M^2 policy is its time-sharing facility while the advantage of the limit allocation policy is its adaptability to workload intensity in sustaining high performance and robustness to processor failures. Using a simple allocation algorithm, the limit scheme is capable of outperforming all other policies. The limit scheme uses the simple FCFS scheduling, which results in a fair and starvation-free mechanism.

The rest of the paper is organized as follows. In Section 2, the proposed processor management schemes are described. In Section 3, applicability of the limit allocation in fault-tolerant processing is discussed. Simulation results are presented in Section 4. Conclusions are drawn in Section 5.

2 Processor Management Alternatives

2.1 Allocation

We use the *first fit* (FF) [11] strategy in this paper for allocation. It is simple and yet very efficient in terms of time complexity and submesh recognition ability. The first fit algorithm tries to find the base of an available submesh for an incoming job. The base of a submesh refers to the lower left corner of the submesh. Whether a node can be the base of any submesh with respect to an incoming job is determined quite efficiently by using the size of the submesh request and the node availability information. With respect to an incoming job J, a set of processors that cannot be the base of any free submesh for accommodating J, called a *coverage set*, can be determined from a collection of busy processors. The policy tries to construct such a coverage set by scanning all the processors and returns the first node which does not belong to the coverage set. The allocation algorithm is given below. For detailed discussion, the reader should refer to [11].

First Fit Allocation Algorithm: First-fit(J_i)

Allocation:

1. Let J_i be the current incoming job.
2. Construct a coverage set C_{J_i} with respect to J_i. Scan all the rows in any order. Scan each row from right to left and mark processors belonging to C_{J_i}.
   Scan all columns from left to right. Starting from the top of each column, mark the processors belonging to C_{J_i}.
3. If a processor does not belong to C_{J_i}, then return it as the base of a free submesh to accommodate J_i. Add the submesh to the allocated submesh list.
4. If no free submesh is found, then enqueue J_i.

Deallocation:

1. Remove the released submesh from the allocated submesh list.

The entire mesh must be scanned at least once to find a base. Therefore, the time complexity of the first fit algorithm is $O(MN)$ for an $M \times N$ mesh.

2.2 Scheduling

In this subsection, an efficient scheduling policy, known as *smallest job first* (SJF), is presented. As the name implies, the smallest submesh request is considered first for allocation. The size of a submesh is defined as the number of processors in it. The advantage of allocating smaller jobs is that it increases the number of allocated jobs, which in turn increases the parallelism.

The SJF policy tries to find a free submesh for an incoming job using a submesh allocation algorithm. If a free submesh is not available, then the job joins the system queue. The system queue is maintained in the FCFS order. Using an appropriate data structure, we can efficiently find the smallest job in the system queue. When a job departs the system, leaving a free submesh, jobs in the system queue are considered for allocation in the SJF order. Large jobs may suffer from starvation under the SJF scheme. To prevent this, we assign a threshold value δt for the maximum queueing delay that a job can tolerate when the job joins the system queue. Whenever the waiting time of a job reaches this threshold value, it gets priority over incoming jobs as well as all jobs in the system queue. No waiting jobs are serviced until it is allocated. A predefined threshold value is useful in imposing deadline for job completion. A heuristic may be used to dynamically derive the value of δt [3]. We use a predefined threshold value in this paper.

SJF Scheduling

Let J_{head} denote the job at the head of the system queue and Deadline$[J_i]$ denote the deadline of job J_i.

Job Arrival:

1. Let J_i be the incoming job.
2. If current_time > Deadline$[J_{head}]$ then
   Deadline$[J_i] \leftarrow$ current_time + δt.
   Enqueue J_i.
   Else
   Call First-fit(J_i) to allocate J_i.
3. If allocation fails, then enqueue J_i.

Job Departure:

1. Let J_i be the departing job.
2. If current_time < Deadline$[J_{head}]$ or First-fit(J_{head}) succeeds then
   Consider all jobs in the system queue in the SJF order. Allocate the jobs by calling First-fit until the allocation fails or the queue is empty.

The most expensive operation in the above algorithm is the first fit allocation. Other steps can be performed in a constant time assuming that the length of the queue that is searched is a small constant in steady state. Therefore, time complexity of the SJF is $O(MN)$ for an $M \times N$ mesh.

2.3 Multitasking and Multiprogramming (M^2)

A processor management policy that combines the spatial allocation and temporal scheduling, called *multitasking and multiprogramming* (M^2) policy, has been recently proposed by Yu [9] and has been analyzed for hypercubes. The key idea of the M^2 policy is to allocate multiple jobs to a subsystem and run them in a time-sharing fashion. After allocated, the processes of each job are executed in synchronous or asynchronous manner. An M^2 policy is defined by using three parameters; *scheduling unit* (process- or job-based), *multiprogramming width* (processor-, subsystem-, or system-wide) and *partitioning mechanism* (static or dynamic).

Scheduling unit in an M^2 scheme could be either a job or a process. A job-based policy forces the processes of a job to be switched in and out at the same time so that all the processes are executed in a synchronous manner. On the other hand, a process-based policy allows each process of a job to behave as an independent unit once a job is allocated, and hence the processes run asynchronously. Multiprogramming width specifies the queue which a process or a job joins after being preempted at the end of a time quantum. With the *system-wide* policy, each process or job returns to the system queue. The *processor-wide* multiprogramming allows increased autonomy to each processor. With this approach, once a process is allocated to a processor, the scheduling of the process is managed by the processor. With the *subsystem-wide* policy, each process or job returns to the subsystem queue. One processor of the subsystem is responsible for managing the subsystem queue. Finally, partitioning mechanism defines how a system is divided to allocate the incoming jobs for multitasking. Static-partitioning divides a system into several subsystems statically. An incoming job is allocated to a subsystem, which might have been already assigned to other jobs. Dynamic-partitioning divides a system dynamically upon job requests. After all jobs or processes of a subsystem are completed, the subsystem is released.

Theoretically there could be twelve possible management schemes using the cross-product of these three parameters. In addition, multiprogramming width and/or partitioning mechanism may not be applicable in some specific implementations. The number of possible representations therefore becomes twenty four including this option. The M^2 is a generic taxonomy, which can be used to describe any shared memory or distributed memory resource management scheme with a three-tuple notation - $M^2_{unit,width,partition}$. We assume FCFS scheduling for all the queues.

Fig. 1 depicts a queueing representation of $M^2_{ps,pr,*}$ and $M^2_{job,pr,*}$ schemes suitable for a distributed memory environment, where process migration between processors of a job is expensive and should be restricted. We analyze a job-based, processor-wide, dynamic M^2 policy ($M^2_{job,pr,dyn}$) for the mesh topology.

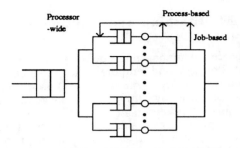

Figure 1: Queueing representation of Process/Job-based Processor-wide M^2.

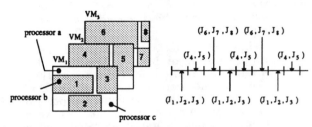

(a) Three virtual meshes multiprogrammed   (b) Timing diagram

Figure 2: Multiprogramming of three virtual meshes.

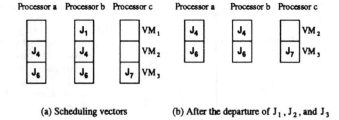

(a) Scheduling vectors   (b) After the departure of J_1, J_2, and J_3

Figure 3: Scheduling vectors for processors in Fig. 2.

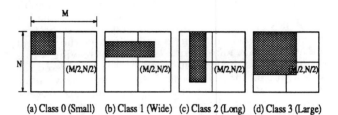

(a) Class 0 (Small)   (b) Class 1 (Wide)   (c) Class 2 (Long)   (d) Class 3 (Large)

Figure 4: Job classification based on form.

We call the M^2 policy *virtual mesh* (VM) scheme. With this scheme, a mesh system is considered to consist of multiple virtual meshes. An incoming job is allocated to one of these virtual meshes using a submesh allocation algorithm. And once allocated, it runs on the allocated virtual mesh until it completes. The virtual meshes are scheduled in a round-robin fashion. That is, jobs allocated to a virtual mesh are scheduled together for a time quantum and run simultaneously. The jobs allocated to the next virtual mesh are executed in the next time quantum, and so on. All processes of a job are switched in and out at the same time, and hence run synchronously. For example, multiprogramming of three virtual meshes is illustrated in Fig. 2. The first VM has three jobs, the second VM has two jobs and the third VM has three jobs. A list of jobs to be executed is associated with each time quantum in Fig. 2.b.

For efficient implementation of the scheduling, each processor maintains a scheduling vector of size D, where D is the degree of multiprogramming (DOM). Note that D is equal to the maximum number of virtual meshes allowed in the system. The ith entry of a processor's scheduling vector contains the process to which the processor is assigned in the ith virtual mesh. The processor executes processes in its scheduling vector one after another in a round-robin fashion. If the processor is not assigned to any process in certain virtual mesh, corresponding entry in its scheduling vector should be empty. Whenever an empty entry is encountered, the processor remains idle for that time quantum.

Consider three processors, **a**, **b**, and **c**, depicted by dark circles in Fig. 2.a. Processor **a** is used in VM_2

and VM_3 but is idle when VM_1 is scheduled (i.e., when jobs allocated to VM_1 are executed). Processor **b** is always busy and processor **c** is busy only when VM_3 is scheduled. Fig. 3.a shows the scheduling vector of each processor. Each entry contains the job the processor is assigned to instead of the process for simple presentation. Note that an entry is empty when the processor is not used in the corresponding virtual mesh. Now suppose jobs J_1, J_2, and J_3 leave the system after completion. Then, all processors will be idle when VM_1 is scheduled. When the last job leaves a virtual mesh, the corresponding entries should be deleted from all scheduling vectors to prevent this unnecessary idling. Fig. 3.b shows the scheduling vectors after J_1, J_2, and J_3 leave the system.

Many variations of the virtual mesh scheme are possible depending on how to distribute the incoming jobs to different virtual meshes. In our algorithm, a job is dispatched to a virtual mesh based on its form. It has been shown that performance is significantly improved when incoming jobs are small [11]. It is because small jobs are more likely to find a free submesh. On the other hand, careful observation indicates that large jobs or jobs with high aspect ratio degrade the system performance when they are allocated. A large job occupies a large number of processors and leaves little room for other jobs. A long or a wide job stretches along the length or width of a mesh and such a job tends to block others. Moreover, when allocated in the middle of a mesh, they divide the system into smaller, less useful submeshes. The class of an $m \times n$ job is defined with respect to an $M \times N$ mesh. A job belongs to class 0 (small job) when $m \leq M/2$ and $n \leq N/2$, and class 1 (wide job) when $m > M/2$ and $n \leq N/2$. When $m \leq M/2$ and $n > N/2$, the job is of class 2 (long) and when $m > M/2$ and $n > N/2$, the job belongs to class 3 (large job). These four classes are depicted in Fig. 4.

The objective of our algorithm is to increase the parallelism among small jobs and to reduce the external fragmentation induced by class 1 and class 2 jobs by allocating each incoming job to a separate virtual submesh based on its class. Therefore, we use a virtual mesh for each of the four job classes in our algorithm (DOM = 4). An incoming job is first attempted for allocation to the class 0 virtual mesh. This step allows the algorithm to work well under light load. If it fails, then the job is dispatched to an appropriate virtual mesh. Each virtual mesh has its own queue to schedule the jobs, dispatched to it by the central scheduler. Each processor has its own queue (scheduling vector) that holds the processes of different jobs. Any scheduling algorithms can be used for each of these queues. In our paper, the FCFS scheduling policy is used. Detailed steps of the M^2 algorithm are given below.

M^2 : Virtual Mesh

Let VM_k denote the virtual mesh for class k.

Job Arrival:

1. Let J_i be an incoming job of class c.
2. Try to allocate J_i to VM_0.
3. If step 2 fails and $c \neq 0$ then
   Allocate J_i to VM_c.
   Update the appropriate scheduling vectors.
4. If J_i is not allocatable, then it joins the queue for VM_c.

Job Departure:

1. Let J_i be a job of class c departing VM_k ($k = 0$ or $k = c$).
2. Deallocate J_i and update the corresponding scheduling vectors.
3. If VM_k is empty, then delete the kth entry from each scheduling vector.
4. Otherwise, try to allocate jobs in VM_k queue in FCFS fashion.

The most expensive step in the above algorithm is the allocation process that has complexity $O(MN)$ for an $M \times N$ mesh. The scheduling vector update can be done using a broadcasting scheme in constant time. Therefore, the time complexity of above algorithm is $O(MN)$.

The job classification scheme used in the $M^2_{job,pr,dyn}$ policy requires the DOM to be four. Higher DOM can be implemented by assigning more than one VM to a job class depending on the anticipated job sizes. Similarly, if the system cannot support four VMs, we can merge class 1, class 2 and class 3 jobs into one or two classes as required.

2.4 Limit Allocation

With the limit allocation, an incoming job is checked to see if it can be allocated to a free submesh. If such a free submesh is not available in the system, the job size is reduced by half by folding it. The smaller submesh request is then checked again for allocatability. This process of allocatability check and

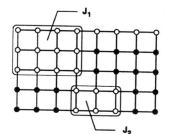

Job	Size	Reduced size
J_1	8 x 3	4 x 3
J_2	6 x 4	3 x 2
J_3	2 x 5	rejected

Figure 5: Allocation of three jobs using limit-Δk with $\Delta k = 2$.

job size reduction is repeated until the job becomes small enough to be allocated or it cannot be scaled down any further. Folding a job can be done either along the length or along the width depending on the length of each side. The longer side is always folded into half. If a job is folded once, it is called limit-1, and if it is folded twice it is called limit-2, etc. When the length of a side being folded is not a multiple of 2, the reduced length is rounded up.

Reduction in job size is reflected by increasing the job execution time. Since the submesh size is reduced by half in each folding step, the service demand should increase at best by a factor of 2 for each folding operation [5]. Thus, the original service demand of a job is linearly increased. For example, a limit-2 reduction results in a four fold increase in the original service time.

The allocation scheme described above is a greedy algorithm in the sense that it always attempts to allocate a request to the largest available submesh So, in the worst case, the algorithm continues scaling down a request until its size is reduced to 1 x 1. In conjunction with the allocation algorithm that has to be called for each reduced size, the overhead involved could be quite high. To make the algorithm more efficient, we restrict the maximum allowed number of folding operations to a certain constant, denoted by Δk. This variation of limit allocation is called limit-Δk. Experiments show that limit-Δk algorithm for a small Δk ($= 2$) easily outperforms other schemes. Fig. 5 shows an example of the limit-Δk allocation mechanism with $\Delta k = 2$. Initially, an 8×5 mesh has three allocated submeshes (represented by dark circles), $(0, 0, 2, 1)$[1], $(6, 0, 7, 1)$ and $(4, 2, 7, 3)$. The next three incoming jobs require 8×3, 6×4, and 2×5 submeshes, respectively. None of these jobs can be allocated without reducing its size. The first job, J_1, is scaled down to 4×3 and then allocated to $(0, 2, 3, 4)$. After being folded twice, J_2 is scaled down to 3×2 and allocated to $(3, 0, 5, 1)$. However, J_3 joins the system queue after being scaled down twice, because it cannot be allocated even after its size is reduced by a factor of four.

Detailed steps of the limit-Δk allocation scheme is

---

[1] A submesh is identified by a quadruple (x, y, x', y'), where $<x, y>$ and $<x', y'>$ are the lower-left and upper-right corners of the submesh.

presented below. The deallocation algorithm is not described here because it is exactly the same as that of the first fit deallocation.

Limit-Δk allocation

1. Let the size of job J_i be $m \times n$.
2. $c \leftarrow 0$.
3. While J_i is not allocatable and $c < \Delta k$ do
   If $m > n$ then $m \leftarrow \lceil m/2 \rceil$.
   else $n \leftarrow \lceil n/2 \rceil$.
   $c \leftarrow c + 1$.
4. If the job could not be allocated in Δk steps then enqueue the original J_i.

Since the limit-Δk tries to allocate a job only a constant number of times (Δk), its time complexity is the same as the first fit algorithm.

3 Application of Limit Allocation in Fault-Tolerant Processing

One advantage of partitionable multiprocessor systems like mesh is that they are less vulnerable to system failure than uniprocessor systems. In a mesh, a faulty node (or processor) can be isolated and an incoming job can be allocated to a fault-free submesh. However, an incoming job, which cannot be allocated due to the processor failure, should be rejected. Even though a mesh system can run in a degraded mode with a few faulty nodes, its performance is significantly affected due to additional external fragmentation caused by the faulty nodes. The external fragmentation problem becomes more severe as the number of faulty processors increases. When the faulty nodes are scattered over the entire mesh, most incoming large jobs are likely to be rejected and only the small jobs can be allocated to the system. This could result in low processor utilization.

A good allocation scheme reduces the number of rejected jobs and hence achieves high processor utilization in faulty environment. Such an allocation algorithm becomes less susceptible to external fragmentation and successfully allocates most of the incoming jobs in a fault prone environment. The limit allocation algorithm discussed in the section 2.4 does meet such requirements quite well. Because of the system fragmentation, jobs are likely to be reduced to smaller submeshes before they are allocated. This increased allocatability should result in lower job rejection ratio and higher processor utilization. All other conventional allocation algorithms cannot adapt to a faulty situation.

4 Simulation Results

Simulation is conducted to evaluate the performance of the proposed processor management schemes. The first fit (FF) policy with the FCFS scheduling also has been simulated for comparison. The model simulated is a 32×16 mesh. Mean response time (including queueing delay) and processor utilization are the performance measures of interest in this study. Rejection ratio, defined as the percentage

of jobs rejected by the system, and processor utilization of each scheme in a faulty environment also have been measured to demonstrate the usefulness of the limit allocation strategy.

The workload is characterized by the distribution of job interarrival time, distribution of job size, and distribution of job service demand. The job interarrival time is assumed to be exponentially distributed. The arrival rate is calculated based on the system load (ρ). The system load is defined as $\rho = \frac{\lambda}{\mu}$, where μ is a service rate and λ is an arrival rate. Therefore, $\lambda = \rho \cdot \mu$. That is, the arrival rate is calculated from the value of the system load and job service rate. In the simulation conducted, we have obtained the performance measures by varying the system load (ρ). The job service demand follows an exponential or a bimodal hyperexponential distribution[2]. Service demand distribution probability α for hyperexponential distribution in our study is taken as 0.95, and the coefficient of variation (C_x) for the residence time is set to 3.0. The mean job service demand is assumed to be 2 time units. Each side of a submesh is computed from a uniform distribution that is limited by the mesh size.

For the SJF scheduling, we predefined the value of threshold δt and set it to 10 time units. For the M^2 scheme, the DOM is set to 4, time quantum for multiprogramming is 0.1 time unit and the context switching time is 0.001 time unit. For the limit allocation, we restrict our experiment to limit-2.

The simulation is event-driven. Each run of the simulation continues until 2000 jobs leave the system. Using independent replication, all the experiment are conducted until the 95% confidence level of the results is within 5% of the mean.

Fig. 6 shows the variation of mean response time and processor utilization of the three processor management schemes analyzed in this paper (SJF, $M^2_{job,pr,dyn}$, and limit-2) as well as the FCFS scheduling policy with respect to system load. Processor allocation scheme is the FF strategy for all schemes. Fig. 7 shows the response time of the four schemes with respect to system load when the job service time follows a bimodal hyperexponential distribution. As Fig. 6 indicates, all of the three proposed schemes outperform the FCFS policy. The average response time is reduced by 50% to 90% under heavy load compared to the FCFS policy. We have noticed that a smaller threshold value δt shifts the response time of the SJF policy toward the FCFS policy. An increase in the δt value beyond 10 time units does not affect the re-

---

[2]Bimodal hyper exponential distribution function is a combination of two exponential distribution functions - one is with a large mean ($\frac{1}{\lambda_l}$) and the other is with a small mean ($\frac{1}{\lambda_s}$). λ_l and λ_s can be determined by three parameters - mean ($\frac{1}{\lambda_m}$), coefficient of variation (C_x) and α, where $\frac{1}{\lambda_l} = \frac{1}{\lambda_m} + \frac{1}{\lambda_m}\sqrt{\frac{(C_x^2-1)\alpha}{2(1-\alpha)}}$ and $\frac{1}{\lambda_s} = \frac{1}{\lambda_m} - \frac{1}{\lambda_m}\sqrt{\frac{(C_x^2-1)(1-\alpha)}{2\alpha}}$. With a probability of α, the total service demand is exponentially distributed with a mean $\frac{1}{\lambda_s}$, and with a probability of $(1-\alpha)$, it is exponentially distributed with a mean $\frac{1}{\lambda_l}$.

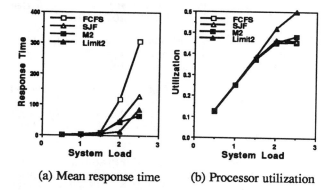

(a) Mean response time　　(b) Processor utilization

Figure 6: Performance of a 32×16 mesh with exponential service demand.

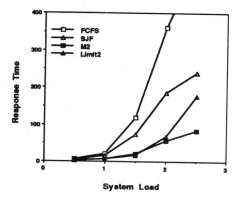

Figure 7: Response time of a 32×16 mesh with hyperexponential service demand.

sponse time curve for the specified workload environment. These results are not reported here due to space limit. Similar performance trend can be observed in Fig. 7.

In Fig. 6, the limit-2 policy outperforms all other schemes (the FCFS policy by 90%) for moderate workload. For higher workload, the M^2 and limit-2 policies have similar behavior. This is because many jobs cannot be allocated after two-fold reduction with the limit-2 scheme. On the other hand, the M^2 policy uses time sharing to accommodate more jobs. For hyperexponential distribution in Fig. 7, the M^2 scheme has the least response time at high load; attributed to the advantage of time sharing. However, the response time of the limit policy can be further reduced by allowing further reduction in job size. It is interesting to note that even though we have increased the job service time linearly with the submesh size reduction, the overall response time has dropped significantly. This is because the job waiting time is the main factor that affects response time behavior. The waiting time can be controlled by the limit scheme.

Fig. 6.b indicates that the limit-2 achieves the

highest processor utilization under moderate to heavy load. It is because more jobs can be allocated by reducing the job size with limit allocation. This keeps many processors busy compared to other policies. The M^2 scheme although has very good response time characteristics, does not improve system utilization like the limit policy. This is because the submesh recognition ability of the M^2 scheme is as that of the FF policy. Therefore, less jobs are allocated to a virtual mesh compared to the limit scheme using temporal partitioning. However, time sharing due to multiprogramming improves the response time behavior. Similarly, the utilization of the SJF policy is not as high as it should have been compared to the improvement in response time. This is due to the fact that we allocate more number of smaller jobs to the system instead of a few large jobs. Thus, the overall utilization does not improve considerably.

Fig. 8 shows the rejection ratio of incoming jobs and processor utilization with respect to the number of faulty nodes in a 32×16 mesh. The mean service demand is 2 time units and the system load is fixed at 2.0. The service time distribution is exponential. The other parameters for the M^2 scheme and SJF are the same as before. An incoming job is rejected when it cannot be allocated due to faulty processors. So, the rejection ratio increases as the number of faulty processors increases as shown in Fig. 8.a. For all the schemes except limit-2, the rejection ratios are identical, because none of these schemes have control over the size of incoming jobs. Rejection ratio reaches as high as 70% for all the other schemes when 10% of the processors are faulty compared to only 25% for the limit-2 policy. As the number of faults increases, only small jobs are processed by the system. This reduces the processor utilization as shown in Fig. 8.b (All other policies have the same utilization.). On the other hand, limit-2 shows far better performance than any other scheme because it is less susceptible to external fragmentation caused by the faulty processors. This highlights the robustness of the limit allocation in addition to its high performance and low complexity.

5　Conclusions

All prior research on designing efficient resource management schemes for mesh-connected multiprocessors has mostly focused on improving the submesh recognition ability of the underlying processor allocation algorithm. However, this approach alone cannot improve system performance considerably due to the blocking property of the FCFS scheduling. This paper, therefore, examines and proposes other alternatives for mesh-connected systems. First, it is shown that the average job response time can be reduced almost by half if the SJF scheduling policy is used. Next, a time-sharing technique, called multitasking and multiprogramming (M^2), is proposed. The M^2 policy is quite generic in nature and can result in many different variations by changing three design parameters; scheduling unit, multiprogramming width, and partitioning mechanism. In this paper, we discuss one implementation of the M^2 policy ,$M^2_{job,pr,dyn}$. This

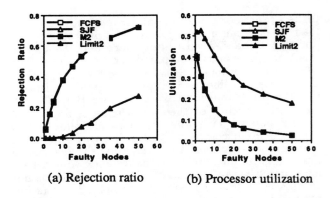

(a) Rejection ratio   (b) Processor utilization

Figure 8: Comparison of proposed schemes with faulty processors.

technique allows allocation of jobs to different virtual meshes dynamically. Jobs are allocated based on their form. All the jobs of a virtual mesh are executed in an assigned time quantum before switching to the next virtual mesh. Simulation results show that significant improvement in job response time is achieved compared to the FCFS and SJF techniques with this time-sharing mechanism.

The third approach, called limit allocation, is another simple and efficient scheme that can be implemented in a mesh OS. This method limits the job size to a smaller submesh if the required submesh cannot be allocated. It can use the simplest available allocation policy and FCFS scheduling and has the potential to outperform all other techniques. The simulation results for the limit-2 policy support this claim. Moreover, job size reduction provides another design attribute - robustness to processor faults. This scheme satisfies many desirable design attributes; high performance, low complexity, fairness, starvation freedom, and robustness. Also, the implementation is not complex since reducing the degree of parallelism of a job is a simple process. The compiler and OS should be able to support this feature.

References

[1] P. J. Chuang and N. F. Tzeng, "An Efficient Submesh Allocation Strategy for Mesh Computer Systems," *Proc. Int'l Conf. on Distributed Computing Systems*, pp. 256-263, May 1991.

[2] J. Ding and L. N. Bhuyan, "An Adaptive Submesh Allocation Strategy for Two-Dimensional Mesh Connected Systems," *Proc. Int'l Conf. on Parallel Processing*, Vol. II, pp. 193-200, Aug. 1993.

[3] P. Mohapatra, C. Yu, C. R. Das and J. Kim, "A Lazy Scheduling Scheme for Improving Hypercube Performance," *Proc. Int'l Conf. on Parallel Processing*, Vol. I, pp. 110-117, Aug. 1993.

[4] P. Krueger, T. H. Lai and V. A. Radiya, "Processor Allocation vs. Job Scheduling on Hypercube Computers," *Proc. Int'l Conf. on Distributed Computing systems*, pp. 394-401, 1991.

[5] F. T. Leighton, *Introduction to Parallel Algorithms and Architecture: Arrays · Trees · Hypercubes*, Morgan Kaufmann Publishers Inc., 1992.

[6] K. Li and K. H. Cheng, "A Two Dimensional Buddy System for Dynamic Resource Allocation in A Partitionable Mesh Connected System," *Proc. ACM Computer Science Conf.*, pp. 22-28, Feb. 1990.

[7] D. Das Sharma and D. K. Pradhan, "A Fast and Efficient Strategy for Submesh Allocation in Mesh-Connected Parallel Computers," *Proc. 5th IEEE Symp. on Parallel and Distributed Processing*, pp. 682-689, Dec. 1993.

[8] D. Das Sharma and D. K. Pradhan, "Job Scheduling in Mesh Multicomputers," *Proc. Int'l Conf. on Parallel Processing*, Vol. II, pp. 251-258, Aug. 1994.

[9] C. Yu, "Processor Management Policies for Multiprocessors," Ph.D. Dissertation, The Pennsylvania State University, pp. 25-54, May 1994.

[10] C. Yu and C. R. Das, "Limit Allocation: An Efficient Processor Management Scheme for Hypercubes," *Proc. Int'l Conf. on Parallel Processing*, Vol. II, pp. 143-150, Aug. 1994.

[11] Y. Zhu, "Efficient Processor Allocation Strategies for Mesh-Connected Parallel Computers," *Journal of Parallel and Distributed Computing*, Vol. 16, pp. 328-337, Dec. 1992.

IRREGULAR LOOP PATTERNS COMPILATION
ON DISTRIBUTED SHARED MEMORY MULTIPROCESSORS

Mounir Hahad, Thierry Priol and Jocelyne Erhel

INRIA/IRISA - Campus de Beaulieu - 35042 Rennes Cedex - France

Abstract – *This paper addresses irregular loops compilation on Distributed Memory Parallel Computers (DMPCs) that provide a Shared Virtual Memory. Runtime techniques are introduced to distribute irregular loops so that page movements are reduced. Experimental results for the KSR1 are presented.*

1 Introduction

Although message passing is the usual programming model of DMPCs, Shared Virtual Memory (SVM) provides an alternative with a data sharing model on these architectures. Software based SVM (like KOAN [7] for the iPSC/2 or MYOAN[4] for the Paragon XP/S) as well as hardware based ones (ALLCACHE on the KSR1) are built on top of a paging mechanism: accessing a page which is not present in the local memory awakes the SVM engine which will look for the missing page. Experiments on a 3D finite element code showed the effectiveness of SVM as compared to the PARTI library [1]. However, to be efficient, such a model requires appropriate optimization techniques to reduce communication (page movements) between processors.

This paper introduces two techniques that can be applied to numerical algorithms containing indirect data accesses to shared arrays, in order to reduce page traffic in a SVM architecture. Other techniques have been already proposed for DMPCs which do not support a SVM. PARTI [8] is one of the most advanced projects in resolving such problems. It has been grafted to several compilers such as FORTRAN-D [6]. Our proposal is an adaptation of the inspector/executor scheme, introduced in PARTI. It has been implemented within the Fortran-S compiler [2] which is targeted to several SVM DMPCs. Comparing to other approaches, based on message-passing, our techniques require less memory storage overhead.

This paper is organized as follows. The next section will go over the context of our study. Section 3 introduces the very heart of our proposal which is the CIL technique. An analytical model is also given in that section. In section 4, an improving extension to the CIL is introduced. Finally, experiments are discussed in section 5.

2 Problem Statement

Let us first consider a parallel loop with an indirect write access to a shared array, as for instance:

```
ℓ₁:Do i=1,n
     S(L(i)) = ...
   EndDo
```

where S is a shared vector and L represents a permutation σ of the iteration space $\mathcal{I}^n = \{1,\ldots,n\}$. Since L is a permutation vector, ℓ_1 is a parallel loop. As L is unknown at compile time, the access order to the elements of S is unknown. Since S is mapped onto memory pages, the loop has to be distributed carefully to reduce page movements. If two iterations are mapped on different processors and the two elements of the array S are stored in the same page, then, that page is moved back from one processor's local memory to the other's. This phenomenon is called *false sharing*. The larger the number of processors that compete for a page is, the more important is the false sharing phenomenon. Generally, the situation is worse if each processor tries to access that page more than once (ping-pong effect). Depending on L values, ℓ_1 is likely to generate a false sharing that usually leads to an important loss of performance.

A more general loop scheme encountered in matrix assembly codes is:

```
ℓ₂:Do i=1,m
     S(L(i)) = S(L(i))  op ...
   EndDo
```

where *op* is an associative and commutative operation, S is of size n and L is of size m with $m > n$. Hence, L is no more a permutation vector, but a projection of the iteration space. Nevertheless, ℓ_2 can be executed in parallel even if it is not a parallel loop provided that each read access to S and the corresponding write access (within the same iteration) are atomically executed, preventing any other processor from accessing that element of S there in between. This is possible on a shared virtual memory by means of an exclusive page locking mechanism that attributes a page to one and only one processor until it explicitly releases the page lock.

So, in addition to the false sharing problem cited above, an overhead is introduced when getting and releasing a lock and a possible loss of parallelism incurs because of mutual exclusion between processors accessing the same page at the same time. Our technique allows parallel execution of both l_1 and l_2 loops while eliminating false sharing and synchronization needs. Hence, overheads due to page access conflicts and page locking are both avoided.

3 Conditioned Iterations Loop

3.1 Principle

Basically, the idea is that two processors should never access the same memory page, at any time. To achieve this goal, our proposal is built on what might be called a Conditioned Iterations Loop. *A Conditioned Iterations Loop (CIL) is a loop which iterations are wholly contained into a conditional statement:*

```
CIL: Do i=1,m
        if (Cond) then
            ...
        Endif
     EndDo
```

so that if an iteration is executed by several processors, the condition Cond *is true on one and exactly one processor (and False on the others).* The CIL technique is used on a SVM machine to specify, at runtime, a particular mapping of the iteration space to minimize the false sharing. Since the coherency grain is the memory page, the objective is that *iterations writing into the same page must be executed on the same processor.* In this context, a processor *owns* a page if it has an exclusive write access to this page. Since it is not allowed to write on pages it doesn't own, the false sharing problem is then eliminated.

During the execution of an application, page ownership is not a constant function: pages move between processors according to the data access pattern. In our implementation, the user can choose between two ownership functions: the first one returns the actual ownership relation while the second one returns a virtual ownership relation on the basis of a user-defined function.

3.2 Analytical model of the l_1 loop

Relatively to a block distributed loop, the CIL introduces an overhead due to the execution of all the iterations by all the processors (instead of a loop bounds reduction), and the evaluation of the condition at each iteration. This overhead may be relatively too expensive depending on the nature of the computations involved by each iteration. This part of the paper aims at comparing this overhead to other possible and correct compilation schemes for the l_1 loop only. To do so, let us assume that our application is running on a DMPC and a coherent page-based SVM is provided. The problem is to decide which compilation technique performs the best execution time among the most common ones:

1P : execution of the ℓ_1 loop on a single processor (no parallelism but no overhead as well);

BS : block-scheduling: block partitioning of the iteration space among the processors (+ page lock);

CIL : execution of the ℓ_1 loop according to a regular block distribution of S pages.

On a sequential computer the execution time of the ℓ_1-loop is:

$$T_{seq} = N(T_{rhs} + 2T_{mem}) \qquad (1)$$

where T_{rhs} stands for the right hand side evaluation time and T_{mem} is the cost of one memory access. Indeed, if we assume that i is stored in a register, only two memory accesses are necessary by iteration, one access to read $L(i)$ and one to store $S(L(i))$.

In our context, we suppose the pages of S to be block distributed among P processors and L to be a local array. Then, the execution time for 1P includes page write faults and page migration (swap) if the local memory is not large enough to store S. Hence,

$$T_1 \geq N(T_{rhs} + 2T_{mem}) + \left(\left\lceil \frac{N}{s} \right\rceil - \left\lceil \frac{N/s}{P} \right\rceil \right) . T_{pf}$$
$$+ \sup \left(\left\lceil \frac{\left\lceil \frac{N}{s} \right\rceil - \left\lceil \frac{N/s}{P} \right\rceil - E}{U_{mig}} \right\rceil , 0 \right) . T_{mig} \qquad (2)$$

where the parameters stand for:

N: loop bound
T_{rhs}: right hand side evaluation time
T_{mem}: one memory access delay
s: one page size (number of S elements in a page)
P: number of processors
T_{pf}: page fault service time
E: initially available local memory for S pages
T_{mig}: page migration service time
U_{mig}: number of pages migrated per swap operation

The execution time for BS is:

$$T_{BS} = \left\lceil \frac{N}{P} \right\rceil (T_{rhs} + 2Tmem) + \alpha.T_{pf} + \beta.T_{mig} \qquad (3)$$

where α stands for the number of page faults and β the number of page migrations. These two parameters depend to a great extent on the runtime values of L.

As to the CIL, the execution time includes a preceding phase that computes the upper and lower bounds of chunks of S virtually attributed to each processor depending on the window of S defined in ℓ_1-loop (in case S has more than N elements). Thus, it is:

$$T_{CIL} = C_0 + NC_1 + (T_{rhs} + 2T_{mem}) \times$$
$$\inf \left(\left\lceil \frac{N/s}{P} \right\rceil . s, \left\lfloor \frac{N/s}{P} \right\rfloor . s + N\text{mod}(P.s) \right) \qquad (4)$$

where C_1 is the time necessary to evaluate Cond. Recall that there is no page fault (nor page migration) in this case. On the other hand, the CIL may introduce a slight load imbalance depending on the page size.

Parameterization In order to reduce the number of parameters that rule our analytical model, we drop the page migration problem. In other words, we assume that each processor has enough local memory to store its whole working set. Although this assumption does not introduce any restriction for BS and CIL, it seems to be quite unrealistic for 1P indeed. Fortunately, in our experiments, an asymptotical behavior has been reached still the problem sizes we used fit in a single processor's memory. Experiments held on a Kendall Square KSR1 machine led to the parameterization depicted in table 1.

$C_0(\mu s)$	$C_1(\mu s)$	$T_{mem}(\mu s)$	s	$T_{pf}(\mu s)$
4.62	0.87	0.9	16	7.5 - 30

Table 1: Analytical model parameterization

Experimental validation We experimentally evaluate the three compilation schemes. In addition, experience is useful to know how close is our simple analytical model to the reality. For purposes of experience, a random permutation σ is generated and stored into L for different values of N. Then, the ℓ_1 loop is executed according to the CIL and BS schemes with the same permutation. Notice that since there is no page migration, the permutation is not important with respect to the execution time of 1P. The experiments held on the KSR1 showed that the first level cache (which is not taken into account in our model) plays an important role. This is why the measured CIL speedup is not so close to its corresponding prediction for the cost-less right hand side. Furthermore, the BS scheme seems to perform better than the CIL one for these cost-less right hand sides. Regarding the expensive right hand sides, the prediction was right.

4 Optimizing CIL

As already mentioned, CIL suffers from the fact that all the processors investigate all the iterations. This operation prevents from scalability and sometimes from efficiency. To alleviate this problem, we use a learning technique that takes advantage from the first iteration to enhance the efficiency of the following iterations. The model loop is:

```
Do t=1, ntimes
    Do i=1,m
       S(L(i)) = S(L(i)) + ...
    EndDo
EndDo
```

The inner loop can be processed by CIL and the outer loop can be used to collect information about the data access pattern that does not change from one iteration to another (the LEARN loop). So, the model loop cited above is transformed into the following code:

```
Do i=1,m
    if (I own the page containing S(L(i)))
       S(L(i)) = S(L(i)) + ...
       j = j+1
       MyIteration(j) = i
    endif
EndDo
Do t=2, ntimes
    Do ij=1,j
       i = MyIteration(ij)
       S(L(i)) = S(L(i)) + ...
    EndDo
EndDo
```

The code showed above is executed by all the processors and MyIteration is a private array. Thanks to this array, the processors do not need to examine all the iterations anymore: this is done only once at the first iteration of the outermost loop (the learning loop). Consequently, the time overhead is reduced while a storage overhead of size $o(\text{m}/number\ of\ processors)$ is needed. In fact, that overhead may be more or less large depending on the actual/virtual page ownership distribution and the actual values in L. If the user has some knowledge about these two parameters (for a given problem), he can insert a directive to allocate a more suitable additional storage.

5 Experiments

For our experiments, we used the *Fortran-S* [2] code generator which provides directives to express parallelism. The output is a Fortran SPMD code targeted to the desired parallel SVM machine. At present, the supported machines are the Intel iPSC/2, the Kendall Square KSR1 and the Intel Paragon XP/S. A set of new directives has been added to implement our runtime techniques. They are described in [5].

To show the impact of the learning technique on the CIL implementation, we carry out some experiments on the KSR1. l_2 is used as a loop model with a right hand side of $1\mu s$ delay at each iteration. The indirection array L is built on the basis of data arising from irregular triangular meshes. The mesh problem we used in our experiments has roughly 200K nodes and 400K triangles. Both a renumbered version (200k-r) with a greedy algorithm that enhances spatial locality [3], and a randomly mixed version (200k-m) of the problem are used.

Figure 1 shows timing results we obtained. The first level cache (subcache) of the KSR1 plays an important role. With roughly the same number of subpage faults, 1P takes much more time to execute when the data are mixed than when they are renumbered for locality. The miss ratios (number of subcache misses / total number of accesses) with 1P is 7.4% for the renumbered version and 56% for the randomly mixed version.

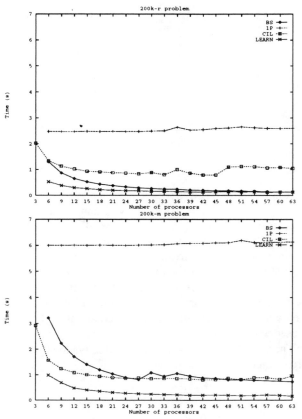

Figure 1: Experimental times of the ℓ_2 loop

Thanks to a relatively cost-less subpage fault service time, BS performs better on the KSR1 than CIL, as long as there are only few page faults. As the mixed data cause a large number of subpage faults (two orders of magnitude higher than the renumbered ones), CIL becomes more advantageous than BS, yet each processor scans the whole iteration space. Particularly interesting is that LEARN remains the most profitable technique with no more than 10 iterations. This obviously suggests that the learning is well written off.

6 Conclusion

In this paper, we were interested in compiling irregular access loop patterns on distributed shared memory multiprocessors. Special attention is given to loops that are encountered in matrix assembly codes arising from unstructured meshes. To execute efficiently these loops on a distributed shared memory multiprocessor, we introduce the CIL scheme which maps the iteration space on the processors according to page ownership criteria. An analytical model is presented and experiments are carried out on a

KSR1. Furthermore, a learning technique is introduced to improve CIL performance. A complete study on an iPSC/2 and a KSR1 is provided in [5]. The experimental results show that the latter technique out-does more conventional compilation schemes. One of the drawbacks of the CIL technique is the load imbalance if too many elements in L focus on the same area of S. However, recent experiments we carried out on real applications show that the efficiency improvement thanks to CIL is worth the load imbalance. Furthermore, it is needless to say that the learning technique becomes expensive in terms of memory storage requirements if there are too many indirect write access statements per iteration. Our future work aims at reducing that storage overhead and extending the CIL to a whole code to allow page-affinity mapping over multiple code sections.

References

[1] R. Berrendorf, M. Gerndt, Z. Lahjomri, and T. Priol. A comparison of shared virtual memory and message passing programming techniques based on a finite element application. In *LNCS 854, proc. of CONPAR-VAPP*, pages 461-472, Austria, Sep.1994.

[2] F. Bodin, L. Kervella, and T. Priol. Fortran-S : a fortran interface for shared virtual memory architectures. In *Supercomputing'93*, pages 274-283, Nov. 1993.

[3] M.O. Bristeau, J. Erhel, P. Feat, R. Glowinski, and J. Périaux. Solving the helmholtz equation at high wave numbers on a parallel computer with a shared virtual memory. *International journal of supercomputer applications and high performance computing*, (9.1), 1995.

[4] G. Cabillic, T. Priol, and I. Puaut. *MYOAN: an Implementation of the KOAN Shared Virtual Memory on the Intel Paragon*. Technical Report 2258, INRIA, Apr. 1994.

[5] M. Hahad, T. Priol, and J. Erhel. *Irregular Loop Patterns Compilation on Distributed Shared Memory Multiprocessors*. Technical Report 2361, INRIA, Campus de Beaulieu - 35042 Rennes Cedex - France, Sep. 1994.

[6] R. von Hanxleden. *Compiler Support for Machine-Independent Parallelization of Irregular Problems*. PhD thesis, Rice University, Houston, TX, Dec. 1994.

[7] Z. Lahjomri and T. Priol. KOAN : a shared virtual memory for an ipsc/2 hypercube. In *LNCS 634, proc. of CONPAR/VAPP*, pages 441-452, France, Sep. 1992.

[8] J. Saltz, K. Crowley, R. Mirchandaney, and H. Berryman. Run-time scheduling and execution of loops on message passing machines. *Journal of Parallel and Distributed Computing*, 8:303-312, 1990.

Performance Extrapolation of Parallel Programs

Allen D. Malony[1]
Dept. of Computer and Information Science
Univ. of Oregon, Eugene OR 97403, USA
malony@cs.uoregon.edu

Kesavan Shanmugam
Convex Computer Corp.
3000 Waterview Pkwy., P.O.Box 833851
Richardson, TX 75083-3851
kesavans@convex.com

Abstract --*Performance Extrapolation is the process of evaluating the performance of a parallel program in a target execution environment using performance information obtained for the same program in a different execution environment. Performance extrapolation techniques are suited for rapid performance tuning of parallel programs, particularly when the target environment is unavailable. This paper describes one such technique that was developed for data-parallel C++ programs written in the pC++ language. The technique uses high-level event tracing of a n-thread pC++ program run on a uniprocessor machine together with trace-driven simulation to predict the performance of the program run on an n-processor machine. Our results show that even with high-level events, performance extrapolation techniques are effective in isolating critical factors affecting a program's performance and for evaluating the influence of architectural and system parameters.*

1 Introduction

One of the foremost challenges for a parallel programmer is to achieve the best possible performance for an application on a parallel machine. The most common motivation for developing high-performing programs is that parallel machines are expensive resources that must be utilized to their maximum potential to justify their costs. However, the process of *performance debugging* (the iterative application of performance *diagnosis* [6] and *tuning*) invariably requires access to the target parallel platform, since the majority of the parallel performance tools are based on the measurement and analysis of actual program execution. As a consequence, during performance debugging, a parallel system is typically running less than optimal codes, often with the additional overhead of execution monitoring. The dependence on actual machine access for performance debugging also restricts the parallel programmer to consider optimization issues only for physically available systems. For parallel programs intended to be portable to a variety of parallel platforms, and scalable across different machine

and problem size configurations, undertaking performance debugging for all potential cases is usually not possible.

In this paper, we describe a performance prediction technique that combines high-level modeling with dynamic execution simulation to facilitate rapid performance debugging. The technique is one example of a general prediction methodology called **Performance Extrapolation** that estimates the performance of a parallel program in a target execution environment by using the performance data obtained from running the program in a different execution environment. Our goal is to demonstrate that performance extrapolation is a viable process for parallel program performance debugging that can be applied effectively in situations where standard measurement techniques are restrictive or costly. Section 2 describes the basic concept of performance extrapolation. Section 3 presents the performance extrapolation tool, *ExtraP*, that we have developed for data-parallel *C++* programs written in the *pC++* language. The experiments applying *ExtraP* to *pC++* benchmark codes are briefly discussed in Section 4. The paper concludes with remarks about future work.

2 Performance Extrapolation

An **execution environment** is a collection of compiler, runtime system, and architectural features that interact to influence the performance of a parallel program. A complete execution environment would include the following important factors.

1. *Architecture*: Interconnection network topology, memory hierarchy, number of processors, and CPU architecture.

2. *Compiler*: Optimization strategies, calling conventions, storage management policies, and source code transformation algorithms.

3. *Runtime system*: Message passing conventions, data distribution policies, thread management, and scheduling policies.

The performance of a parallel program should always be specified with respect to a particular execution environment. An execution environment provides a reference framework for interpreting a program's performance on a

---
1 This research is supported by ARPA under Rome Labs contract AF 30602-92-C-0135 and Fort Huachuca contract ARMY DABT63-94-C-0029. The research is also supported by a NSF National Young Investigator (NYI) award.

single run as well as comparing the performance of different versions of the same program. A given program can also be evaluated under different execution environments and the environment best suited to that program chosen.

We define a **performance metric** as a measure of the quality of a parallel program. Performance metrics are important because they assist the user during performance debugging to identify performance bottlenecks and investigate how to resolve them. Performance metrics are obtained by analyzing the static and dynamic **performance information** about a parallel program's execution. In some cases, the performance metric is trivially derived from the performance information. Sometimes more detailed analysis of the performance information is required.

The importance of execution environment is evident in the definition of performance extrapolation. **Performance extrapolation** is the process of obtaining the performance information PI_1 of a parallel program for an execution environment E_1 and using PI_1 to predict the performance information PI_2^p (the superscript p indicates a predicted quantity) of the same program in a different execution environment E_2. The performance information PI_2^p is then used to compute the predicted performance metrics of the program in E_2, PM_2^p. This process can be considered as a translation or extrapolation of PI_1 to PI_2^p using the knowledge about E_1 and its similarities to and differences from E_2. PM_1 and PM_2, the performance metrics derived from measured performance information, PI_1 and PI_2, in the environments E_1 and E_2, respectively, can be compared with the predicted performance metrics for validation of the extrapolation technique.

3   A Performance Extrapolation Technique for pC++

We have developed a performance extrapolation technique that allows performance information and metrics to be predicted for data-parallel programs written in the pC++ language [1,3,4,8]. In particular, we investigated the problem of extrapolating from a 1-processor execution of a n-thread parallel program to a n-processor execution for an environment where certain architectural and system parameters are configurable.

In general, our approach is to execute a n-thread pC++ program on a single processor using a non-preemptive threads package. Important high-level events are recorded during the program run in a trace file. The events are then sorted on a per thread basis, adjusting their timestamps to reflect concurrent execution. The resulting set of trace files look as if they were obtained from a n-thread, n-processor run, except that they lack certain features of a real parallel execution. A trace-driven simulation using these trace files attempts to model those features and predict the events as they would have occurred in a real n-processor execution environment. The extrapolated trace files are then used to obtain various performance metrics related to the pC++ program.

3.1 pC++: The Language, Compiler, and Runtime System

pC++ is a language extension to C++ that supports an *object-parallel* execution model [1,3,4]. Under this model, a *collection* of objects can be distributed across a set of threads, in much the same way as arrays are distributed in HPF [2,7]. The objects which make up the collection are called *elements* of the collection. The collection inherits certain member functions of its elements, so that when such a member function is called, it is called for every element in the collection. This parallel method invocation for collections of elements is the main source of parallelism in pC++.

The compiler accomplishes a parallel method invocation by generating code so that each thread calls the method for all its local elements. At the end of each parallel method invocation, the threads are synchronized by a global barrier. In addition, when a thread wants to access an element which it does not own, it generates a *remote element* request to be serviced by the thread that owns the element. The runtime system provides facilities for creating the threads, synchronizing the threads, and accessing remote elements [1].

3.2 Instrumentation and Trace Translation

For the purposes of our performance extrapolation work, we modified the pC++ runtime system so that all n threads of a parallel program are executed on a single processor (in a virtual parallel manner) using a non-preemptive threads package [5]. The elements of a collection are allocated in a global space accessible by all the threads. When a thread requires access to a remote element, it gets it directly from the global space. Thus, under this runtime system, remote accesses are indistinguishable from local accesses (in terms of timing characteristics) and thread switches happen only at barrier entry and exit points.

Since the only interactions between threads in the pC++ programming model occur during barrier synchronizations and remote element accesses, the runtime system was instrumented to record all such interactions. The result of a 1-processor performance measurement is a trace file which contains barrier entry, barrier exit, and remote access events from all the threads in the pC++ program. These high-level events form the basis for the extrapolation and the time between the events reflect the thread computation times.

The trace translation algorithm takes in the trace file produced by the *n*-thread, 1-processor run of a *pC++* program and creates *n* trace files each containing events from one thread. The timestamps are adjusted to reflect the ideal parallel execution of the threads' computation on a *n*-processor machine. This is accomplished by retaining the time between two consecutive events for a thread and by enforcing the semantics of the barrier synchronization events. Notice that the trace translation algorithm relies on the fact that the threads are scheduled only at synchronization boundaries (i.e., the threads are not preempted until they encounter the barrier). That is the reason why a non-preemptive threads package must be used for the *n*-thread, 1-processor run. The trace translation algorithm is easily modified to handle the overhead for recording the events, flushing the event buffer, and switching the threads.

The translated trace files capture the program execution times between events under the assumptions of instant remote accesses, instant barrier synchronization (threads exit a barrier as soon as the last thread comes in), and unperturbed thread computation. Although these assumptions are idealized, the claim that we make is that *pC++* performance extrapolation can now be done for an *n*-processor execution by using the high-level events in the traces to drive simulations where models attempt to capture the execution realities of these performance factors (i.e., cost of remote accesses, synchronization overheads, processor performance) in the target environment.

3.3 Simulation Architecture and Models

The trace-driven simulation is the heart of the *pC++* performance extrapolation. The simulation system consists of three main components:
- Processor model
- Remote data access model
- Barrier model

3.3.1 Processor model

If the performance of the target system's processor is different from the measured machine, the difference must be addressed during extrapolation. For *pC++* extrapolation, we use a simple ratio of processor speeds to scale the computation time between events appropriately for the target machine. Such a ratio can be easily obtained by measuring the MFLOPS (or other processor performance) ratings of the target machine and the machine on which the experiments are performed. In addition to processor speed scaling, the processor model represents certain operational aspects of the *pC++* runtime system. One important aspect is the policy about how remote data accesses are serviced and what message handling functions are performed for the cho-

sen policy. The following remote data access policies are currently supported: *No interrupt*, *Interrupt*, and *Poll*. The runtime system also determines how threads are assigned to processors, affecting data locality and processor sharing.

3.3.2 Remote data access model

The simulation models each remote access in the program as a remote request for data from one thread to the thread that "owns" the data (*pC++* follows the "owner computes" model [4,7]). The owner thread services the request and returns the data to the requesting thread. This is equivalent to how the *pC++* system operates in distributed memory environments. The remote data access model includes parameters to represent: communication start-up overheads, communication bandwidth, message types and sizes, message construction overhead, network topology, and network contention. The contention models we developed include parameters based on the intensity of concurrent use of shared system resources (e.g., the interconnection network) during the simulation. The contention models were analytical expressions of remote access delay involving the contention factors calculated from the simulation state.

3.3.3 Barrier model

The current barrier model is based on a linear, master-slave barrier synchronization algorithm. Thread 0 acts as the master thread while all the other threads are slaves. Every slave thread entering a barrier sends a message to the master thread and waits for a release message from the master thread to continue to the next data-parallel phase.The master thread waits for messages from all the slaves and then sends release messages to all of them. For distributed memory systems, the *pC++* runtime system must continue to service remote data access messages that arrive at a processor even when the threads that run on that processor have reached the barrier. This is also true in the simulation. The linear barrier model delivers an upper bound on barrier synchronization times. We can easily substitute other barrier algorithms (e.g. logarithmic) if a more accurate simulation of barrier operation is required.

4 Experimental Results

To evaluate the concept of performance extrapolation and, in particular, the efficacy of the *ExtraP* tool, we performed several extrapolation experiments on codes in the *pC++* benchmark suite[2]. These codes represent a wide range of

---

2 The complete presentation of the experimental results can be found in the full paper at: **http://www.cs.uoregon.edu/paracomp/papers/papers.html**.

execution behaviors, reflecting different degrees of computation and communication. First, we wanted to establish that the extrapolation methodology could be applied in an actual parallel programming context where performance debugging is an important component [8]. Second, we wanted to verify that modifying simulation parameters of interest resulted in observable and expected effects in extrapolated benchmark performance behavior.

Initially, to observe processor scaling effects, we selected a single parameter combination and ran *ExtraP* on all benchmarks and processor numbers. The results clearly show the range of performance found in the *pC++* benchmark suite and the speedup values are representative of what we have observed in different real environments. We were then able to test the performance outcome of changes in the execution environment, including different data distributions.

Our second experiment was designed to observe the effects of extrapolating processor speed. The expected slowdown and speedup performance effects were clearly seen in all benchmark extrapolation results, but different speedup behaviors arose depending on the effect of changes in the ratio of computation to communication. The ability to extrapolate target system parameters such as processor speed and communication start-up and to see such performance effects is important when "what if" questions are posed, especially about systems that do not physically exist.

The last experiment with the benchmarks demonstrated *ExtraP*'s ability to simulate different runtime system policies for servicing remote data accesses. Two policies are supported in the actual *pC++* system: polling and interrupt. The choice of which policy to implement is usually based on whether the target machine environment supports message interrupts (e.g., active messages on the CM-5 9). The extrapolation process allowed us to again observe the potential performance effects of changes in the target execution environment, this time at the runtime system level, and to possibly use that information to make application-specific runtime system optimizations.

5 Future Work and Conclusions

Performance extrapolation uses performance information from one execution environment to predict the performance in another target environment. We have described a particular performance extrapolation technique that uses the execution time and events from a n-thread, 1-processor run to predict the execution time of a n-thread, n-processor run of the same program. Our experience suggests that performance extrapolation is a viable technique to be use in a performance debugging system, particularly for language environments, like that of *pC++*, where simple, restricted execution semantics make programs more amenable to per-

formance prediction. Two important results of our research are that access to actual target platforms are not always required for accurate performance evaluation of parallel programs, and that low-level system simulations are not the only alternative required for detailed performance prediction. Performance extrapolation based on high-level events, like implemented in *ExtraP*, can support both diagnosis and tuning in a performance debugging system.

This work can be extended in a several ways. The simulation can be extended to handle multithreaded processors (i.e., the scheduling of more than one thread on a processor). This will extrapolate the performance from a n-thread, 1-processor run to a n-thread, m-processor run, where $m \leq n$. We are currently modifying *ExtraP* to support multithreading. Another direction is to apply this work to other language systems, like HPF. The focus of our present work is to integrate the *ExtraP* tool into the *pC++* program analysis environment.

6 References

[1] F. Bodin et al., Implementing a Parallel C++ Runtime System for Scalable Parallel Systems, Proc. Supercomputing 93, pp. 588-597, Nov. 1993.

[2] Z. Bozkus et al., Compiling Distribution Directives in a Fortran 90D Compiler, Tech. Report SCCS-388, Northeast Parallel Architectures Center, July 1992.

[3] D. Gannon, F. Bodin, S. Srinivas, N. Sundaresan and S. Narayana, Sage++, An Object Oriented Toolkit for Program Transformations, Tech. Report, Dept. of Computer Science, Indiana Univ., 1993.

[4] D. Gannon and J. K. Lee, Object Oriented Parallelism: *pC++* Ideas and Experiments, Proc. Japan Society of Parallel Processing, pp. 13-23, 1991.

[5] D. C. Grunwald, A Users Guide to AWESIME: An Object Oriented Parallel Programming and Simulation System, TR 552-91, Dept. of Computer Science, Univ. of Colorado at Boulder, Nov. 1991.

[6] R. Helm, A. Malony and S. F. Fickas, Capturing and Automating Performance Diagnosis: The Poirot Approach, IPPS'95, April 1995.

[7] High Performance Fortran Forum, High Performance Fortran Language Specification version 1.0, TR92225, CRPC, Rice University, Jan. 1993.

[8] A. Malony et al., Performance Analysis of *pC++*: A Portable Data-Parallel Programming System for Scalable Parallel Computers, IPPS'94, pp. 75-85, April 1994.

[9] On-line information on CM-5, CM-5 Technical Summary, http://www.think.com/Prod-Serv/Products/cmmd.html, November 1992.

*GRAPH: A Tool for Visualizing Communication and Optimizing Layout in Data-Parallel Programs *

SANDRA G. DYKES   XIAODONG ZHANG   YI SHEN
CLINTON L. JEFFERY   DEVIN W. DEAN

High Performance Computing and Software Laboratory
The University of Texas at San Antonio
San Antonio, Texas 78249
sdykes@dragon.cs.utsa.edu

Abstract

GRAPH is an event-driven visualization and optimization tool for data-parallel communication, currently implemented for the C language on the CM-5. Its goal is to help data-parallel programmers reduce their communication costs by providing information about communication bottlenecks and suggesting system directives for an optimized data layout. To accomplish this, *GRAPH provides facilities for

1. visualizing a trace of internode communication,

2. pinpointing source code statements where communication bottlenecks occur, and

3. optimizing data layout for the program's communication pattern.

This paper presents the motivation for *GRAPH, its design concept and the initial implementation. We include two case studies illustrating how performance can be improved by applying *GRAPH's recommended data layout directive. Elapsed execution times were reduced by 31% for a folded convolution algorithm, and by 17% for a Gaussian elimination.

*This work is supported in part by the National Science Foundation under grants CCR-9102854 and CCR-9400719, by the U.S. Air Force under research agreement FD-204092-64157, and by the Air Force Office of Scientific Research under grant AFOSR-95-1-0215 and by two Fellowships from the Southwestern Bell Foundation. Part of the experiments were conducted on the CM-5 machines in Los Alamos National Laboratory and in the National Center for Supercomputing Applications at the University of Illinois.

1 Introduction

Two factors are necessary to achieve good parallel performance: the program design must be efficient and the data distribution must match communication patterns. Optimization of program design and optimization of data layout usually are not independent. Source code modifications can alter communication so that it no longer matches the data layout. Alternatively, some communication bottlenecks can be alleviated by better distributing the data rather than by rewriting the source code. Here the question is: which approach is easier and faster for the applications programmer? If a tool were available that automatically computed an optimized data layout for the given source code, then the answer must be it is easier to change the data layout.

One of our goals in the *GRAPH project is to investigate methods for building layout optimization tools and to study their effectiveness in improving performance. We are focusing this effort on data-parallel languages. In message-passing languages, the data distribution and internode communication are handled explicitly by the applications programmer. However, in data-parallel languages the data distribution and amount of internode communication are normally hidden from the programmer. A tool to optimize data layout and visualize internode communication is certainly helpful to the message-passing programmer, but is needed far more by the data-parallel programmer.

This paper presents the motivation for *GRAPH, its design concept and the initial implementation. Section 2 discusses related visualization and optimization tools for data-parallel languages, and explains why *GRAPH's integrated approach is important. Section 3 provides a brief background covering the virtual processor concept in data-parallel languages, particularly as implemented on the CM-5. The *GRAPH Monitor, Visualizer and Optimizer modules are described in Sections 4 through 6. In Section 7 we present two case studies demonstrating how *GRAPH's recommended data layout directive can improve performance. Section 8 concludes with a discussion of results and current work.

2 Motivation

In data-parallel languages, the system software normally assumes responsibility for mapping data onto distributed memory. The applications programmer need not specify nor even be aware of the run-time data layout. When communication occurs, the system software computes data locations: the applications program is unaware of which communication is on-node and which is off-node.

2.1 Pros and cons of data abstraction

Data abstraction is both a strength and a weakness of data-parallel languages. By isolating applications code from machine architecture, data-parallel languages free the programmer to focus on a natural representation of the problem. For example, image processing applications can view data elements as pixels and directly map pixel neighborhood functions onto grid communication statements. Similarly, simulations of chemical reactions can structure data as a 3D grid of molecules in which molecular interactions translate directly into grid communications. Because data-parallel layout and communication are so easily molded into the application domain, data-parallel languages potentially have shorter learning curves, faster development times and less debugging than message-passing languages.

Unfortunately, the logical corollary of the benefits of system abstraction is system dependency: the efficiency of a data-parallel program can not exceed the efficiency of the automation resources it employs. For example, if the data layout generated by system software does not match a program's communication pattern then performance will suffer. Thus the strength of the data-parallel paradigm is also its weakness. By relying on system software in the critical area of data layout and parallel communication, the programmer unknowingly may trade performance for ease of use. For this reason it is important for data-parallel systems to have a tool which a) visualizes physical communication and b) optimizes data layout for a specific source code-architecture combination.

2.2 Related work on communication visualization

Parallel programming paradigms are usually viewed as either shared-memory, message-passing, or data-parallel. Performance monitoring and visualization tools exist for each paradigm, such as Para-Graph for message-passing [9], MIN-Graph for shared-memory [23], and Prism for data-parallel [22] to name but a few. In some cases a performance tool supports several paradigms. For example, the Paradyn tool [16] developed at the University of Wisconsin-Madison supports data-parallel Fortran on the CM-5 as well as message-passing programs on the CM-5, Sun workstation and PVM. Paradyn uses dynamic instrumentation and offers automatic bottleneck searching. Currently Paradyn is limited to probe insertions at procedure boundarys.

The Prism programming environment on CM-5 provides Motif-based data visualizers, a parallel debugger, editing capabilities and performance data for C* and CM-Fortran data-parallel programs. Prism gathers its performance data from preprocessor insertion of system calls to the CM_timer facility. Unfortunately, we have encountered applications where Prism's performance data was diametrically opposed to measurements made directly with CM_timers and were unable to resolve the discrepancy [18].

2.3 Related work on layout optimization

The problem of finding the optimal data layout for distributed memory systems has been shown to be NP-complete [14, 15]. Numerous heuristics have been proposed to optimize data distributions, usually with the goal of incorporating automatic layout optimization into parallelizing compilers [1, 8, 17]. Post-mortem optimizations have the luxury of more time and more precise data than "on-the-fly" optimizations so they can in general produce a higher quality solution, especially for dynamically allocated parallel data. If the data layout problem is represented as a graph, general optimization heuristics such as simulated annealing [12] or greedy algorithms [13] can be applied.

Performance data can be provided by simulation rather than actual program monitoring. One important use of such models has been as an aid to selecting data layouts [2, 5].

2.4 Uniqueness of *GRAPH - Combining visualization and optimization

If a program is communication-bound, performance can potentially be improved by modifying either the source code (the goal of performance monitors) or the data layout (the goal of layout optimizers). *GRAPH is different from other performance tools because it combines these two approaches. The Visualizer pinpoints which source code statement corresponds to the largest off-node communication and shows the distribution of messages across nodes. This information helps a programmer with source code modification. Alternatively, if the programmer decides to change data layout rather than source code then *GRAPH's Optimizer can be used to suggest data layout directives.

Using both visualization and layout optimization together provides a more powerful approach. For example, a programmer can try the suggested layout directives, view the result and compare to the unoptimized version. If two source code versions are being compared, *GRAPH's Optimizer can suggest optimized layouts for each version and *GRAPH's Visualizer can be used to compare the two optimized source code versions.

2.5 The *GRAPH project

Performance data from a program monitor can be used both for visualizing communication and for optimizing data distribution. It therefore seems natural to integrate these two areas into a single tool. One important contribution of *GRAPH is this integration.

The goal of *GRAPH is to help data-parallel programmers reduce their communication costs by providing information about communication bottlenecks and suggesting system directives for an optimized data layout. To accomplish this, *GRAPH provides facilities for

1. visualizing a trace of internode communication,

2. pinpointing source code statements where communication bottlenecks occur, and

3. optimizing data layout for the program's communication pattern.

The current version reported in this paper considers regular and irregular point-to-point communication, but does not include broadcasts or reductions, and is limited to C* source code on the CM-5. To allow for expansions to other parallel languages and architectures, *GRAPH is split into three independent modules:

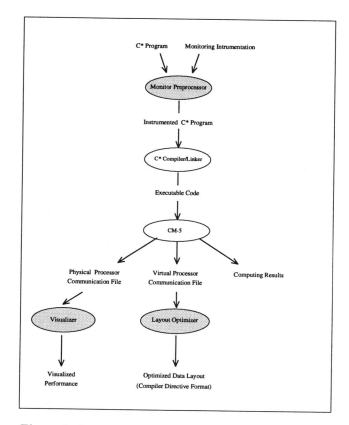

Figure 1: *GRAPH tool for visualization and optimization of data-parallel communication. Shaded areas represent *GRAPH modules.

- Monitor

- Visualizer

- Layout Optimizer.

Figure 1 shows an overview of the *GRAPH tool and functions of its three modules. The Monitor is the only module that is language specific. The Visualizer merely requires that communication data be stored in a particular file format, permitting visualization of data from any parallel language including CM Fortran. Visualization of communication data from another architecture requires changes in the Visualizer architecture view, but does not affect any other view. Similarly, new forms of visualization can be added without changing either the Monitor or the Optimizer.

The Optimizer also reads a language-independent data file and produces data distribution arguments that can be used either by the LAYOUT directive of CM Fortran or the allocate_detailed_shape() function of C*. The current version of the Optimizer supports any data-parallel language that supplies directives or functions to compute data layout from weighting the data dimensions. A more general Optimizer module is planned for future versions.

3 Virtual processor communication channels

Each element in a parallel data set such as a C* shape or a CM Fortran parallel array is viewed as having its own *Virtual Processor* (VP). The mapping of virtual processors onto physical processors defines the data distribution. VPs assigned to a physical processor have their data elements stored in that processor's memory and their execution implemented in vector loops on that processor. On most CM-5 installations, nodes contain four vector processing units. Therefore, point-to-point data communication on CM-5 can be between [20, 21] :

- VPs on the same vector unit
- VPs on the same node, but different vector units
- VPs on different nodes

Data locality is of overwhelming importance for most applications. Communication costs are lowered substantially by placing communicating VPs on the same vector unit or even on the same node.

3.1 Control of memory layout

By default, the CM-5 run-time system assigns the data distribution [20]. C* programs can influence or completely specify memory layout by calling the function *allocate_detailed_shape()*. CM Fortran programs specify memory layout with the LAYOUT directive. The *allocate_detailed_shape()* function allows specific layout details to be dictated or allows the programmer to specify weights for each data dimension (*axis*). When an axis receives a larger relative weight, more elements along that axis are stored in the same local memory. This reduces communication cost along that axis.

4 *GRAPH Monitor

The *GRAPH Monitor inserts data collection code into the applications program to gather a physical communication trace for the Visualizer and virtual communication statistics for the Optimizer. Tracing can be turned on or off at desired locations in the source code.

4.1 Collecting the physical processor trace

At instrumented communication steps, the active virtual processors 1) collect their individual communication data $< VP_{Dest}, Message\_bytes >$ and 2) compute the physical processor id number of their destination:

$$VP_{Dest} \rightarrow PP_{Dest}.$$

Typically many VPs have the same source and destination processors, so the trace must record the sum of their message bytes. This is most efficiently done using a parallel-to-parallel reduction [20]. After the reduction, the data is condensed by storing only events with non-zero message bytes, with events defined by the tuple

$$< Step, PP_{Source}, PP_{Dest}, Message\_bytes > .$$

Each non-zero event is stored in the memory of its source processor, PP_{Source}. Upon exit, the trace data is written to a Physical Processor Communication File for use by the Visualizer module. In our case study, we found this method of recording trace data requires a negligible amount of execution time and has relatively small storage requirements while providing the desired trace granularity.

4.2 Collecting the virtual processor statistics

In concert with the physical processor trace, the Monitor code collects virtual processor communication statistics:

$$< VP_{Source}, VP_{Dest}, \sum_{Steps} Message\_Bytes > .$$

Statistics are stored on the source VP, and upon exit are written to a *Virtual Processor Communication File* for use by the Optimizer module.

5 *GRAPH Visualizer

The Visualizer reads trace data from a *Physical Communication File* and presents it in an interactive X-windows display. *GRAPH's visualization package is being developed in the Icon programming language because of Icon's support for graphics and interface programming [6, 11].

5.1 Visualizer views

Although data was collected for physical processors, the current Visualizer views display only internode communication. Examples of the Visualizer views are shown in Figure 2. Views are organized hierarchically, starting with a view that maps communication data onto an architectural representation of the interconnection network.

5.2 Locating the communication bottleneck

To locate a communication bottleneck, the user views the *Step* histograms to find the step with the largest total communication. Alternatively, the user can focus on the step with the largest "hot node"; that is, the step with the largest single node send or receive. For the example in Figure 2, the largest total and single node communication occurs at Step 27. After finding the bottleneck step, the user can examine the *Send Bytes* and *Receive Bytes* windows to see the distribution of communication across the nodes. This determines if the bottleneck is localized at a small number of nodes or is distributed across the network. Most useful is the source code window. This pinpoints the source code statement which corresponds to the communication bottleneck. At this point the user can decide to either accept the current performance, rewrite the source code, or use the Optimizer module to potentially improve data layout.

5.3 Comparing data layouts or source code versions

*GRAPH reports the application's sum of internode communication bytes, and sum of bytes sent or received by the "hottest" node at each step. Both numbers can be used to compare the total amount of internode communication for different data layouts or different source code versions.

6 *GRAPH Optimizer

The Optimizer reads the virtual communication file and produces an optimized data layout in the form of arguments (axis weights) for the C* *allocate_detailed_shape()* function. Computing the optimal placement of virtual processors onto the distributed memory is an NP-complete problem. This type of problem has been shown to be well-suited to simulated annealing [12]. However, data distribution directives in both C* and CM Fortran support a much simpler heuristic. The C* *allocate_detailed_shape()* function and the CM Fortran LAYOUT directive contain arguments for weighting the data dimensions. Assigning a larger weight to one dimension results in more contiguous storage of data from that dimension, thereby reducing its communication cost. For example, if communication in a 3D grid occurs primarily in the z direction, then the z data axis should receive the largest weight. We compute axis weights by summing VP communication along each data axis.

Optimization using axis weights is too simple a solution to dismiss and has performed well in limited tests. However, it is important to have a finer-tuned and more general form of optimization. For this we are currently adding a second optimization module based on the advanced simulated annealing package *Very Fast Simulated Reannealing* [3, 10].

7 Case studies

7.1 Folded convolution

The folded convolution is an efficient data-parallel algorithm for 2D image convolutions [4]. Unlike a naive convolution, the communication in the folded algorithm is asymmetric with respect to rows and columns. Figure 3 shows the two stages of the algorithm. In the first stage, single elements are passed along the row axis while in the second stage arrays are passed along the column axis. For the experiments reported below, the convolution window size was 15×15 and the image size was 256×256.

7.1.1 Default data layout

Figure 2 shows *GRAPH's visualization of the folded convolution when the default data layout was used. On 32 nodes, the system chose a subgrid layout of 32×16, which slightly favors communication along the row axis (i.e,. between elements with the same column number). As shown in the figure, maximum communication occurred at *Step 27* and was evenly distributed across nodes. The source code statement corresponding to this step was a *from_torus()* call at line 601.

Average elapsed time for the unoptimized layout was 0.135 seconds. Total number of bytes for VP communication was 66 MB, of which 7.8 MB required off-node messages. The Optimizer suggested setting the row:column weight ratio to 8:1.

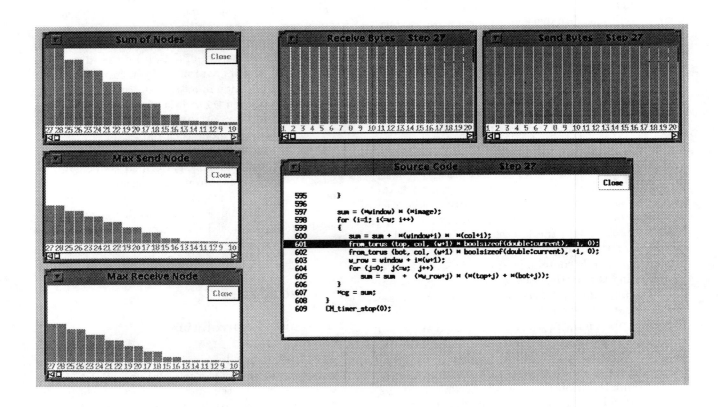

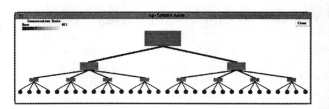

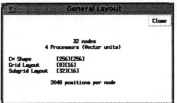

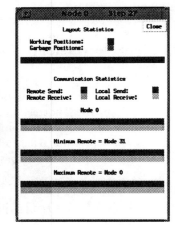

Figure 2: *GRAPH Visualizer views for the folded convolution with default data layout at Step 27 (maximum communication step). **Sum of Nodes** window displays the total off-node message bytes at each communication step. The **Source Code** window highlights the source code statement corresponding to the selected step. **Send Bytes** and **Receive Bytes** windows display off-node message bytes for each node at the selected step. Flat histograms in these windows indicate communication at this step is evenly distributed across nodes.

Convolution Folding

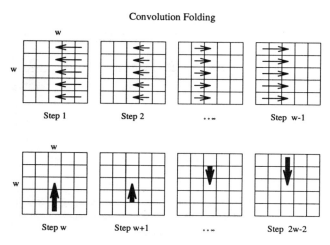

Figure 3: Case Study Algorithm: the *Folded* Convolution. Skinny arrows represent single element messages and fat arrows represent array messages.

7.1.2 Optimized data layout

Using the *allocate_detailed_shape()* function with the Optimizer's suggested weight ratios produced a subgrid layout of 512 × 1. As a result, communication in the second stage of the algorithm has become completely on-node. Visualization of the optimized version (omitted here for lack of space) showed no off-node communication at *Step 27* or at any step corresponding to source code line 601. The step with maximum communication in the optimized layout version corresponds to a C* communication syntax statement at source code line 594.

Table 1 summarizes the results. Changing the data layout reduced the number of off-node message bytes to 58 MB. Most importantly, data layout optimization reduced the average elapsed execution time to 0.09 seconds, a 31% improvement.

7.2 Gaussian elimination

For Gaussian elimination with partial pivoting, there are three communication phases: broadcast, reduction and column (or row) swapping when a pivot occurs. Because *GRAPH currently collects only grid communication data, the broadcast and reduction communication do not affect the optimization. Optimization and communication measurements reported in Table 1 refer to the partial pivoting portion of the algorithm. The test algorithm implements pivoting with column swaps, causing all VP communication to occur along the row axis.

7.2.1 Default data layout

With a problem matrix size of 128 × 128 and a partition size of 32 nodes, the CM run-time system chose a subgrid layout of 16 × 8. This caused the most of the pivot communication to occur off-node (see Table 1).

The C* language provides several logically equivalent communication syntax and functions. However, performance of these alternatives is not equivalent, Using the *GRAPH Visualizer we located the source code statement corresponding to maximum communication cost, and replaced this statement with all possible alternatives to find the fastest C* alternative for the default layout. This optimized syntax was used for both the default layout and optimized layout versions.

7.2.2 Optimized data layout

For pivoting via column swap, *GRAPH's Optimizer suggested a weight ratio of 0:1. Using a layout directive with this ratio, the CM run-time system allocated a serial data layout, 512 × 1. The optimized layout eliminated off-node messages, reduced the pivot time by 45% and overall elapsed time by 17%.

8 Conclusions

Data distribution is critical to performance of parallel programs. For data-parallel languages, a tool such as *GRAPH can fill the gap between data abstraction and an optimized data distribution. By combining visualization of the data distribution and program communication with an automatic means of generating data distribution directives, *GRAPH provides a two-pronged approach to performance improvement. First, for each communication step *GRAPH's Visualizer displays the corresponding source code line and the distribution of messages across the network. *GRAPH's visualizer also helps find communication bottlenecks by sorting communication steps on the basis of message bytes. Second, for a given source code-architecture combination, *GRAPH's Optimizer suggests arguments for data layout directives to improve the data layout. A programmer can try the suggested layout directives, view the result and compare to the unoptimized version. If two source code version are being compared, *GRAPH's Optimizer can suggest optimized layouts for each version and *GRAPH's Visualizer can be used to compare the two optimized source code versions.

	Axis Weights		Subgrid Size	Off-node Message Bytes	On-node Message Bytes	Elasped Time (uninstrumented)		Improvement	
	row	col				Total	Communication	Total	Communication
Folded Convolution									
Default Layout	1	1	32 × 16	7.80 MB	58.26 MB	0.13 s	0.10 s		
Optimized Layout	8	1	512 × 1	3.67 MB	62.39 MB	0.09 s	0.06 s	31%	40%
Gaussian Elimination									
Default Layout	1	1	16 × 8	144,272	107,696	1.05 s	0.33 s		
Optimized Layout	0	1	1 × 128	0	251,968	0.87 s	0.18 s	17%	45%

Table 1: Algorithm performance before and after layout optimization on a 32 node CM-5. On-node and off-node message bytes are summed across nodes and across communication steps. The folded convolution applied a 15 × 15 window to a 256 × 256 image. The Gaussian Elimination measured a 127 × 127 matrix; communication time is for partial pivot section only.

8.1 Current work

*GRAPH's simple optimization heuristic of calculating data axis weight from virtual processor communication statistics has been shown to substantially improve performance in our two test cases. Because of its simplicity, this method requires negligible execution time and is therefore potentially valuable. One of *GRAPH's goals is to investigate different optimization heuristics and the breath of their applicability. The optimization presented in this paper reports our initial work towards that goal. We are currently investigating another method for layout optimization using a simulated annealing package known as Very Fast Simulated Reannealing [3, 10].

Future development of *GRAPH is aided by its modularity. For example, development of a CM Fortran version of the *GRAPH Monitor requires no changes in either the Visualizer or Optimizer modules. Likewise, the Visualizer and Optimizer modules can be extended independently. Currently, the Visualizer is being extended by the addition of views termed *Novas* [7] that can be used for trace animation.

We hope that *GRAPH's philosophy and design approach will yield flexible and useful tools for improving parallel program performance. We also hope this approach will help retain the advantages of data-parallel language abstractions while mitigating performance problems due to system-assigned data distributions.

References

[1] J. Anderson and M. Lam, "Global Optimizations for Parallelism and Locality on Scalable Parallel Machines", *ACM SIGPLAN Notices, Proceedings of the ACM SIGPLAN '93 Conference on Programming Language Design and Implementation*, Vol. 28, pp. 112-125, June, 1993.

[2] V. Balasundaram, G. Fox, K. Kennedy, and U. Kremer, "A Static Performance Estimator to Guide Partitioning Decisions", *Proceedings of the 3rd ACM Symposium on Principles and Practice of Parallel Programming*, pp. 213-223, 1991.

[3] S. Dykes and B. E. Rosen, "Parallel Very Fast Simulated Reannealing by Temperature Block Partitioning", *Proceedings of the 1994 IEEE International Conference on Systems, Man, and Cybernetics*, Vol 2, pp. 1914-1919, 1994.

[4] S. Dykes and X. Zhang, "Folding Spatial Image Filters on the CM-5", *Proceedings of the 9th International Parallel Processing Symposium*, Apr., 1995.

[5] T. Fahringer and H. P. Zima, "A Static Parameter Based Performance Prediction Tool for Parallel Programs", *Proceedings of the 7th ACM International Conference on Supercomputing*, July, 1993.

[6] R. E. Griswold and M. T. Griswold, *The Icon Programming Language, second edition*, Prentice-Hall, Englewood Cliffs, New Jersey, 1990.

[7] R. E. Griswold and C. L. Jeffery, "Nova: Low-Cost Data Animation Using a Radar Sweep Metaphor", *Proceedings of UIST '94*, pp. 131-132, Nov., 1994.

[8] M. Gupta and P. Banerjee, "Demonstration of Automatic Data Partitioning Techniques for Parallelizing Compilers on Multicomputers", *IEEE Trans. on Parallel and Distrib. Sys.*, Vol. 3, pp. 179-193, Mar., 1992.

[9] M. T. Heath and J. A. Etheridge, "Visualizing the Performance of Parallel Programs", *IEEE Software*, Vol. 8, No. 5, pp. 29-39, Sept., 1991.

[10] L. Ingber, "Very Fast Simulated Re-annealing", *Mathl. Comput. Modeling*, Vol. 12, No.8, pp. 967-973, 1989.

[11] C. L. Jeffery, G. M. Townsend and R. E. Griswold, "Graphics Facilities for the Icon Programming Language; Version 9.0", Technical Report IPD 255, Department of Computer Science, University of Arizona, July, 1994

[12] D.S. Johnson, C. R. Aragon, L. A. McGeoch and C. Schevon, "Optimization by Simulated Annealing: An Experimental Evaluation", *Operations Research*, Vol. 39, No. 3, pp. 378-406, May-June, 1991.

[13] B. Kernighan and S. Lin, "An Efficient Heuristic Procedure for Partitioning Graphs", *Bell Systems Technical Journal 49*, Vol. 49, pp. 291-307, 1970.

[14] U. Kremer, "NP-Completeness of Dynamic Remapping", *Proceedings of the Fourth International Workshop on Compilers for Parallel Computers*, pp. 135-141, Dec., 1993.

[15] M. Mace, *Memory Storage Patterns in Parallel Processing*, Kluwer Academic, 1987.

[16] B. Miller, J. M. Cargille, R. Bruce Irvin, K. Kunchithapadam, M. D. Callaghan, J. K. Hollingsworth, K. L. Karavanic, T. Newhall, *The Paradyn Parallel Performance Measurement Tools*, Technical Report, Computer Sciences Department, University of Wisconsin-Madison, 1994.

[17] R. Ponnusamy, J. Saltz, R. Das, C. Koelbel and A. Choudhary, "Embedding Data Mappers with Distributed Memory Machine Compilers", *ACM SIGPLAN Notices, Workshop on Languages, Compilers, and Run-Time Environments for Distributed Memory Multiprocessors*, Vol. 28, pp. 52-55, Jan, 1993.

[18] Thinking Machines Corporation consultants, *private communication*, 1994.

[19] Thinking Machines Corporation, *CM-5 C* Performance Guide, V.7.1*, 1993.

[20] Thinking Machines Corporation, *CM-5 C* Programming Guide*, 1994.

[21] Thinking Machines Corporation, *CM-5 CM Fortran Programming Guide*, 1994.

[22] Thinking Machines Corporation, *Prism User's Guide*, 1994.

[23] X. Zhang, N. Nalluri and X. Qin, "MIN-Graph: A Tool for Monitoring and Visualizing MIN-Based Multiprocessor Performance", *J. Parallel Distrib. Comput.*, Vol. 18, No. 2, pp. 231-241, 1993.

The Integration of Event- and State-Based Debugging in Ariadne *

Joydip Kundu
Department of Computer Science
University of Massachusetts
Amherst, MA 01003

Janice E. Cuny
Department of Computer Science
University of Oregon
Eugene, OR 97403

Abstract. Parallel programs often have complex behaviors that require multi-level debugging strategies. Here, we propose an integration of event- and state-based strategies that initially uses event-based behavioral modeling to narrow the focus of attention and then uses state-based techniques to relate observed errors to specific code segments. Our debugger supports this strategy, allowing the results of behavioral analysis to be used in specifying consistent, global breakpoints. These breakpoints differ from those provided by other debuggers in that they are meaningful in the context of the ongoing event-based analysis. In addition, the breakpoints that users select with our debugger would be difficult to set using existing state-based approaches. We demonstrate here that the combination of event- and state-based debugging is significantly more powerful than either strategy alone.

1   Introduction

The complex behavior of parallel programs often requires a multi-level debugging strategy. Such a strategy might, for example, employ event-based techniques [1, 3, 7, 9] at the highest level where gross patterns of process interactions are investigated. As debugging proceeds and the focus of attention narrows, the behavior of progressively smaller parts of the program could be analyzed in progressively finer detail, using a combination of event- and state-based techniques. Finally, at the lowest level, when the error has been isolated to specific sections of sequential code, traditional state-based techniques could be used. This strategy effectively addresses many of the issues in parallel debugging. The initial use of event-based techniques focuses the user's attention on manageable portions of the state space and (as we will see below) provides the basis for establishing consistent, meaningful, global breakpoints. Event-based techniques can incorporate replay mechanisms that support reproducible execution [13] and logical time transformations to filter out perturbations due to asynchrony

[8, 12]. State-based techniques, on the other hand, allow the user to directly examine an execution to an arbitrary level of detail and often make it easier to relate errors to source code constructs.

A number of event- and state-based debuggers already exist. Here we discuss the integration of two – Ariadne [2] and *ipd* [10] – into a single environment. The key to this integration is the use of user-defined abstract events in establishing consistent, global breakpoints: the states we provide are *meaningful in the context of the user's ongoing event-based analysis*. Often these breakpoints would often be difficult, if not impossible, to set using existing state-based techniques.

In Section 2, we give a brief overview of the integrated environment; in Section 3, we demonstrate its use.

2   Setting Breakpoints with Abstract Events

To effectively use break/examine techniques on massively parallel programs, it is necessary to stop those programs in meaningful states, that is, states that make sense within the context of the current debugging session. Standard mechanisms that set breakpoints in terms of program locations or combinations of local state predicates are problematic: How are the breakpoints to be specified across multiple processes? Must they be globally consistent or are local breakpoints sufficient? If local, how and when should the remaining processes be stopped in order to provide a meaningful state?

One possibility is to allow breakpoints only at barrier synchronizations [17] but this is overly constraining in multithreaded language environments. Alternatively, local breakpoints can be extended dynamically to global breakpoints [6, 16, 18]. Unfortunately, debuggers that support this typically allow processes not directly involved in the breakpoint to continue executing for some time, potentially obliterating information relevant to the source of the error. We choose instead to take advantage of replay mechanisms: we use the results of *post mortem* modeling to insert directives that will, during replay, halt each process at the latest point at which it could have influenced the event(s) triggering the breakpoint.

Manabe and Imase [15] also use this approach, allowing breakpoints to be to specified as conjunctions and disjunctions of local predicates. It may, however, be difficult to determine the appropriate predicates: errors observable only as incorrect output are often hard to relate back to individual, local predicates in specific processes. Instead, our Ariadne/*ipd* debugger allows the user to specify breakpoints in terms of the same abstract events that he/she is already using for behavioral modeling.

During event-based behavioral modeling, user-specified models of intended program behavior are matched against the actual program behavior as captured in event traces. Within Ariadne, abstract events are sets of primitive events grouped by logical, temporal relationships. Breakpoints can be set immediately before or after a matched or partially matched abstract event. Once set, the debugger annotates the trace and then reexecutes the program to the latest *consistent* state preceding those annotations. Often, this allows the programmer to easily stop his/her program at states that would be difficult to specify using local state predicates. For example, the program could be stopped at the first point in an execution where two processes see a sequence of broadcast operations in a different order, or at a point where all of the processes reached by a specific spreading activation have sent replies, or at the beginning of the first phase in which some processes communicate over different distances.

A matched (or partially matched) abstract event can be used to partition the processes of a computation into three equivalence classes: the *core set* containing the processes that executed parts of the abstract event; the *influence set* containing the processes that could have causally affected a component of the abstract event; and the *other set* containing the remaining processes. In order to stop a computation, the state most relevant to the abstract event is restored as follows:

1. processes in the *other* set are stopped at their earliest state,

2. processes in the *influence* set are stopped immediately after the latest event that could causally influence the abstract event, and

3. processes in the *core* set are stopped immediately before (after) the earliest (latest) local event that

is a part of the abstract event, depending on the user request.

In some cases, this procedure results in a state that is not *consistent*, that is, a state that could not have occurred in an actual execution. We define consistency as in [4]. Whenever a user's breakpointing request would result in an inconsistent state, our debugger automatically constructs an earlier, consistent state using a greedy algorithm.

3 Tracking Elusive States with Event-Based Breakpointing

In this section, we demonstrate the use of our integrated debugging environment on a parallel program for querying a knowledge base [5]. The knowledge is stored hierarchically as objects and classes of objects related by the *is-a* relation. To avoid redundancy, object attributes are stored as high as possible in the hierarchy and are then inherited at the lower levels. Thus, for example, "warm-blooded" is an attribute of "mammals" and "dog" *is-a* "mammal," so we can infer that any "dog" is "warm-blooded." To query such a database, the attributes of an object are found by following its *is-a* links outward in a spreading activation; as attribute values are encountered, they are returned to the originating node. The process is complicated by exceptions: "birds fly" but "penguins are birds" and "penguins don't fly." As processing proceeds, returned exceptions may invalidate earlier responses. Thus, the originating node must wait until all of the activity resulting from a query has terminated before replying. To detect termination, a *mass* attribute is attached to each message. The originating node sends out a total mass of 1, each activated node sends as much *mass* as it receives, and the originating node waits until it has received back a *mass* of 1. Multiple queries can be active in the knowledge base simultaneously.

The program uses five different message types. It was instrumented so that these message types appear as labels on interprocess read and writes within the trace. **plus** and **minus** messages spread activity through the network. **plus** messages request attribute values and **minus** messages indicate that existing values are overridden by an exception. **found**, **ignore**, and **mass** messages return answers to the originating node. **found** signals a valid attribute value, **ignore** notes an exception to a previously reported value, and **mass** indicates that no attribute value was found.

We tested our program and discovered that it did not operate correctly even on simple queries. It never found any attributes at all. To debug it, we first coarsely modeled its intended behavior. Ariadne uses a simple modeling language based on regular expressions.

For clarity, responses to the originating node were modeled by the writing of those messages as

`ch response = W_found or W_ignore or W_mass;`

and the receiving of those messages as

`ch gather = R_found or R_ignore or R_mass;`

where `W` and `R` indicate write and read operations respectively and `ch` is a keyword indicating that this is the definition of a local pattern of interaction. The behavior of an activated node was modeled as

`ch scatter = (( W <: @ R response )* :> )*;`

where the `@` symbol moves the pattern matching to the process that is the destination of the immediately preceding write. A complete description of the modeling language is beyond the scope of this paper but can be found in other references [2].

The behavior of the originating node was more complex because it must also collect responses. It was described using the above patterns as

`scatter gather+`

This behavior was bound to concurrent executions on a process set (here, `procset`), with

`pch query = scatter gather+ onsome procset;`

The entire series of queries over the lifetime of the program was defined as

`ptch Query = query*;`

To match this expected behavior against the execution trace, we used

`match infer = Query;`

which assigned the result of that match to the identifier `infer`.

For this example, the intended patterns of behavior were found in the execution trace and the match succeeded. Ariadne's feedback is shown in Figure 1 for a trace that contained a single originating node sending 5 different queries. As can be seen, Ariadne imposes a tree-hierarchy on the matched events that is based on the user-defined model. Internal nodes represent user-defined abstract events and leaves (not shown in the figure) represent primitive events. This tree is called a *match tree*. For scalability, Ariadne compresses it. Here compression is shown by the stacked boxes below the `query` level (indicating that the subtrees are not shown) and the slider between the two `query` nodes (indicating that 3 of their siblings are not shown). Again, the complete description of this figure is be-

yond the scope of this paper [11].

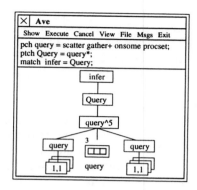

Figure 1: Ave feedback showing coarse grain pattern match in knowledge-base example.

This first step in debugging did not show any obvious error in query handling. It is possible that the interprocess communication patterns were, in fact, correct, but it is also possible that our model was too coarse to detect the error. To rule out the latter, we examined the matched behavior more thoroughly using Ariadne's query facility which provides a spreadsheet-like interface for computing incremental attributes of a match tree [11]. With a few simple queries, we were able to ascertain the following:

1. The number of **found** messages received by the originating node was one more than the number of **ignore** messages received. (Thus, it appeared that a valid attribute value was found but somehow discarded.)

2. A node that sent a **mass** message did not send any other message to the originator for this query. (Thus, the originating node was not receiving extraneous **mass** messages that might cause it to reply prematurely.)

3. **ignore** messages were sent only after a node received a **minus** message. (Thus, extra **ignore** messages were not overwriting good values.)

After this examination, we were confident that the interprocess communication patterns of the program were correct. Using event-based analysis we had not been able to detect any errors. We had, however, considerably narrowed our focus of attention: since the interprocess communication appeared correct, it was likely that the error was within the sequential calculation of mass or attribute values. Event-based debugging would not be useful in tracking this kind of an error, thus, at this point we switched to the *ipd* state-based debugger.

To use state-based debugging, we must be able to easily specify breakpoints that are consistent and meaningful to the programmer. What state would that be in our example? Mass and attribute values are processed both at the activated nodes and at the originating node. The obvious place to stop the program would be in the middle of a query, at a point where all of the activated nodes have just finished sending their replies and the originating node has not yet processed their replies. This state is difficult to specify in standard, state-based manner. Multiple queries can be present in the knowledge base and different nodes may handle the queries in different orders (if at all). We do not know in advance which nodes are activated by a particular query or even which queries will be of interest. Which processes should be stopped for a particular query? How will we know which query to stop them on?

The specification of this breakpoint in Ariadne, however, is simple: the user scans the match tree, selects any matched **query** node, and requests that a breakpoint be set just after its **scatter** subpattern.

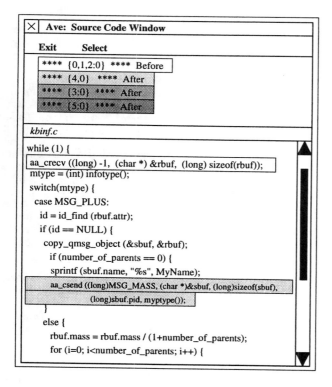

Figure 2: Ave: Source Code Window, showing processes stopped at different points.

Having done this, we replayed this execution of the program, stopping at the designated breakpoint as shown in the Ave Source Code Window in Figure 2. Highlights in the figure indicate the source lines at which processes have stopped. From the output at the top of the screen, we can see that processes 0, 1, and 2 are stopped at the line indicated by the clear box, process 4 is waiting at the send operation indicated by the lightly shaded box, and processes 4 and 5 are waiting at source lines not currently visible in the window.

To continue debugging, we next choose to selectively execute the originating node (process 0) by setting a local breakpoint after each message receive with

```
(0:0) > break kbinf.c{}#412
(0:0) > continue
```

We examined the messages as they were received. The first **found** message, for example, was examined with

```
(0:0) > display main()rbuf

  ** kbinf.c{}main()rbuf **
  ***** (0:0) *****
  rbuf.mtype = found
  rbuf.sender = 3
  rbuf.element = Penguin
  rbuf.attribute = Locomotion
  rbuf.value = NULL
```

Notice that while the message was of type **found**, it did not contain a **value** (that is, **rbuf.value** was NULL). In fact, none of the **found** messages we saw contained any attribute values. Thus the problem was not in the originating node but at the responding, activated nodes (which were apparently not sending values in their messages). These nodes were already stopped immediately after they had sent those messages. A look at the preceding code section revealed that the programmer forgot to copy the result of the query to the send buffer.

4 Conclusion

In this paper, we have presented a multi-level debugging strategy which uses event-based behavioral debugging at the highest levels of abstraction and then switches to stated-based mechanisms as the focus of attention narrows. This strategy effectively addresses many of the issues in parallel debugging. The initial use of event-based techniques focuses the user's attention on manageable portions of the state space; it can provide the basis for establishing consistent, global breakpoints using replay mechanisms; and it can support logical time transformations that filter out perturbations due to asynchrony. State-based techniques, on the other hand, allow the user to directly examine an execution to an arbitrary level of detail and often make it easier to relate errors to source code constructs.

Here, we discussed the integration of two existing debuggers, Ariadne and *ipd*. The key to their integration was the use of user-defined abstract events in establishing *consistent, global breakpoints that were meaningful in terms of the user's ongoing event-based analysis of his/her code*. These breakpoints would be difficult or impossible to set with standard breakpoint debuggers.

We are currently extending the implementation to a number of different platforms, focusing on an integration of Ariadne and the Breezy breakpoint debugger [17].

Acknowledgements

George Forman, Alfred Hough, Calvin Lin, Lawrence Snyder, and David Stemple contributed to the initial design of Ariadne.

References

[1] P. C. Bates. *Debugging Programs in a Distributed System Environment*. PhD Thesis, University of Massachusetts, 1986.

[2] J. E. Cuny, G. Forman, A. Hough, J. Kundu, C. Lin, L. Snyder, and D. Stemple. The Ariadne debugger: scalable application of event-based abstraction. *SIGPLAN Notices*, Vol. 28, No. 12, pages 85–95, 1994.

[3] R. J. Fowler, T. J. Leblanc, and J. M. Mellor-Crummey. An integrated approach to parallel program debugging and performance analysis on large-scale multiprocessors. *SIGPLAN Notices*, Vol. 24, No. 1, pages 163–173, 1989.

[4] A. P. Goldberg, A. Gopal, A, Lowry, and R. Strom. Restoring consistent global states of distributed computations. In *Proceedings of the ACM/ONR Workshop on Parallel and Distributed Debugging*, pages 144–154, 1991.

[5] M. Greenberg and J. Cuny. Parallelism in knowledge-based systems with inheritance. In *Proceedings of the 1988 International Conference on Parallel Processing*, pages 141-145, 1988.

[6] D. Haban, and W. Weigel. Global events and global breakpoints in distributed systems. *21st Annual Hawaii International Conference on System Sciences*, pages 166-174, 1988.

[7] A. A. Hough. *Debugging Parallel Programs Using Abstract Visualizations*. PhD Thesis, University of Massachusetts, 1991.

[8] A. A. Hough and J. E. Cuny. Perspective views: A technique for enchancing visualizations of parallel programs. In *1990 International Conference on Parallel Processing*, pages II 124–132, 1990.

[9] W. Hseush and G. E. Kaiser. Modeling concurrency in parallel debugging. In *Proceedings of the Second ACM SIGPLAN Symposium on Principles and Practice of Parallel Programming*, pages 11–20, 1990.

[10] Intel Supercomputer Systems Division. Paragon Interactive Parallel Debugger Reference Manual. Paragon Documentation, 1994.

[11] J. Kundu and J. E. Cuny. A scalable, visual interface for debugging with event-based behavioral abstraction. In *Proceedings of New Frontiers on Massively Parallel Processing*, pages 472–479, 1995.

[12] R. J. LeBlanc and A. D. Robbins. Event-driven monitoring of distributed programs. In *Proceedings of the 5th International Conference on Distributed Computing Systems*, pages 515–522, 1985.

[13] T. J. LeBlanc and J. M. Mellor-Crummey. Debugging parallel programs with instant replay. *IEEE Transactions on Computers*, Vol C-36, No. 4, pages 471–482, 1987.

[14] T. J. LeBlanc, J. M. Mellor-Crummey, and R. J. Fowler. Analyzing parallel program executions using multiple views. *Journal of Parallel and Distributed Computing*, Vol 9, pages 203–217, 1990.

[15] Y. Manabe, and M. Imase. Global conditions in debugging distributed programs. *Journal of Parallel and Distributed Computing*, Vol. 15, pages 62-69, 1992.

[16] K. Mani Chandy, and L. Lamport. Distributed snapshots: determining global states of distributed systems. *ACM Transactions on Computer Systems*, Vol 3, No. 1, pages 63-74, 1985.

[17] D. Brown, S. Hackstadt, A. Malony, B. Mohr. Program Analysis Environments for Parallel Language Systems: The TAU Environment. *Proceedings of the 2nd Workshop on Environments and Tools For Parallel Scientific Computing*, pages 162-171, 1994.

[18] B. Miller, and J. -D. Choi. Breakpoints and halting in distributed programs. *Proceedings of 8th International Conference on Distributed Computing Systems*, pages 316-323, 1988.

graze: A TOOL FOR PERFORMANCE VISUALIZATION AND ANALYSIS*

Lantz Moore Debra A. Hensgen David Charley
Venkatram Krishnaswamy   Dale E. Martin
Timothy J. McBrayer   Philip A. Wilsey
University of Cincinnati
Distributed Operating Systems Lab
Cincinnati, OH 45220-0030
lantz_moore@ece.uc.edu

Abstract — *We present* **graze**, *a framework for the collection, visualization, and analysis of performance data. Our framework allows a designer to focus on different aspects of an application in execution. Its purpose is to provide a simple, flexible, and robust environment for performance analysis. Facilitating user-defined events and intervals,* **graze** *permits flexible monitoring of parallel and distributed programs, as well as generic visualization. The generic visualization is easily extensible by the user.*

INTRODUCTION

This paper presents **graze**, a portable, flexible framework for the collection, visualization, and analysis of the performance data of applications with multiple threads of control. It details how we were able to use **graze** to uncover performance problems of a mature VHDL simulation system implemented with the Time Warp paradigm.

Until very recently, profiling was the sole means of system optimization. Profiling pinpointed the function or section of code where the system was spending the majority of its time. The programmer would analyze this code, determining whether it could be replaced with a more efficient sequential or parallel algorithm, and then repeat the whole process. Unfortunately, the profiling technique doesn't scale well past a single thread of control. In profiling a multithreaded application, information such as the time spent by a thread waiting for synchronization or the length of a threads incoming message queue, can be

*This work is partially supported by the US Air Force Wright Laboratory Avionics Directorate under contract number F33615-93-C-1301.

lost or simply doesn't exist in the profiling data.

RELATED WORK

In the past several years, several tools that transcend the capability of profiling have been produced. Example performance visualization tools are ParaGraph [1], Traceview [3], Quantify, and Paradyn [5]. PSpec [6] is a tool that provides performance assertion checking. ParaGraph and PSpec will be discussed briefly below.

ParaGraph is a performance optimization tool for use with PICL [7]. ParaGraph is tightly tied to the message paradigm, and can be used to gain great insight into communication overhead. However, it lacks the ability to adequately illustrate generic performance data.

PSpec is used to describe certain aspects of an application by identifying "interesting" events and intervals that may occur. PSpec can be used to check that a certain event happens at least x number of times, or that a certain interval never takes more than y microseconds. Since PSpec employs a user defined set of events and intervals, it can analyze arbitrary performance data. However, it may not always be possible to glean all the specified information from the collected data, even though PSpec has interfaces to several data collection facilities.

THE graze FRAMEWORK

The **graze** framework is comprised of three main components: a specification language, a data collection facility, and a generic data visualization facility. The specification language allows a programmer to define application-specific events and intervals of interest. A set of data collection routines is automatically produced from the programmers specification. The programmer manually inserts these probes into the application, and then compiles and executes the application. The programmer can then analyze the resulting log files using the generic data visualization facility. A detailed description of the specification language, and the data collection and generic visualization facilities are given below.

Specification Language

The **graze** specification language is similar to the language used by PSpec. In this specification language, there are two main constructs: an event, and an interval. The user specifies what events and intervals are of interest. In the **graze** framework, this specification is used to generate the data collection routines as well as guide the generic visualization. In the following discussion, a message passing specification will be used as a running example.

Events denote points of interest during the execution of an application. To distinguish events of the same type, or to provide more detailed data, events may have attributes. Intervals are bounded by events. The bounding events of an interval can simply match on event type, or a boolean expression involving the bounding events can be used. To further distinguish events and intervals, they can be given visual representations. Events can be displayed as any of several different shapes, and the bounding events of an interval can be connected with a line or rectangle. The following example specifies the events corresponding to sending and receiving a message, as well as the intervening interval:

```
event Send = Plus;
event Recv = Box;
interval Transit
  [ s:Send -> r:Recv ] = Line;
```

The interval `Transit` starts with a `Send` and ends with the next `Recv`. It should be noted that more information is needed to identify the interval correctly; namely, the source identifier of the message, and a sequence number. Adding this information, the above example would look like:

```
event Send(src, seq) = Plus;
event Recv(src, seq) = Box;
interval Transit
  [ s:Send -> r:Recv ]
  { s.src == r.src &&
    s.seq == r.seq; } = Line;
```

Every event has two implicit attributes: an owner and a time-stamp. The owner is the identifier of the

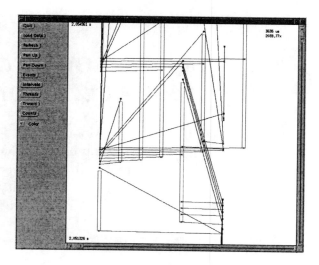

Figure 1: **gorge** visualizing message data from a simulation containing eight threads of control. Time increases along the Y-Axis.

thread that produced the event. The resolution of the time-stamp depends on the hardware.

Data Collection

Using the specification provided by the user, the **graze** framework automatically generates a custom data collection facility for an application. For each event in the specification, there is a corresponding function. The function name is simply the event name with the suffix "`_Stamp`" appended to it. If the event has explicit attributes then the function has formal parameters matching those attributes. This collection of probes is put into a header file that the programmer can include where needed.

The programmer simply instruments the application by making calls to the various probes, filling in the parameters as needed. The probes are designed so that they can be turned on and off by recompiling. When the probes are on, and the application is executed, a set of log files is produced. These log files are used by the data visualization facility.

Generic Data Visualization

After an instrumented program has been executed, there are several ways in which to visualize the data. We have built two tools for generic data visualization: **gorge** and **nibble**. **gorge** uses a space-time graph to display the collected data exactly as it is defined in the programmers specification, while **nibble** allows the programmer to graph information pertaining to specific events and intervals. Both of these tools were developed using **cud**,

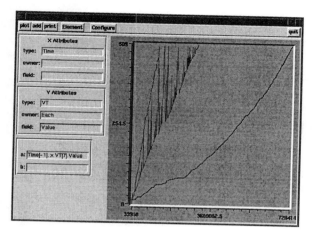

Figure 2: **nibble** graphing an event parameter for three threads. Time increases along the X-Axis.

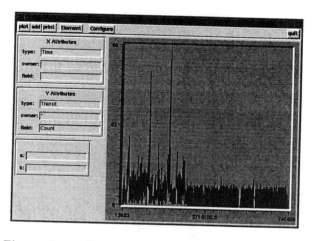

Figure 3: **nibble** graphing the total messages queued in the system.

C and TCL/TK interfaces that allow a programmer to write a completely specialized display tool.

The **gorge** tool uses a space-time graph to display the collected data exactly as is defined in the programmers specification. The user can control the amount of information displayed by zooming in or out on the graph, or displaying any combination of events, intervals, and threads. Figure 1 shows an example screen from **gorge**.

The **nibble** tool is a generalized graphing tool. An event's explicit attribute or the total number of occurrences can be graphed versus time (figure 2). An interval's elapsed time, aggregate elapsed time, or total number of occurrences can be graphed versus time. We treat these graphs as curves, so we also allow them to be smoothed, added, subtracted, and even integrated (figure 3).

EXPERIENCE

We have successfully applied the **graze** framework by analyzing a mature VHDL simulator [4]. The simulator consists of two main parts: a simulation kernel and a communication fabric. A simulation consists of several cooperating autonomous threads of control executing in parallel. These threads execute at their own pace, while sending messages amongst one another. The VAST simulator uses the Time Warp [2] synchronization protocol to preserve causality.

We first instrumented the communication fabric. Being primarily interested in the latency of message delivery notification to threads waiting for a message, we were surprised to find a very peculiar pattern to the message traffic. Figure 4 shows filtered message traffic. The nearly horizontal lines denote

one thread sending a message to another. No messages to or from thread zero, a special centralized controller, are displayed.

In the class of simulation that produced this log file, we know that the simulation is complete when the simulation objects finish passing messages amongst themselves. However, as evidenced by figure 4 the simulation was taking twice the time needed to complete. In our simulator, termination is a by-product of garbage collection. While Time Warp processes periodically save state, they need not keep the saved states forever. There is a lower bound on virtual time, called Global Virtual Time, below which saved state can safely be garbage collected. GVT is defined as the minimum across all simulation objects of the objects Local Virtual Time, any unprocessed events, and any unacknowledged messages. If a simulation object has no unprocessed events, nor unacknowledged messages, it reports that it is idle to thread zero. If all simulation objects report idle, then the simulation is complete. It was for this reason that we needed to track GVT, as well as the value of each simulation object's LVT.

The result of this instrumentation of virtual time is shown in figure 5; GVT, the low gently curving line, seemed to be held back. On further examination of the code that the simulation objects used to form their GVT estimate, it was discovered that the estimate was being calculated far too conservatively. We noted over a factor of 2 speedup on most simulations after fixing this performance bug.

We were also able to quantify the approximate overhead due to causality errors. When a simulation object must process an event with a time-stamp lower than the objects LVT, it must rollback to a

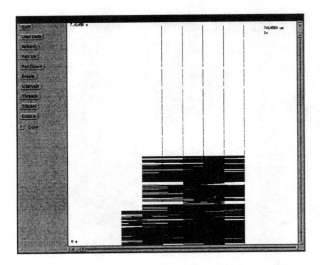

Figure 4: **gorge** visualizing message traffic.

state that has a time-stamp less than or equal to that of the offending event. It is possible for an object to process very far into the future, called lookahead computation, and then get rolled back. The time spent in lookahead computation that gets rolled back is called wasted lookahead computation. Specifying and graphing the intervals corresponding to lookahead and wasted lookahead, we noted that 60% of the total lookahead computation was wasted lookahead computation.

FUTURE ENHANCEMENTS

We envision many enhancements to the **graze** framework. Integrating with PSpec is of high priority; **graze** has much to gain from this marriage, in that assertions make wonderful filters. Automatic instrumentation of object code would dramatically lessen the burden on the programmer. Integrating processor information into the data collection facility would provide the programmer with even more detailed information.

CONCLUSIONS

In conclusion, we have presented **graze**, a framework for data collection, visualization, and analysis. The **graze** framework differs from other performance visualization and analysis tools in that it is totally user specified, in both data collection and visualization. We have shown **graze** useful in analyzing a mature and complex application by uncovering numerous performance bugs. Taking advantage of **graze** could provide a great opportunity for programmers to easily instrument, analyze, demonstrate, and improve their applications.

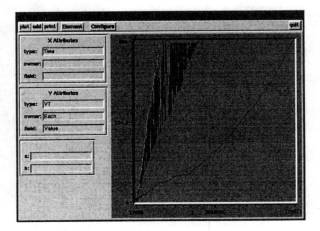

Figure 5: **nibble** graphing the virtual time of several threads.

References

[1] M.T. Heath and J.A. Etheridge. Visualizing the performance of parallel programs. *IEEE Software*, 8(5):29–39, September 1991.

[2] D.R. Jefferson. Virtual time. *ACM Transactions on Programming Languages and Systems*, 7(3):405–425, July 1985.

[3] A.D. Malony, D.H. Hammerslag, and D.J. Jablonowski. Traceview: A trace visualization tool. *IEEE Software*, 8(5):19–28, September 1991.

[4] T. McBrayer, V. Krishnaswamy, L. Moore, X. Liu, J. Carter, D. Charley, P. A. Wilsey, D. A. Hensgen, H. W. Carter, P. Chawla, J. Collier, and S. Bilik. VAST: Time warp simulation of VHDL on SMP workstations. In *VHDL Users' Group Fall 1994 Conference*, Washington, DC, November 1994.

[5] Barton P. Miller, Mark D. Callaghan, Johathan M. Cargille, Jeffery K. Hollingsworth, R. Bruce Irvin, Karen L. Karavanic, Krishna Kunchithapadam, and Tia Newhall. The paradyn parallel performance measurement tools. Technical Report 1256, University of Wisconsin–Madison, 1994.

[6] S.E. Perl and W.E. Weihl. Performance assertion checking. *Operating Systems Review*, 27(5):134–145, December. 1993.

[7] P.H. Worley. A new picl trace file format. Technical Report ORNL/TM-12125, Oak Ridge National Laboratory, 1992.

PROGRESS: A TOOLKIT FOR INTERACTIVE PROGRAM STEERING

Jeffrey Vetter[a]   Karsten Schwan

College of Computing
Georgia Institute of Technology
Atlanta, GA 30332-0280
{vetter, schwan}@cc.gatech.edu

Abstract -- *Interactive program steering permits researchers to monitor and guide their applications during runtime. Progress is a toolkit for developing steerable applications. Users instrument their applications with library calls and then steer parallel applications with Progress' runtime system. Progress provides steerable objects which encapsulate program abstractions for monitoring and steering during program execution. This toolkit provides sensors, probes, actuators, function hooks, complex actions, and synchronization points. Progress has been applied to several large-scale parallel application programs, including a molecular dynamics code and an N bodies simulation. It is currently being used with a complex global atmospheric modeling code.*

1. INTRODUCTION

If high performance computing continues to remain 'non-interactive' [10], end-users and program developers alike will not capitalize on new techniques for interactive data visualization and program animation [8], interactive debugging and monitoring [5,12], and on-line program adaptation [11]. Program steering provides end-users with the capability to monitor and guide their applications during runtime. The Progress toolkit presented and evaluated in this paper provides facilities for increased interactivity [2,6] for existing high-performance multiprocessor programs.

Progress explores the requirements and opportunities of selective and application-dependent runtime monitoring and program steering. The hypothesis we explore in this research is "program steering should be dynamically initiated, enabled and disabled, changed in scope and functionality, and used selectively."

Interactive program steering concerns data collection, data interpretation, and steering. Data collection has a rich history of research in program monitoring. Data interpretation also has an extensive past with visualization [8] and other techniques such as filtering and clustering [1]. Steering, however, is rather immature. Progress addresses this immaturity by

---

(a) Vetter is financially supported by a NASA Graduate Student Researchers Program grant.

defining a framework for steering.

Section 2 describes the construction of the Progress steering system. Section 3 evaluates Progress with an application. Section 4 reviews related work. Section 5 concludes this paper with a review of research goals and future research.

2. PROGRESS STEERING TOOLKIT

Progress is an acronym for *Prog*ram and *Re*source *S*teering *S*ystem. The goal of ProgReSS is "to provide application developers with a set of tools and facilities for creating steerable parallel applications." Progress is not intended as an advanced remote visualization system, or a parallel debugger. The Progress steering toolkit has several concepts and stages. First, the application developer must understand the steering object model to properly instrument the target application. The Progress object model encapsulates components of the application that the user might wish to steer or monitor. Second, the application developer actually instruments the application with calls to the Progress library to create and maintain these steering objects throughout the application's lifetime. The developer understands the application's operation sufficiently to allow external changes to application state. And finally, the end-user controls the application with the steering runtime system.

Our design requirements for Progress include four fundamental notions. First, basic steering should be possible with many dynamic applications. Second, the steering library should provide functionality for steering beyond updates to simple variables (e.g., the steering system should support complex steering operations). Third, the user interface should allow end-users the capability to explore their executing programs. Finally, the steering system must provide mechanisms for protecting application integrity while allowing changes to application state that do not degrade application performance. Steering tools should provide a level of abstraction that is effective for an end-user and useful for steering. User-defined types including structures and arrays must be accessible to allow complex steering operations on arrays and structures of data. Moreover, for performance reasons and in contrast to debugging, steering should not allow access to every program variable and the ability to arbitrarily interrupt program flow. Such functionality is not required for steering which concentrates on observing and changing parameters of a correct application.

The end user, not the application developer, controls the steering system through the steering user interface. From the steering interface, the user should be able to selectively monitor and modify steering objects within the application. This interface must provide various types of steering actions consistently on a continuous range of steerable programs. Operations on steering objects should be consistent between different applications.

An object model allows the steering system (and the user) to concentrate on only those components of the application that the developer explicitly instrumented. This abstraction is necessary to limit the amounts of information visible to the end user and the quantity of data processed by the monitoring and steering system. Progress does not support inheritance and other characteristics of object-oriented languages; Progress' object model encapsulates and identifies components of the program for monitoring and steering (this concept is similar to LeBlanc's instant replay mechanism [9], where the user identifies 'replayable' objects within the code). An object becomes known to the steering server when the object is registered.

Once a steering object is registered, the steering server manipulates the object with several different operations. These operations allow a variety of synchronous and asynchronous accesses to the application through steering objects. Synchronous and asynchronous accesses are differentiated because, in some cases, asynchronous accesses to steering objects may produce inconsistent views of the object or actually invalidate the application results! To perform synchronous operations, the developer must instrument the application more thoroughly. The basic operations are probe read, probe write, sense, and actuate. Both probes and sensors are investigated in [12] as part of an application specific monitoring system. In Progress, these sensors and probes monitor steering objects instead of language specific application components.

A probe is the simplest of the steering operations because once an object is registered, the steering server just reads the object's data. No additional instrumentation is necessary. The steering server performs probes without respect to the application's control flow; therefore, probes are asynchronous. Probes are particularly useful for inspecting stalled programs or updating non-critical variables in the application.

A sensor captures an object's state within the application and forwards it to the monitoring system as an event. When the application encounters a 'Sense' call in its thread of execution, it copies the object state to an event record, and places the record into a buffer destined for the monitoring system. Because sense is executed within the control flow of the application, sense is synchronous. The steering server can enable or disable sensors for each particular steering object.

Actuators perform steering actions. When an application thread executes this call, actuate checks its buffer to determine if any changes are intended for its steering object. These changes are 'programmed' for a certain steering object by the steering server. Actuate is synchronous because the modification is performed on the application by one of its own threads (as opposed to the steering server) when the application encounters an actuate instrumentation point.

Additionally, Progress provides a set of specialized objects [14]. These operations are synch points, function execution, and scripts.

2.1 Runtime System

Progress' runtime system in Figure 1 is composed of a server and a client. With the server on the same machine as the application, the server can access and control the application with low latency [5]. The client is remote. This architecture is advantageous because the server exists throughout the application's lifetime while the client does not. The client is transitory and can connect to the server many times throughout the server's (application's) existence. Because the client may use visualizations for data interpretation, the end-user may choose to run the client on a high-performance graphics system.

The steering server is a separate thread that executes in the same memory space as the application. The server architecture allows the application to execute

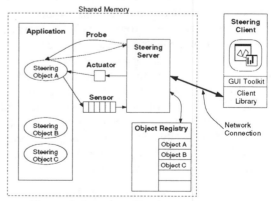

Figure 1. Progress Runtime Architecture

normally. The server thread has three basic tasks: interact with the steering client, gather monitoring output from the application, and steer the application via steering objects. The server continuously maintains a registry of the application's steering objects. Each object that the application registers is stored in the registry. Registry records contain enough information for the server to access objects at anytime and interpret information generated by the application. The registry is also used to route information from the application to the client and visa-versa.

The server monitors the application by receiving sensor events and probing objects. The server does simple analysis to filter irrelevant events out of the stream. The server controls monitoring so that the end-user can selectively observe different steering objects throughout the application's lifetime. Based on commands from the end user, the server enables and disables sensors and probes objects to gather information.

The server steers the application through steering objects using the steering operations detailed earlier. Several actions can trigger a steering operation. The user can manually request a steering operation on a steering object, or the server can execute a steering operation in response to an event received from the application.

The client is a user interface application that communicates with the steering server through a steering client library. The client has three main tasks: interact with the user, communicate with the steering server, keep relatively consistent state information about all the steering objects. The client receives all of its information from the steering server; it receives periodic updates to its registry from the server. The client has a shadow registry of the server's object registry. Client interfaces are built on a client library that actually communicates with the steering server. Customized interfaces are built using this same client library.

3. EVALUATION

To evaluate Progress, we use N-body because it is a well-understood and concise example with which we can describe the functionality and performance of our toolkit. Additional functionality of Progress is evaluated by demonstrating new techniques of interacting with the executing application. Performance is important because this toolkit cannot prohibitively degrade performance of the application.

The numerical N-body of gravitation [3] simulates dynamical behavior of large stars with only gravitational forces acting between them. This simulator uses a straightforward N^2 algorithm for calculating pairwise gravitational forces and updating the velocities and positions of the bodies.

Integrating N-body with the Progress system requires three distinct steps. First, the developer adds appropriate calls to create and initialize the steering server. Second, registration calls for all steerable objects are inserted into the source code. Third, all synchronous objects have instrumentation code inserted into the application to identify points where these objects can be safely accessed.

Prior to the steering integration, user interaction was limited to file I/O. While this type of interaction could be customized to provide interactivity, there are no generic methods or tools to facilitate selective, application-specific interaction. Progress allows far more interaction with the simulation than file I/O. With Progress, the user interactively explores intermediate results of the simulation. For example, the Sun's velocity can be traced during the simulation using a sensor to determine if the parameters are realistic. If the user notices an error in the parameters, then the user can stall the application and inspect other stars. Furthermore, if the user decides to take corrective action, the user can update parameters with a probe write and allow the application to continue, or the user can modify the parameter with an actuator without stalling the application. Users may also execute functions to add new bodies and delete existing bodies to the executing simulation with the aid of Progress.

The utmost concern of Progress' designers was the performance degradation due to instrumentation with the Progress library. Obviously, the user can chose to degrade performance by controlling the program; however, this option remains with the user. Performance of the sample N-body application degraded 4% with the addition of the Progress instrumentation with no interactivity. The same application performance suffered penalties of 26% and 30% for gprof and standard debugging options, respectively. These measurements were gathered on an SGI 8-node multiprocessor with the standard SGI compiler and multiprocessor library. Additional performance measurements are presented in [5].

4. RELATED WORK

Several steering systems exist [4] as well as research in dynamic applications and adaptable systems. Because Progress focuses on interactive systems, we limit our review to systems that allow interactivity. Functionality and generality of these systems vary but they are consistent with Progress' goals of performing interactive program steering. Program directing is investigated in [13]. DYNA3D and AVS (Application Visualization System from AVS Inc.) are combined with customized interactive steering code to produce a time-accurate, unsteady finite-element simulation in [7].

Both VASE [6] and Progress provide a user interface for interacting with a remotely executing application. They both also provide a technique for abstracting uninteresting details from the steering process. VASE, however, concentrates of abstracting blocks of code and control flow, whereas Progress focuses on abstracting important data and operations with steering objects.

5. CONCLUSIONS

Progress is a prototype steering toolkit for the specific purpose of evaluating the necessary components of a general steering system and the essential functionality required by interactive steering. Two improvements to the existing Progress system are essential. First, the object registry system must allow complex user defined types including arrays and structures. Existing technology forces our toolkit to define these user types are runtime, rather than compile time. Second, visualizations of steering objects at runtime are necessary to interpret the massive amounts of information that the user might select [8].

6. REFERENCES

[1] P. Bates. "Debugging Heterogeneous Distributed Systems Using Event-Based Models of Behavior." ACM/ONR Workshop on Parallel and Distributed Debugging (1988), pp. 11-22.

[2] G. Eisenhauer, W. Gu, K. Schwan, and N. Mallavarupu. "Falcon -- toward interactive parallel programs: The on-line steering of a molecular dynamics application." In *Proceedings of The Third International Symposium on High-Performance Distributed Computing*, San Francisco, CA, (August, 1994).

[3] L. Greengard. "The Numerical Solution of the N-Body Problem." *Computers in Physics*, (March/April, 1990), pp. 142-152.

[4] W. Gu, J. Vetter, and K. Schwan. "An Annotated Bibliography of Interactive Program Steering." *ACM SIGPLAN Notices*, 29(9):140-148, (September 1994).

[5] W. Gu, G. Eisenhauer, E. Kraemer, K. Schwan, J. Stasko, J. Vetter, and N. Mallavarupu. "Falcon: On-line Monitoring and Steering of Large-Scale Parallel Programs." *Proceedings of FRONTIERS'95*, (February 1995).

[6] D. Jablonowski, J. Bruner, B. Bliss, and R. Haber. "VASE: The Visualization and Application Steering Environment." *Proceedings of Supercomputing 93*, (November, 1993), pp. 560--569.

[7] D. Kerlick and E. Kirby. "Towards Interactive Steering, Visualization and Animation of Unsteady Finite Element Simulations." *Proceedings of Visualization 93*, (October, 1993).

[8] E. Kraemer, J. T. Stasko. "The Visualization of Parallel Systems: An Overview." *Journal of Parallel and Distributed Computing*, (May, 1993).

[9] T. J. LeBlanc, J. M. Mellor-Crummey. "Debugging Parallel Programs with Instant Replay." *IEEE Transactions on Computers*, C-36(4):471-481, (April 1987).

[10] B. H. McCormick, T. A. DeFanti, M. D. Brown. "Visualization in Scientific Computing." *ACM SIGGRAPH Computer Graphics*, 21(6):, (November, 1988).

[11] B. Mukherjee and K. Schwan. "Improving Performance by Use of Adaptive Objects: Experimentation with a Configurable Multiprocessor Thread Package." *Proc. of Second International Symposium on High Performance Distributed Computing (HPDC-2)*, (July, 1993), pp. 59-66.

[12] D. Ogle, K. Schwan, and R. Snodgrass. "Application Dependent Dynamic Monitoring of Distributed and Parallel Systems." *IEEE Transactions on Parallel and Distributed Systems*, 4(7):762-778, (July, 1993).

[13] R. Sosic. "Dynascope: A Tool for Program Directing." In Proceedings of SIGPLAN'92 Conference on Programming Language Design and Implementation, SIGPLAN Notices, 27(7):12-21, (July, 1992).

[14] J. Vetter, K. Schwan. "Progress: a toolkit for interactive program steering.", College of Computing, Georgia Institute of Technology, Tech Report GIT-CC-95-16, (March, 1995).

An Incremental Parallel Scheduling Approach to Solving Dynamic and Irregular Problems

WEI SHU AND MIN-YOU WU

Department of Computer Science

State University of New York, Buffalo, NY 14260

{shu,wu}@cs.buffalo.edu

Abstract —Global parallel scheduling is a new approach for runtime load balancing. In parallel scheduling, all processors are cooperated together to schedule work. Parallel scheduling accurately balances the load by using global load information. As an alternative strategy to the commonly used dynamic scheduling, it provides a high-quality, low-overhead load balancing. This paper presents a parallel scheduling algorithm for tree structured interconnection networks.

1. Introduction

Application problem structures can be classified into two types: problems with a predictable structure, also called *static problems*, and problems with an unpredictable structure, called *dynamic problems*. There are two basic scheduling strategies: *static scheduling* and *dynamic scheduling*. The static scheduling distributes the work load before runtime, and can be applied to static problems. Most existing static scheduling algorithms are sequential, executed on a single processor system. Dynamic scheduling performs scheduling activities concurrently at runtime, which applies to dynamic problems. Although dynamic scheduling can apply to static problems as well, we usually use static scheduling for static problems because static scheduling provides a more balanced load distribution than dynamic scheduling.

Static scheduling utilizes the knowledge of problem characteristics to reach a global optimal, or nearly-optimal, solution with well-balanced load. It has recently attracted considerable attention among the research community. The quality of scheduling heavily relies on the accuracy of weight estimation. The requirement of large memory space to store the task graph restricts the scalability of static scheduling. In addition, it is not able to balance the load for dynamic problems.

Dynamic scheduling has certain advantages. It is a general approach suitable for a wide range of applications. It can adjust load distribution based on runtime system load information. However, most runtime scheduling algorithms utilize neither the characteristics information of application problems, nor global load information for load balancing decision. Efforts to collect load information for a scheduling decision certainly compete the resource with the underlying computation during runtime. System stability usually sacrifices both quality and quickness of load balancing.

It is possible to design a scheduling strategy that combines the advantages of static and dynamic scheduling. This scheduling strategy should be able to generate well-balanced load without incurring large overhead. With advanced parallel scheduling techniques, this ideal scheduling becomes feasible. In a parallel scheduling, all processors cooperate together to schedule work. Some parallel scheduling algorithms have been introduced in [11, 3, 18]. Parallel scheduling is stable because of its synchronous operation. It uses global load information stored at every processor and is able to accurately balance the load. Parallel scheduling opens a new direction for runtime load balancing. As an alternative strategy to the commonly used dynamic scheduling, it provides a high-quality, low-overhead scheduling.

In this paper, we present a new method, called *Runtime Incremental Parallel Scheduling (RIPS)*. RIPS is a runtime version of global parallel scheduling. In RIPS, the system scheduling activity alternates with the underlying computation work during *runtime*. Tasks are *incrementally* generated and scheduled in *parallel*. The RIPS system paradigm is shown in Figure 1. A RIPS system starts with a system phase which schedules initial tasks. It is followed by a user computation phase to execute the scheduled tasks, and possibly generate new tasks. In the second system phase, the old tasks that have not been executed will be scheduled together with the newly generated tasks. This process will repeat iteratively until the entire computation is completed. Note that we assume the Single Program Multiple Data (SPMD) programming model, therefore, we rely on a uniform code image accessible at each processor. In addition, we assume that the job to be executed is computation-intensive and can be partitioned into medium grain-size tasks to ensure the cost to migrate a task is less than the execution time of the task.

RIPS is a general approach. It can be used for a single job on a dedicated machine or a multiprogramming environment. It can be applied to both shared memory and distributed memory machines. Algorithms for scheduling a single job on a dedicated distributed memory machine is presented in this paper.

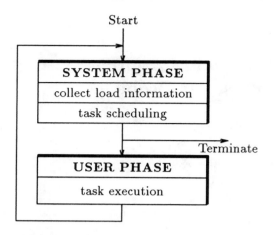

Figure 1: Runtime Incremental Parallel Scheduling.

2. Previous Works

RIPS and the static scheduling share some common ideas [6, 19, 21]. Both of them utilize the systemwise information and perform scheduling globally to achieve high-quality of load balancing. They also clearly separate the time to conduct scheduling and the time to perform computation. But, RIPS is different from the static scheduling in three aspects. First, the scheduling activity is performed at runtime. Therefore, it can deal with the dynamic problems. Second, the possible load imbalance caused by inaccurate grain size estimation can be corrected by the next turn of scheduling. Third, it eliminates the requirement of large memory space to store task graphs, as scheduling is conducted in an incremental fashion. It then leads to a better scalability for massively parallel machines and large size applications.

RIPS is similar to the dynamic scheduling in a certain degree [13, 5, 10]. Both methods schedule tasks at runtime instead of compile-time. Their scheduling decisions, in principle, depend on and adapt to the runtime system information. However, there exist substantial differences, making them appear as two separate categories. First, the system functions and user computation are mixed together in dynamic scheduling, but there is a clear cutoff between system and user phases in RIPS, which potentially offers easy management and low overhead. Second, placement of a task in dynamic scheduling is basically an individual action by a processor, based on partial system information. Whereas in RIPS, the scheduling activity is always an aggregate operation, based on global system information.

In the following, we describe three dynamic scheduling algorithms which will be compared to RIPS. A randomized allocation strategy dictates that each processor, when it generates a new task, should send it to a randomly chosen processor [1, 4, 8]. The major advantages of this strategy are its simplicity and topology independent. No local load information needs to be maintained, nor is any load information sent to other processors. Statistical analysis shows that randomized allocation has a respectable performance. However, a few factors may degrade the performance of the randomized allocation. First, the grain sizes of tasks may vary. Even if each processor processes approximately the same number of tasks, the load on each processor may be uneven. Second, the lack of locality leads to large overhead and communication traffic. Only $1/N$ subtasks stay on the creating processor, where N is the number of processors in the system. Thus most messages between tasks have to cross processor boundaries. The average distance traveled by messages is the same as the average interprocessor distance of the system.

In the gradient model [10], instead of trying to allocate a newly generated task to other processors, the task is queued at the generating processor and waits for some processor to request it. A separate, asynchronous process on each processor is responsible for balancing the load. This process periodically updates the *state* function and *proximity* on each processor. The *state* of a processor is decided by two parameters, the *low water mark* and *high water mark*. If the load is below the *low water mark*, the state is *idle*. If the load is above the *high water mark*, the state is *abundant*. Otherwise, it is *neutral*. The proximity of a processor represents an estimate of the shortest distance to an idle processor. An idle processor has a proximity of zero. For all other processors, the proximity is one more than the smallest proximity among the nearest neighbors. If the calculated proximity is larger than the network diameter, it is in the saturation state and the proximity is set to be *network_diameter+1* to avoid unbounded increase in proximity values. If the calculated proximity is different from the old value, it is broadcast to all the neighbors. Based on the state function and the proximity, this strategy is able to balance the load between processors. When the state is abundant and not in the saturation state, the processor sends a task from its local queue to the neighbor with the least proximity.

The receiver-initiated diffusion (RID) algorithm is a near-neighbor diffusion approach [18]. Load information is exchanged between neighbor processors. Whenever a load L_i increases to $(1/u)L_i$ or drops to uL_i, update messages are sent to its neighbors to update the load information, where u is called the *load update factor*. Load balancing is initiated by any processor whose load drops below a threshold L_{LOW}. Then, the average load of the processor itself and its neighbors, $\bar{L}_i$, is calculated. If a processor's load is below the average load by more than another threshold, $L_{threshold}$, a weight h_k is assigned to each neighbor k:

$$h_k = \begin{cases} L_k - \bar{L}_i, & if \ L_k > \bar{L}_i \\ 0 & otherwise \end{cases} \qquad H_i = \sum_{k=1}^{K} h_k$$

The amount of load requested by processor i from neighbor k is computed as:

$$\delta_k = (\bar{L}_i - L_i)\frac{h_k}{H_i}.$$

Load requests are sent to appropriate neighbors. Upon receipt of a load request, a processor will fillful the request only up to an amount equal to half of its current load.

There is another category of scheduling which carries out computation and scheduling alternately. It is sometimes referred to as *prescheduling* which is more closely related to RIPS. Prescheduling utilizes partial load information for load balancing. Fox *et al.* first adapted prescheduling to application problems with geographical structures [7, 11]. The project PARTI automates prescheduling for nonuniform problems [2]. The dimension exchange method (DEM) is a parallel scheduling algorithm applied to application problems without geographical structure [3]. It balances load for independent tasks with an equal grain size. The method has been extended by Willebeek-LeMair and Reeves [18] so that the algorithm can run incrementally to correct the unbalanced load due to varied task grain sizes. The DEM scheduling algorithm produces redundant communications. It is designed specifically for the hypercube topology and implemented much less efficiently on a simpler topology such as a tree or a mesh [18]. RIPS uses optimal parallel scheduling algorithms, which is also applied to non-geographically structured problems. RIPS minimizes the number of communications, as well as the data movement. Furthermore, RIPS is a general method and applies to different topologies, such as the tree, mesh, and k-ary hypercube. In this paper, we present a basic algorithm for the tree topology.

3. Incremental Parallel Scheduling

RIPS can be presented in its two major components: incremental scheduling and parallel scheduling. The incremental scheduling policy decides when to transfer a user phase to a system phase and which tasks are selected for scheduling. The parallel scheduling algorithm is applied in the system phase to collect system load information and to balance the load.

The transfer policy in RIPS includes two sub-policies: a local policy and a global policy [15]. Each individual processor determines if it is ready to transfer to the next system phase based on its local condition. Then all processors cooperate together to determine the transfer from the user phase to the system phase based on the global condition.

The local policies used in RIPS include *eager scheduling* and *lazy scheduling*. In the eager scheduling, every task must be scheduled before it can be executed. In the lazy scheduling, scheduling is postponed as much as possible so that some tasks could be executed without being scheduled. Two global policies are the *ALL* policy and the *ANY* policy. The ALL policy states that the transfer from a user phase to the next system phase will be initiated only when all the processors satisfy their local conditions. Whereas,

with the ANY policy, as long as one processor has met its local condition, the transfer is initiated. The details of these policies can be found in [15]. The ANY-Lazy policy has shown to be the best of all four combinations. In this paper, we use the ANY-Lazy policy.

Next, we discuss *parallel scheduling*. Parallel scheduling executes the scheduling algorithm *in parallel*. All processors cooperate together to collect load information and to exchange work load in parallel. With parallel scheduling, it is possible to obtain a high-quality scheduling and scalability simultaneously. Furthermore, parallel scheduling is stable because it is a synchronous approach.

Different network topologies need different algorithms. The algorithms for the mesh and hypercube topologies can be found in [20]. Here we present a parallel scheduling algorithm for the tree topology, in which the communication network is tree-structured and each node of the tree represents a processor. The algorithm, called *Tree Walking Algorithm (TWA)*, is shown in Figure 2. The objective of parallel scheduling is to schedule works so that each processor has the same work load. This objective requires an estimation of task execution time. The estimation can be application-specific, leading to a less general approach. Sometimes, such an estimation is difficult to obtain. Therefore, in TWA, each task is presumed to require the equal execution time and the objective of the algorithm becomes to schedule tasks so that each processor has the same number of tasks. Inaccuracy caused by grain-size variation can be corrected in the next system phase.

The other objective of this algorithm is to minimize communication overhead of load balancing. Ideally, the tasks with less communication/computation ratios should have a higher priority to migrate. Although the communication/computation ratio is difficult to predict, it has been observed that the communication/computation ratio does not change substantially in a single application. Thus, we can use the number of task migrated instead of the actual communication cost as the objective function.

In this algorithm, the first step will be executed only at the system setup time, but not in every system phase. Steps 2 and 3 collect the system load information. In step 2, the total number of tasks is counted with a global reduction operation. At the same time, each node records the number of tasks in its subtree and its children's subtrees, if any. In step 3, the root calculates the average number of tasks per processor and then broadcasts the number to every processor so that each processor knows if its subtree is over-loaded or under-loaded in step 4. If the number of tasks cannot be evenly divided by the number of processors, the remaining R tasks are evenly distributed to the first R processors so that they have one more task than others. In step 5, the work load is exchanged so that at the end of the system phase, each processor has the almost same number of tasks. This algorithm is deadlock-free because of the tree topology.

Tree Walking Algorithm (TWA)

1. Assign each node an *order i* according to the pre-order traversal; and N_i, the number of nodes of its subtree, where $i = 0, 1, ..., N - 1$, and $N = N_0$ is the number of nodes in the system. This step executes only at the system setup time.

2. Let w_i be the number of tasks in node i. Perform a *sum* reduction of w_i and let W_i denote the sum of w_i in the subtree of node i. Each node also keeps records of W_i of its children (if any).

3. The root calculates $w_{avg} = \lfloor W_0/N \rfloor$ and $R = W_0 \bmod N$, and broadcasts w_{avg} and R to all other nodes.

4. Each node calculates its quota q_i that indicates how many tasks are to be scheduled to the node:

$$q_i = \begin{cases} w_{avg} + 1 & if \ i < R \\ w_{avg} & if \ i \geq R \end{cases}$$

Each subtree, rooted at node i, calculates its quota

$$Q_i = \sum_{all \ nodes \ in \ subtree \ i} q_j$$

that indicates how many tasks are to be scheduled to the subtree. Q_i can be calculated directly as follows:

$$Q_i = w_{avg} * N_i + r_i$$

where

$$r_i = \begin{cases} 0 & if \ i \geq R \\ N_i & if \ i \leq R - N_i \\ R - i & if \ R - N_i < i < R \end{cases}$$

Each node keeps records of Q_i and Q_j, where node j is node i's child (if any).

5. Each node i waits for the incoming message from its parent if $W_i < Q_i$, and the incoming message from each of its children j if $W_j > Q_j$.

Each node i, after it has received all incoming messages, sends $(W_i - Q_i)$ tasks to its parent if $W_i > Q_i$; and for each child j, if $W_j < Q_j$, sends $(Q_j - W_j)$ tasks to child j.

Figure 2: A Parallel Tree Walking Algorithm

Example 1:

An example is shown in Figure 3. The nodes in the tree are numbered by preorder traversal. At the beginning of scheduling, each node has w_i tasks ready to be scheduled. Values of W_i are calculated at step 2. The root calculates the value of w_{avg} and R:

$$w_{avg} = 4, \ R = 5$$

Then, each node calculates the value of Q_i at step 4. The values of w_i, N_i, W_i, and Q_i are as shown below:

i	w_i	N_i	W_i	Q_i
0	1	9	41	41
1	4	3	20	15
2	5	1	5	5
3	11	1	11	5
4	7	2	9	9
5	2	1	2	4
6	3	3	11	12
7	3	1	3	4
8	5	1	5	4

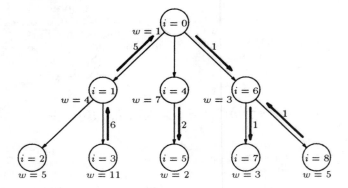

Figure 3: Example for the Tree Walking Algorithm.

The numbers of tasks to be exchanged between nodes are shown in Figure 3. The load exchange takes 4 communication steps to finish:

1) node 3 to node 1, node 8 to node 6, node 4 to node 5
2) node 1 to node 0
3) node 0 to node 6
4) node 6 to node 7

At the end of scheduling, nodes 0–4 have 5 tasks each and nodes 5–8 have 4 tasks each.

In the following, we prove that the algorithm can

- achieve a well-balanced load;
- minimize communication; and
- maximize locality.

Theorem 1: The difference of the number of tasks in each processor is at most one after executing the Tree Walking Algorithm.

Proof: Executing the algorithm, the number of tasks in processor i is

$$w_i' = w_i + (Q_i - W_i) + \sum_{j \text{ is child of } i} (W_j - Q_j).$$

Because

$$\sum_{j \text{ is child of } i} W_j = W_i - w_i,$$

$$w_i' = w_i + (Q_i - W_i) + (W_i - w_i) - \sum_{j \text{ is child of } i} Q_j$$

$$= Q_i - \sum_{j \text{ is child of } i} Q_j = q_i.$$

Since q_i is either w_{avg} or $w_{avg} + 1$, the difference in the number of tasks in each processor is at most one. □

When the total number of tasks is evenly divided by N, each processor has w_{avg} tasks. If the number of tasks cannot be evenly divided by N, each of the first R processors have $(w_{avg} + 1)$ tasks and the rest of processors have w_{avg} tasks. In the following, we assume that the number of tasks T is evenly divided by N, the number of processors. When T is not evenly divided by N, the algorithms are nearly-optimal.

Theorem 2: The Tree Walking Algorithm minimizes the total number of communications and the total number of task-hops $\sum_j e_j$, where e_j is the number of tasks transmitted through edge j.

Proof: In each subtree, if $Q_i \neq W_i$, then there is at least one communication between the subtree and its parent, which is the minimum number of communications. If $Q_i = W_i$, then there is no communication between the subtree and its parent.

In each subtree, if $Q_i \geq W_i$, then it needs to receive from its parent $Q_i - W_i$ tasks, which is the minimum number of tasks to be transmitted to the subtree. Similarly, if $Q_i < W_i$, then it needs to send to its parent $W_i - Q_i$ tasks, which is the minimum number of tasks to be transmitted from the subtree. Therefore, the total number of task-hops $\sum_j e_j$ is minimized. □

This algorithm also maximizes locality. *Local tasks* are the tasks that are not migrated to other processors, and *non-local tasks* are those that are migrated to other processors. Maximum locality implies the maximum number of local tasks and the minimum number of non-local tasks. The following lemma gives the minimum number of non-local tasks.

Lemma 1: To reach a balanced load, the minimum number of non-local tasks is

$$\sum_i \max(w_{avg} - w_i, 0).$$

Proof: Each processor with $w_i < w_{avg}$ must receive $(w_{avg} - w_i)$ tasks from other processors for a balanced load. Therefore, a total of $\sum_i \max(w_{avg} - w_i, 0)$ tasks must be transmitted between processors. □

Another measure of locality is $\sum_k d_k$, where d_k is the distance that non-local task k has traveled. The following theorem shows that the TWA algorithm maximizes locality in both measures.

Theorem 3: The number of non-local tasks in the TWA algorithm is

$$\sum_i \max(w_{avg} - w_i, 0),$$

and the total number of distances $\sum_k d_k$ traveled by non-local tasks is minimized.

Proof: In TWA, each processor receives tasks before sending tasks. At any time when executing the TWA algorithm, the number of tasks in each processor is not less than $\min(w_i, w_{avg})$. Thus, in all processors, at least $\sum_i \min(w_i, w_{avg})$ tasks are local. Therefore, the number of non-local tasks is no more than

$$N \times w_{avg} - \sum_i \min(w_i, w_{avg}) = \sum_i (w_{avg} - \min(w_i, w_{avg}))$$

$$= \sum_i \max(w_{avg} - w_i, 0).$$

As stated in Lemma 1, TWA minimizes the number of non-local tasks.

The total distance traveled by all tasks is equal to the total number of task-hops:

$$\sum_k d_k = \sum_j e_j,$$

which has been proved to be minimum by Theorem 2. □

In this algorithm, steps 2 and 3 spend $2m$ communication steps, where m is the depth of the tree. The communication steps in step 5 is that from a leaf node to another leaf node, which is at most $2m$. Therefore, the total communication steps of this algorithm is at most $4m$. With a balanced tree, $m = logN$, and the number of communication steps of this algorithm is $O(logN)$.

In step 5 of the Tree Walking Algorithm, each node must receive all incoming messages before sending out messages. This algorithm can be further optimized by relaxing this constraint, shown in Figure 4, where only the fifth step in the TWA has been modified and rewritten. We called this algorithm the Modified Tree Walking Algorithm (MTWA) which is shown in Figure 4. In this algorithm, a node is able to send some messages out before it has received all incoming messages. The communication time and processor idle time can be reduced. It takes only two communication steps for Example 1:

1) node 3 to node 1,   node 0 to node 6,
    node 6 to node 7,   node 8 to node 6,
    node 4 to node 5

2) node 1 to node 0

Modified Tree Walking Algorithm (MTWA)

The first four steps are the same as TWA algorithm.

5. For each node i, construct two sets: one is *InMsg* set including all nodes that are expected to send tasks to node i, and the other is *OutMsg* set consisting of all destination nodes to which node i needs to send tasks. The two sets are constructed as follows:

If $W_i < Q_i$, add its parent node to the *InMsg* set; if $W_i > Q_i$, add tuple (*its parent node*, $W_i - Q_i$) into the *OutMsg* set.

For each of its child node j, if $W_j > Q_j$, add its parent node to the *InMsg* set; if $W_j < Q_j$, add tuple (*its parent node*, $Q_j - W_j$) into the *OutMsg* set.

Sort the tuples in the *OutMsg* set according to the number of tasks in an increasing order to form a *OutMsg list*.

While neither the *InMsg* set nor the *OutMsg* list is empty

while there is an incoming message arrived

receive the message and update the *InMsg* set

if there are adequate tasks to meet the first request in the *OutMsg* list

send the message to the destination node and update the *OutMsg* list

Figure 4: Modified Tree Walking Algorithm

The Modified Tree Walking Algorithm may have a negative impact in locality. In TWA, a node can keep the maximum possible number of local tasks and send non-local tasks to other nodes. In MTWA, a node may send local tasks to other nodes and then receive tasks from others. Therefore, the decision on use of TWA or MTWA is a trade-off between scheduling time and locality.

4. Experimental Study

The RIPS system has been implemented on a 32-node TMC CM-5 machine. The network that connects the processors in the CM-5 is a 4-ary *fat tree*. Different from the general tree topology used in the Tree Walking Algorithm, CM-5 has all processors as leaf nodes. A mapping algorithm is used to map the scheduling tree to the fat tree topology [14].

The system has been tested with three application problems. The first one, the exhaustive search of the N-Queens problem, has an irregular and dynamic structure. The number of tasks generated and the computation amount in each task are unpredictable. The second one, iterative deepening A* (IDA*) search is a good example of parallel search techniques [9]. The sample problem is the 15-puzzle with three different configurations. The grain size may vary substantially, since it dynamically depends on the currently estimated cost. Also, synchronization at each iteration reduces the effective parallelism. Performance of this problem is therefore not as good as others. The third one, a molecular dynamics program named GROMOS, is a real application problem [17, 16]. The test data for GROMOS is the bovine superoxide dismutase molecule (SOD), which has 6968 atoms [12]. The cutoff radius is predefined to 8 Å, 12 Å, and 16 Å. GROMOS has a more predictable structure. The number of tasks is known with the given input data, but the computation density in each task varies.

The Modified Tree Walking Algorithm (MTWA) does not improve performance substantially compared to the Tree Walking Algorithm (TWA). While MTWA reduces the communication steps, its locality is not as good as TWA. In general, non-local tasks in MTWA are about 40 percent more than that in TWA. We did not do an extensive comparison since both algorithms perform about the same. Therefore, only the performance with the Tree Walking Algorithm is shown below.

In Table 1, we compare RIPS to three other dynamic load balancing strategies: random allocation, gradient model, and RID. The comparison is done with (1) the number of tasks that are sent to other processors, which is a measure of locality; (2) the overhead time, which includes all system overhead; (3) the idle time, which is a measure of load imbalance; (4) the execution time; and (5) the efficiency. Here, the efficiency is defined as $\mu = \frac{T_s}{T_p * N}$, where N is the number of processors, T_s is the sequential execution time, and T_p is the parallel execution time. The randomized allocation, although its locality is not good, can balance the load fairly well. The gradient model does not show good performance for the N-Queens problem. However, it performs fairly on the less irregular, highly parallel GROMOS program. Generally speaking, it cannot balance the load well, since the load is spread slowly. In addition, the system overhead is large since information and tasks are frequently exchanged. In RID, three parameters, L_{LOW}, $L_{threshold}$, and u are adjusted to 2, 1, and 0.4, respectively. RID shows a better performance than the randomized allocation in most cases. However, it does not perform well for IDA* because of its low parallelism. It is known that a receiver-initiated approach does not do well in a lightly loaded system [5]. When the problem size becomes large, such as the configuration #3, RID's performance is improved. In RIPS, the Tree Walking Algorithm can balance the load very well and the incremental scheduling is able to correct the load imbalance. Many people may expect large overhead

Table 1: Comparison of Four Scheduling Algorithms

		# of tasks	# of non-local tasks	overhead (seconds)	idle time (seconds)	exec. time (seconds)	efficiency
Exhaustive search 13-Queen	Random	7579	7342	0.40	0.08	1.50	68%
	Gradient Model	7579	4255	0.97	0.88	2.87	36%
	RID	7579	2597	0.41	0.10	1.53	67%
	RIPS	7579	314	0.29	0.05	1.36	75%
Exhaustive search 14-Queen	Random	11166	10832	0.62	0.25	7.14	88%
	Gradient Model	11166	6305	1.28	5.45	13.0	48%
	RID	11166	4218	0.57	0.27	7.11	88%
	RIPS	11166	645	0.52	0.08	6.87	91%
Exhaustive search 15-Queen	Random	15941	15459	1.10	1.50	43.4	94%
	Gradient Model	15941	9058	2.10	30.5	73.4	56%
	RID	15941	7103	0.78	1.42	43.0	95%
	RIPS	15941	925	0.97	0.23	42.0	97%
IDA* search config. #1	Random	2895	2804	0.34	0.63	2.31	58%
	Gradient Model	2895	2010	0.85	1.70	3.89	34%
	RID	2895	619	0.95	2.57	4.86	28%
	RIPS	2895	203	0.53	0.13	2.04	66%
IDA* search config. #2	Random	3382	3277	0.65	1.97	9.02	71%
	Gradient Model	3382	2184	2.34	7.25	16.0	40%
	RID	3382	382	3.20	14.8	23.4	27%
	RIPS	3382	257	1.17	0.41	7.98	80%
IDA* search config. #3	Random	29046	28138	1.33	4.65	33.5	82%
	Gradient Model	29046	19285	2.28	22.6	52.3	52%
	RID	29046	3038	2.35	6.83	36.7	75%
	RIPS	29046	1138	1.04	0.52	30.2	91%
GROMOS (8 Å)	Random	4986	4831	0.26	0.56	4.10	80%
	Gradient Model	4986	2311	0.52	0.73	4.53	72%
	RID	4986	528	0.43	0.36	4.07	81%
	RIPS	4986	494	0.60	0.11	3.99	82%
GROMOS (12 Å)	Random	4986	4833	0.25	1.76	11.9	83%
	Gradient Model	4986	2184	0.51	1.40	11.8	84%
	RID	4986	547	1.01	0.90	11.8	84%
	RIPS	4986	556	1.16	0.36	11.4	87%
GROMOS (16 Å)	Random	4986	4832	0.26	3.94	25.0	83%
	Gradient Model	4986	2363	0.54	5.26	26.6	78%
	RID	4986	485	1.79	2.01	24.6	84%
	RIPS	4986	582	0.95	0.65	22.4	93%

from this accurate load balancing algorithm. A surprising observation is that the overhead of RIPS is slightly larger than that of the randomized allocation and much smaller than that of other dynamic scheduling algorithms, such as the gradient model. It is partly due to the fact that many tasks are packed together for transmission, reducing communication overhead, whereas, in dynamic scheduling, tasks are distributed individually.

5. Concluding Remarks

It has been widely believed that a scheduling method that collects load information from all processors in the system is neither practical, nor scalable. This research has demonstrated a scalable scheduling algorithm that uses the global load information to optimize load balancing. At the same time, this algorithm minimizes the number of tasks to be scheduled and the number of communications. Furthermore, in a dynamic system, when it intends to quickly and accurately balance the load, the system could become unstable. In RIPS, a synchronous approach eliminates the stability problem and is able to balance the load quickly and accurately.

RIPS combines the advantages of static scheduling and dynamic scheduling, adapting to dynamic problems and producing high-quality scheduling. It balances the load very well and effectively reduces the processor idle time. Tasks are packed together to be sent to other processors so that the number of messages is significantly reduced. Its overhead is comparable to the low-overhead randomized allocation. It applies to a wide range of applications, from slightly irregular ones to highly irregular ones.

Acknowledgments

We are very grateful to Reinhard Hanxleden for the GROMOS program, Terry Clark for the SOD data, and Marc Feeley for the elegant N-Queens program. This research was partially supported by NSF grant CCR-9109114.

References

[1] W. C. Athas and C. L. Seitz. Multicomputers: Message-passing concurrent computers. *IEEE Computer*, 21(8):9–24, August 1988.

[2] H. Berryman, J. Saltz, and J. Scroggs. Execution time support for adaptive scientific algorithms on distributed memory machines. *Concurrency: Practice and Experience, accepted for publication*, 1991.

[3] G. Cybenko. Dynamic load balancing for distributed memory multiprocessors. *J. of Parallel Distrib. Comput.*, 7:279–301, 1989.

[4] D. L. Eager, E. D. Lazowska, and J. Zahorjan. Adaptive load sharing in homogeneous distributed systems. *IEEE Trans. Software Eng.*, SE-12(5):662–674, May 1986.

[5] D. L. Eager, E. D. Lazowska, and J. Zahorjan. A comparison of receiver-initiated and sender-initiated adaptive load sharing. *Performance Eval.*, 6(1):53–68, March 1986.

[6] H. El-Rewini and T. G. Lewis. Scheduling parallel program tasks onto arbitrary target machines. *Journal of Parallel and Distributed Computing*, June 1990.

[7] G. C. Fox, M. A. Johnson, G. A. Lyzenga, S. W. Otto, J. K. Salmon, and D. W. Walker. *Solving Problems on Concurrent Processors*, volume I. Prentice-Hall, 1988.

[8] R. M. Karp and Y. Zhang. A randomized parallel branch-and-bound procedure. *Journal of ACM*, 40:765–789, 1993.

[9] R. E. Korf. Depth-first iterative-deepening: An optimal admissible tree search. *Artificial Intelligence*, 27(1):97–109, September 1985.

[10] F. C. H. Lin and R. M. Keller. The gradient model load balancing method. *IEEE Trans. Software Engineering*, 13:32–38, January 1987.

[11] J.K. Salmon. Parallel hierarchical N-body methods. Technical report, Tech. Report, CRPC-90-14, Center for Research in Parallel Computing, Caltech, 1990., 1990.

[12] J. Shen and J. A. McCammon. Molecular dynamics simulation of superoxide interacting with superoxide dismutase. *Chemical Physics*, 158:191–198, 1991.

[13] Niranjan G. Shivaratri, Phillip Krieger, and Mukesh Singhal. Load distributing for locally distributed systems. *IEEE Computer*, 25(12):33–44, December 1992.

[14] W. Shu and M. Y. Wu. Runtime Incremental Parallel Scheduling on distributed memory computers. Technical Report 94-25, Dept. of Computer Science, State University of New York at Buffalo, June 1994.

[15] W. Shu and M. Y. Wu. Runtime Incremental Parallel Scheduling (RIPS) for large-scale parallel computers. In *Proceedings of the 5th Symposium on the Frontiers of Massively Parallel Computation*, pages 456–463, 1995.

[16] Reinhard v. Hanxleden and Ken Kennedy. Relaxing SIMD control flow constraints using loop transformations. Technical Report CRPC-TR92207, Center for Research on Parallel Computation, Rice University, April 1992.

[17] W. F. van Gunsteren and H. J. C. Berendsen. GROMOS: GROningen MOlecular Simulation software. Technical report, Laboratory of Physical Chemistry, University of Groningen, Nijenborgh, The Netherlands, 1988.

[18] Marc Willebeek-LeMair and Anthony P. Reeves. Strategies for dynamic load balancing on highly parallel computers. *IEEE Trans. Parallel and Distributed System*, 9(4):979–993, September 1993.

[19] M. Y. Wu and D. D. Gajski. Hypertool: A programming aid for message-passing systems. *IEEE Trans. Parallel and Distributed Systems*, 1(3):330–343, July 1990.

[20] M. Y. Wu and W. Shu. Runtime parallel scheduling algorithms. Technical report, Dept. of Computer Science, State University of New York at Buffalo, January 1995.

[21] T. Yang and A. Gerasoulis. PYRROS: Static task scheduling and code generation for message-passing multiprocessors. *The 6th ACM Int'l Conf. on Supercomputing*, July 1992.

Mapping Iterative Task Graphs on Distributed Memory Machines

Tao Yang   Cong Fu
Dept. of Computer Science
University of California
Santa Barbara, CA 93106
{tyang,cfu}@cs.ucsb.edu

Apostolos Gerasoulis
Dept. of Computer Science
Rutgers University
New Brunswick, NJ 08903
gerasoulis@cs.rutgers.edu

Vivek Sarkar
ADTI, 555 Bailey Ave.
IBM Software Solutions Division
San Jose, CA 95141
vivek_sarkar@vnet.ibm.com

Abstract

This paper addresses the problem of scheduling iterative task graphs on distributed memory architectures with nonzero communication overhead. The proposed algorithm incorporates techniques of software pipelining, graph unfolding and directed acyclic graph scheduling. The goal of optimization is to minimize overall parallel time, which is achieved by balancing processor loads, exploring task parallelism within and across iterations, overlapping communication and computation, and eliminating unnecessary communication. This paper gives a method to execute static schedules, studies the sensitivity of run-time performance when weights are not estimated accurately at compile-time, and presents experimental results to demonstrate the effectiveness of this approach.

1   Introduction

Many scientific applications can be viewed as the repeated execution of a set of computational tasks and can be modeled by iterative task graphs (ITGs). Mapping weighted iterative task graphs on message-passing architectures requires the exploration of both task and loop parallelism. Task parallelism has been addressed in the context of DAG (Directed Acyclic Graphs) scheduling[4, 16, 20, 24, 27]. Using graph scheduling algorithms for iterative computation is not feasible since the number of iterations may be too large or may not even be known at compile-time. Loop scheduling has been studied extensively in the previous work e.g. [6, 17, 21]. Software pipelining [2, 11, 15, 19, 22] is an important technique proposed for instruction-level loop scheduling on VLIW and superscalar architectures. Loop unrolling or graph unfolding techniques [2, 18] have also been developed to allow a compiler to explore more paral-

lelism. Our work has been motivated by the above research but takes into consideration the characteristics of asynchronous parallelism and the impact of communication. We will examine how task graph and loop scheduling techniques can be combined together for mapping iterative task computation on message-passing architectures.

In addition to load balancing, the important optimization needed for distributed memory systems is the elimination of unnecessary communication and overlapping of computation with communication. It is known that communication in such systems suffers high startup cost and *data locality* must be explored. In the task model, exploring data locality means to localize data communication between tasks by assigning them in the same processor [20].

The main focus of our work is to guide task mapping based on the theory of graph scheduling and minimize the performance difference between the proposed solution and an optimal solution. The goal of this paper is to briefly present scheduling algorithms for ITGs and then discuss how the scheduling results can be used in executing an ITG. The proposed algorithm contains the following parts: graph characterization, unfolding, graph transformation and DAG scheduling, construction of the ITG schedule, and schedule optimization. The loop unfolding factor is derived so that the scheduling performance can be guaranteed compared to the optimal solution. To make use of the scheduling result, we also present a method that executes the derived ITG schedule on distributed memory machines. We analyze the performance of run-time execution of static schedules when the estimation of task weights at static-time is not accurate. We conduct several experiments to verify our approach.

This paper is organized as follows. Section 2 gives the problem definition and assumptions. Section 3 gives a brief description of a scheduling algorithm and Section 4 provides the properties of this algorithm.

Section 5 presents the schedule executing scheme and examines the run-time performance of static schedules. Section 6 presents the experimental results.

2 Definitions and Backgrounds

Formally, an ITG G contains v tasks $T_1, T_2, \cdots, T_v$, and e data dependence edges. This graph is repeatedly executed for N iterations. Let the instance of task T_i at iteration k be T_i^k ($1 \leq k \leq N$). Let (T_i, T_j) be a dependence edge from T_i to T_j in G, labeled $d_{i,j}$. $d_{i,j}$ is a non-negative integer called dependence distance. Thus task instance $T_j^{k+d_{i,j}}$ depends on T_i^k and it cannot start to execute until T_i^k has completed its computation and the data produced by T_i^k is available at the local memory. Control dependence is not considered in this paper. Let $Children(T_x, G)$ be a set of tasks in G that depend on task T_x. Let $Parent(T_x, G)$ be a set of tasks in G that T_x depends. The expanded graph $E(G, N)$ contains all $N * v$ task instances T_i^k ($1 \leq i \leq v$, $1 \leq k \leq N$), and these tasks and their dependencies in $E(G, N)$ constitute a DAG.

Each task T_x in an ITG has a computation cost τ_x and there is a communication cost $c_{x,y}$ for sending a message from task T_x in one processor to task T_y in another processor.

The scheduling problem for an ITG is to assign the instances of tasks to the given p processors and determine the execution order of tasks within each processor. The parallel time is the completion time of the last task instance and the goal of scheduling is to minimize the parallel time. Fig. 1(a) shows an ITG where edges are labeled with dependence distances. Fig. 1(b) shows another ITG. For these two examples, all task weights are 20 and all edge weights are 10 between any two processors. The expanded graph for Fig. 1(b) is in (c). A schedule for this graph when $N = 3$ is depicted in Fig. 1(d). The parallel time is 290.

Let f be the unfolding factor. When a graph is unfolded f times, the number of tasks increases by a factor of f and the number of iterations for the new ITG is $\lfloor N/f \rfloor$. For example, Fig. 1(b) is an ITG obtained by unfolding the ITG of Fig. 1(a) by a factor of 2. The number of tasks is doubled from (a) to (b). If an ITG is unfolded N times, then the resulting task graph is the expanded graph of G, $E(G, N)$.

In general, the scheduling problem for ITGs is NP-hard. There could be several approaches for addressing this problem. 1) Applying DAG algorithms for an expanded graph has a high complexity and it is not feasible if N is unknown. 2) Applying a DAG algorithm for one iteration can only exploit parallelism within one iteration but not across iterations. 3) The

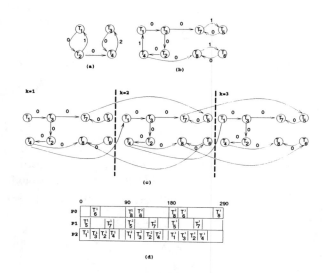

Figure 1: (a) An ITG. (b) Another ITG. (c) The expanded graph of (b) when $N = 3$. (d) A schedule for (b).

idea of exploring parallelism within and across iterations has been proposed for instruction-level software pipelining [2, 15]. In [11], a near-optimal scheduling algorithm for pipelining with no communication delay is proposed. We need to extend this result to incorporate communication optimization with load balancing since communication is a major overhead in a message-passing machine. The optimal solution for a special class of ITGs has been studied in [5]. Graph unfolding techniques [2, 18] can increase the number of tasks within each iteration and thus more parallelism could be explored. Our goal is to only examine tasks from few iterations (limited by the unfolding factor f) but explore parallelism of the entire iteration space. We need to derive a small unfolding factor for general graphs so that the complexity of scheduling is not too high. [9] presented an approach to optimal unfolding for a special case.

The problem of scheduling with communication is much harder than that without communication [20]. Our algorithm needs to eliminate unnecessary communication and overlap communication with computation. The above communication optimization should be consistent with the goal of load balancing. We assume that communication between two tasks is zero if these two tasks are mapped to the same processor. Thus placing two tasks together in one processor to take advantage of data locality could eliminate some communication overhead although it might reduce parallelism. A trade-off between parallelization and communication locality must be addressed.

We will use the concept of integer-periodic scheduling [19] as in instruction-level software pipelining and structure a schedule as follows: all instances of task T_i are assigned to the same processor. If task T_i^k is executed at time t, then task T_i^{k+1} is executed after T_i^k at time $t + \beta$ where β is the iteration interval. In order to execute a schedule on a parallel machine, we need to construct a code that follows the scheduling result. It can be seen from Fig. 1(d) that the execution of each processor possesses certain local periodic patterns. In general, the local schedule of a processor contains three parts: prelude, iterative execution based on periodic pattern and postlude. For example, Processor 0 has a periodic pattern T_6, T_8 at interval $[20, 110)$. It will repeat at $[110, 200)$ and so on. There is no prelude and postlude. However the periodic pattern is not unique. Another pattern is T_8, T_6 at $[90, 180)$. The prelude is T_6^1 and the postlude is T_8^3. Notice that in instruction-level software pipelining [2], a *global* periodic pattern (a pattern which starts at the same time for all processors) is computed and this is not feasible in message-passing machines with asynchronous communication. In our approach, a local periodic pattern is produced. We will also uniformly express the prelude and postlude within a periodic pattern to avoid explicitly enumerating and sorting tasks in the prelude and postlude. The generated code is MPMD (multiple-program multiple-data) in nature, since each processor has a different local schedule. However, in practice, it is convenient to use a single SPMD program for all processors, and to let the program determine which local schedule should be used based on the processor identification.

In order to derive a schedule which is near optimal, we conduct a comparison of our schedule with an optimal solution with the minimum time for $E(G, N)$. We assume that N is very large and our analysis measures the asymptotic performance of scheduling. We also assume that the given graph always has a cycle. The acyclic case is discussed in [28].

3 The Scheduling Algorithm

The algorithm for mapping an ITG on p processors contains the six steps.

3.1 Computing performance parameters

Smallest iteration interval. We first estimate the maximum performance this task graph could achieve. The most important factor is

$$\beta^*(G) = \max_{\text{all cycles C in } G} \frac{\tau(C)}{d(C)}.$$

where $d(C)$ is the summation of edge distances in this cycle C and $\tau(C)$ is the summation of task weights in this cycle. This is the well-known optimal rate (the smallest iteration interval) of pipelining when communication cost is zero and there is a sufficient number of processors [19]. To compute this value, we set up the following inequalities for all edges (T_x, T_y) in G: $\alpha_y - \alpha_x + d_{x,y}\beta \geq \tau_x$ where α_y is a non-negative unknown associated with task T_y and β is an approximation of $\beta^*(G)$. These inequalities can be solved in a complexity of $O(\sqrt{ve} \log^2 \frac{Seq(G)}{\epsilon})$ using the shortest path algorithm [10] where constant ϵ is the desired accuracy in finding the minimum β such that $\beta^*(G) \leq \beta \leq \beta^*(G) + \epsilon$. $Seq(G) = \sum_{i=1}^{v} \tau_i$. For Fig.1(a), ϵ is chosen as 0.5 and β is computed as 40.

Granularity. Next we estimate the granularity of the graph as follows. This value will be useful in computing the performance bound of the produced schedule and the unfolding factor.

$$g(G) = \min_{T_x \in G} \{ \min_{T_y \in Parents(T_x, G)} \frac{\tau_x}{c_{x,y}} \}.$$

For example, for Fig.1(a), $g(G) = 2$.

Unfolding factor. Then we compute the unfolding factor f. The analysis in Section 4 shows that such a factor will guarantee that the asymptotic performance of this scheduling is competitive to an optimal solution. The competitive ratio is: $B = S1(1 + \frac{\epsilon}{\max(\beta - \epsilon, Seq(G)/p)}) + 1$. where $S1 = 1 - 1/p + 1/g(G)$. The unfolding factor f is $\lceil \min(L1, L2) \rceil$ where

$$L1 = \frac{S1(\tau^{max} + \epsilon) + \tau^{max}/g(G)}{(B-1)Seq(G)/p - S1\beta},$$

$$L2 = \frac{S1(\tau^{max} + \epsilon) + \tau^{max}/g(G)}{B(\beta - \epsilon) - S1\beta - Seq(G)/p},$$

and τ^{max} is the maximum task weight in G. For example, for Fig.1(a), $B = 2.68$ and $f = 2$.

3.2 Unfolding the ITG

In this step, the graph is unfolded by a factor of f. This graph is called G^f. The new graph needs to be executed only $\lfloor N/f \rfloor$ iterations. We use the unfolding algorithm proposed in [18] to construct G^f.

For example, unfolding the ITG in Fig. 1(a) with $f = 2$ results in an ITG shown in Fig. 1(b). Notice the unfolded graph has the following properties [18]: $\beta^*(G^f) = f\beta^*(G)$.

3.3 Graph transformation

We transform G^f into a DAG so that the DAG scheduling technique could be applied. Define

$DIV(x, y)$ as the largest integer that is less than or equal to x/y, i.e. $\lfloor x/y \rfloor$. Define $MOD(x, y) = x - DIV(x, y) * y$. Let the tasks in G^f be renumbered as $T_1, T_2, \cdots, T_{fv}$. We first compute the minimum integer values for β_f and α_i ($1 \leq i \leq fv$) based on the following inequalities for each dependence edge (T_i, T_j) in G^f: $\alpha_j - \alpha_i + \beta_f d_{i,j} \geq \tau_i$. where α_j is a non-negative unknown associated with task T_j in G^f and β_f is an approximation of $\beta^*(G^f)$. Since we know that $\beta^*(G^f) = f\beta^*(G)$, it can be shown that $f\beta - f\epsilon \leq \beta^*(G^f) \leq f\beta$. Thus the value β_f can be easily found in this range using the binary searching and the shortest path algorithm. The result satisfies $\beta^*(G) \leq \beta_f \leq \beta^*(G) + \epsilon$.

For example, edge (T_4, T_1) in Fig. 1(b) corresponds to the inequality $\alpha_1 - \alpha_4 + 1\beta_f \geq 20$. Based on inequalities of all edges, we can produce a solution $\beta_f = 80$ with $\epsilon = 0.5$, $\alpha_1 = 0$, $\alpha_2 = 40$, $\alpha_3 = 20$, $\alpha_4 = 60$. $\alpha_5 = 0$, $\alpha_6 = 20$, $\alpha_7 = 40$, $\alpha_8 = 80$.

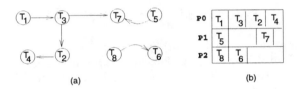

(a)   (b)

Figure 2: (a) The kernel DAG of Fig. 1(b). (b) The schedule for this DAG.

Then we transform the dependence graph G by deleting all edges (T_i, T_j) in the graph G which satisfy: $MOD(\alpha_j, \beta) - MOD(\alpha_i, \beta) < \tau_i$. For example, edge (T_4, T_1) of Fig. 1(b) is deleted because its edge weight satisfies $MOD(\alpha_4, \beta) - MOD(\alpha_1, \beta) < \tau_1$. This transformation was first proposed in [11] for mapping graphs when communication is zero. We can show such a transformation is still valid for our case by carefully designing the ITG schedule in Section 3.5. The resulting graph after edge deletions is a DAG. We call the transformed graph the kernel DAG $K(G^f)$. For example, the kernel DAG of Fig. 1(b) is in Fig. 2(a).

3.4 Mapping the kernel graph

In this stage, a DAG schedule is derived for $K(G^f)$. The starting time γ_i and the processor assignment $Proc(i)$ of each task T_i are provided. In mapping the kernel DAG, we use two algorithms: one is a multi-stage algorithm designed for the PYRROS system [27], another is a one-stage algorithm. We will choose the smaller one between two solutions produced by these algorithms. The result of the DAG scheduling is used for constructing the ITG schedule in Section 3.5. Both

algorithms have a complexity $O((v + e) \log v)^1$.

Our multi-stage approach contains the following mapping operations: 1) Assign tasks to a set of clusters using the DSC Algorithm [25]. The goal of clustering is to identify communication-intensive tasks and Tasks in the same cluster will be executed in the same processor. 2) If the number of clusters is larger than p, we use a simple load balancing heuristic to balance the processor load. 3) We further order the execution of tasks within each processor using the RCP algorithm [26] with a goal of overlapping communication with computation to hide communication latency. [12] conducted a comparison of this approach with a higher complexity ETF method[14] and found that this approach has a much lower complexity but a competitive performance.

The one-stage approach uses the idea of the DSC algorithm [25] but limits the number of clusters to p. The detail is in [28]. We apply this algorithm to the given DAG and also the reversed graph. This algorithm can obtain the optimum for fork, join, coarse grained tree DAGs if there is a sufficient number of processors and has the following bound.

Theorem 1 *Given a DAG R and p processors, the parallel time produced by this algorithm is bounded by*

$$PT(R) \leq (1 - 1/p + 1/g(R))CP(R) + Seq(R)/p$$

where $CP(R)$ is the length of the critical path in this graph R.

The ETF algorithm [14] has a similar performance bound with a complexity $O(pv^2)$ while our algorithm has a lower complexity $O((v + e) \log v)$.

3.5 Constructing an ITG schedule

The DAG scheduling produces processor assignment $Proc(i)$ and the starting time γ_i for each task T_i in $K(G^f)$. Next we construct a schedule for the original ITG based on the DAG scheduling result. The starting time of a task T_i at iteration k is modeled as $ST(T_i^k) = \alpha_{p,i} + \beta_p(k - 1)$ where $\alpha_{p,i}$ is called startup delay and β_p is the length of the iteration interval. Notice that for a task T_i, T_i^1 is executed at time $\alpha_{p,i}$, T_i^2 is executed at time $\alpha_{p,i} + \beta_p$ and so on. The startup delay of task T_i in the ITG schedule is $\alpha_{p,i} = \gamma_i + \beta_p DIV(\alpha_i, \beta_f)$ and $\beta_p = max(PT(K(G^f)), D)$ where D is

$$\max_{(T_x, T_y) \in G^f - K(G^f)} \left\{ \frac{\gamma_x + \tau_x - \gamma_y + c_{x,y}}{DIV(\alpha_y, \beta_f) - DIV(\alpha_x, \beta_f) + d_{x,y}} \right\}.$$

---

[1]The size of the graph increases by f after unfolding. However in practice, we find that f is small, thus f is not included in the complexity term.

The processor assignment of instances of T_i in this ITG schedule is the same as $Proc(i)$ derived by the DAG scheduling algorithm.

For the kernel DAG of Fig. 2(a) and $p = 3$, the DAG scheduling produces the following solution as shown in Fig. 2(b): $Proc(T_6) = Proc(T_8) = 0$. $Proc(T_5) = Proc(T_7) = 1$, $Proc(T_1) = Proc(T_2) = Proc(T_3) = Proc(T_4) = 2$, And $\gamma_1 = \gamma_5 = \gamma_8 = 0$, $\gamma_3 = \gamma_6 = 20$, $\gamma_2 = 40$, $\gamma_7 = 50$, and $\gamma_4 = 60$. $D = 90$ and $PT(K(G^j)) = 80$, thus $\beta_p = 90$. The startup delays for $T_1, T_2, \cdots, T_8$ are 0, 40, 20, 60, 0, 20, 50, 90 respectively.

3.6 Optimization for communication contention

So far we have not considered the processor distance. It has been shown (e.g. [7]) in the modern processor architectures, the distance factor does not affect the cost of communication significantly because of wormhole routing technology. However, this routing scheme requires the exclusive use of channels, and there exists communication contention if network traffic is heavy. Assigning communication-intensive tasks close to each other in terms of processor distances would reduce the chance of contention. In this step, we incorporate the processor distance factor in estimating communication overhead and perform a series of pairwise interchanges to adjust the ITG schedule. The complexity is $O(p^2(v \log v + e))$. A similar strategy is used in PYRROS [27]. It should be noted that communication contention exists in other routing schemes. Modeling contention is hard and more investigation is needed.

4 Analysis

The overall time complexity of this algorithm is $O(\sqrt{v}e \log^2 \frac{Seq(G)}{\epsilon} + p^2(v \log v + e))$. We can show that the derived schedule satisfies all dependence and resource constraints. We also conduct an analysis to compute the unfolding factor and show that this factor gives a guarantee on the asymptotic performance bound of this heuristic algorithm.

Theorem 2 *Let the unfolding factor f be chosen as:* $f = \lceil \min(L1, L2) \rceil$ *where*

$$L1 = \frac{S1(\tau^{max} + \epsilon) + \tau^{max}/g(G)}{(B-1)Seq(G)/p - S1\beta} \quad and$$

$$L2 = \frac{S1(\tau^{max} + \epsilon) + \tau^{max}/g(G)}{B(\beta - \epsilon) - S1\beta - Seq(G)/p}.$$

Let $\epsilon \neq \beta - Seq/p$. Then

$$\lim_{N \to \infty} \frac{PT(G)}{PT_{opt}(G)} \leq B$$

where $B = S1(1 + \frac{\epsilon}{max(\beta - \epsilon, Seq(G)/p)}) + 1$ and $S1 = 1 + 1/g(G) - 1/p$. $PT(G)$ is the solution produced by our algorithm, and $PT_{opt}(G)$ is the optimal solution.

This theorem indicates that since ϵ could be chosen small, the bound is about $S1 + 1 = (1 - 1/p + 1/g(G)) + 1$. When $g(G)$ is not too small, this bound is tight. Notice that our granularity estimation may not be effective when some tasks perform very small computation but receive large messages from others. In practice, the size of data a task communicates is proportional to the amount of computation work it does, we expect our algorithm performs well and the bound is relatively tight. We will verify this in our experiments. The theorem also indicates that program partitioning that produces ITGs should make $g(G)$ not too small. This is consistent to the previous results [13].

5 Run-time execution and sensitivity analysis

5.1 A schedule executing method

We assume that the target is a message-passing parallel machine, each processor has its own private memory and a communication buffer to hold the outgoing and incoming messages. In Section 3, we have derived the starting function value for each task instance. If we execute tasks by sorting the starting time of these tasks at run-time, then the cost of sorting contributes a significant amount of overhead in the overall performance. Our approach is based on the idea of "periodic execution" in software pipelining. The difference is that in instruction-level pipelining, periodic pattern is global usually [2] while in a distributed memory machine, we use local period patterns to utilize asynchronous parallelism. Our code does not have separate prelude and postlude parts, and in this way the code length can be shortened.

The starting time of task instances depends on the iteration number k. We use the following method to find the periodic pattern for processor j from the coefficients in starting time function $ST(T_x) = \alpha_{i,p} + \beta_p(k - 1)$. Let $Tasks(j)$ be all tasks assigned to processor j. Let $L_j = \max_{T_i \in Tasks(j)} \alpha_{p,i}$. We can show that $[L_j, L_j + \beta_p)$ contains the periodic pattern of Processor j. Let $o_i = \lceil \frac{L_j - \alpha_{p,i}}{\beta_p} \rceil + 1$. We can show that o_i is the iteration sequence number of task T_i at $[L_j, L_j + \beta_p)$. The tasks in this interval are executed in the increasing order of value $\alpha_{p,i} + \beta_p(o_i - 1)$. In Fig. 1, processor 0 has two tasks T_6 and T_8. $max(\alpha_{p,6}, \alpha_{p,8}) = max(20, 90) = 90$. A local periodic pattern starts at $[90, 180)$. The iteration sequence numbers for T_6 and T_8 are 2 and 1. Thus the ordering is T_8, T_6. We show the correctness of this method as follows.

Theorem 3 $[L_j, L_j + \beta)$ *is the initial interval of a local periodic pattern for processor j where $L_j = \max_{T_i \in Tasks(j)} \alpha_{i,p}$. Within interval $[L_j, Lj + \beta_p)$, each processor j executes one and only one instance of a task T_i in $Tasks(j)$ and the iteration number is $o_i = \lceil \frac{L_j - \alpha_{i,p}}{\beta_p} \rceil + 1$.*

Based on the above theorem, we give the code structure for executing the schedule at processor j in Fig. 3. When executing each individual task, the task receives the data items it needs, performs computation, and then sends out data produced to its children (using aggregate multicasting if possible). The style is similar to the one used in the PYRROS system [27].

Let $o^{max} = \max_{T_x \in Tasks(j)} o_x$.
Let $o^{min} = \min_{T_x \in Tasks(j)} o_x$.
For $r = -o^{max}$ **to** $N - o^{min}$
    Execute tasks T_i^k in an order of $\alpha_{i,p} + \beta_p(o_i - 1)$
where $k = r + o_i$; If $k > N$ or $k < 1$, skip this task.
endfor

Figure 3: A generic form of code for Processor P_j.

5.2 An analysis of run-time performance

Code in Fig.3 executes the given ITG G N times using the derived periodic pattern. When it executes a task, it starts task computation as soon as this task is ready. Thus if run-time weights are the same as the static time weights, it is possible that the code may execute tasks in an iteration interval smaller than computed β_p. We call the periodic schedule used in Fig.3 to be S, and define $PT(G)$ to be the parallel time estimated by our algorithm using parameter β_p. Let $PT(S, G)$ be the parallel time produced by the code in Fig.3. Then $PT(S, G) \leq PT(G)$.

At run-time, the computation time of each task and communication delay between tasks may not be the same as the one predicted at compile-time. The following parameters affect the performance of a static schedule in a run-time environment: 1) Variation of task computation weight, due to inaccurate estimation of static weight and the cost of data copying. 2)

Variation of communication delay, due to inaccurate estimation of static communication weight and unexpected run-time network contention.

We address the sensitivity of the scheduling performance on weight variation. Let G be the graph for which the scheduling algorithm produces a schedule S. The weights of this graph may change at run-time and we call the new graph as G^r. The run-time method in Fig. 3 uses the mapping and task ordering of S to execute graph G^r, resulting in a schedule S^r in a length of $PT(S, G^r)$. Assume that $PT_{opt}(G^r)$ is the optimal schedule length for G^r. The interesting question is: what is the competitive ratio of $PT(S, G^r)$ over $PT_{opt}(G^r)$.

We assume that at run-time, each communication weight $c_{i,j}$ is changed to $c_{i,j}^r$ and the computation weight τ_i is changed to τ_i^r. They satisfy: $(1-\delta_1)c_{i,j} \leq c_{i,j}^r \leq (1 + \delta_2)c_{i,j}, (1 - \delta_1)\tau_i \leq \tau_i^r \leq (1 + \delta_2)\tau_i$. We have shown in Theorem 2 that the derived schedule satisfies: $PT(G) \leq B \times PT_{opt}(G)$ where $B = (1 - 1/p + 1/g(G))(1 + \frac{\epsilon}{max(\beta-\epsilon, Seq(G)/p)}) + 1$. The following theorem identifies the relationship between $PT(S, G^r)$ and $PT_{opt}(G^r)$. It shows that if the run-time variation of weights is not significant, a static schedule still produces a competitive performance.

Theorem 4

$$\lim_{N \to \infty} \frac{PT(S, G^r)}{PT_{opt}(G^r)} \leq \frac{1 + \delta_2}{1 - \delta_1} \times B.$$

6 Experiments

We have implemented our scheduling scheme and have conducted several experiments to examine the scheduling performance of randomly generated ITGs, and application programs including matrix multiplication, solving a Laplace equation and a banded matrix system, and the SOR method for sparse matrix systems. We have also hand-coded some of schedules using our executing scheme in nCUBE-2 to verify the correctness of the simulation. In this section we report some experimental results.

Overall performance and its sensitivity. The left part of Fig. 4 lists the simulated and actual speedups on nCUBE-2 for solving an Laplace PDE (400x400 grid) and a banded matrix system (matrix dimension 4000, width 35). In the simulation, we compute the asymptotic speedup of an ITG schedule as: $\lim_{N \to \infty} \frac{N * Seq(G)}{max_{T_i \in G} ST(T_i^N)} = Seq(G)/\beta_p$. We can see that the simulated performance is close to the actual performance for the studied programs. The right part of Fig. 4 shows the performance variation after changing the weights of PDE and banded matrix ITGs in a range of $[-10\%, +10\%]$. The variation ratio is $\frac{Speedup_{new}}{Speedup_{old}} - 1$. We can see the performance is sensitive but variation is small, within 5%.

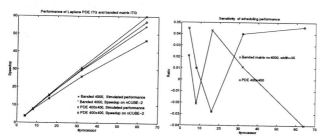

Figure 4: The left is speedups of solving a Laplace equation and a banded matrix system. The right is the performance variation ratio using the inaccurate and accurate weights.

Unfolding and performance bounds. We examine if the computed factor identifies the range of unfolding in improving performance for three groups of randomly generated ITGs: coarse grained graphs with $g(G) \geq 1$; fine grained graphs where $0.4 < g(G) < 1$ and the local granularity (local computation weights vs. communication weights) is also in this range; graphs with mixed local grain values in a range $[0.1,1]$. The number of tasks in each ITG varies from 30 to 60. The results for $p = 32$ are presented in Fig. 5. The X axis is the unfolding percentage which is calculated with respect to the unfolding factor obtained by our scheduling algorithm. Assume that f is the calculated factor. 50% unfolding percentage means that $0.5f$ is used. The Y axis is the speedup improvement ratio after unfolding: $\frac{Speedup_{new}}{Speedup_{old}} - 1$. We can see that unfolding greatly improves the performance. This is because unfolding produces more tasks and gives more flexibility for the DAG scheduling.

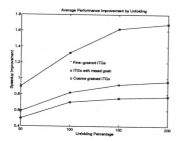

Figure 5: Performance improvement by unfolding.

We have also examined the actual bounds and the estimated upper bounds after unfolding for above

tested cases. The actual bounds are no more than 1.8. We find that if the granularity value is large, our bound formula is very tight. When granularity value is small (say 0.1), the estimated bound is loose. But the actual upper bound is small, which indicates that our unfolding factor effectively guides the algorithm to obtain a near optimal schedule. The unfolding factors computed are small (4-8) in general except for three cases (30-50). We studied those cases and found that B bound is set to be too small. If we increase B by 1, then the average unfolding factor is reduced to 6.

Software pipelining and DAG scheduling. We have discussed that parallelism is not fully utilized if using the DAG scheduling for one iteration without pipelining iterations (called simple DAG scheduling approach). It should be noted that after unfolding an ITG, the simple DAG scheduling approach could also explore parallelism between iterations *for the original ITG* but *not for the new ITG*. We have examined the 50 random ITGs and compared the performance of this simple DAG scheduling approach with our algorithm. Before unfolding, our approach can achieve as high as 50% performance improvement due to the exploration of the across-iteration parallelism especially when the number of processors is large. After unfolding, both approaches have gained more flexibility. But our approach is still a lot better. The left part of Fig 6 shows the average performance improvements of our approach over this simple DAG scheduling approach after unfolding by a factor of 2.

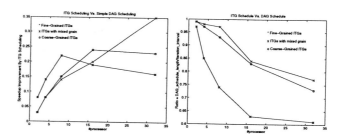

Figure 6: The left is the average improvement ratio by software pipelining after the graphs are unfolded by a factor of 2. The right shows the role of DAG scheduling in our algorithm.

Our pipelining algorithm still uses the DAG scheduling for kernel graphs and the performance of DAG scheduling is critical for constructing the final ITG schedule. Notice that $\beta_p = max(D, PT(K(G^f)))$. The right part of Fig. 6 shows the average ratios $PT(K(G^f))/\beta_p$ for three groups of these randomly generated graphs. It shows that when p is small, each processor has a heavy load and the DAG schedule dominates the ITG performance. When p is large

($p = 32$), the load per processor becomes smaller and the role of communication comes to play. But still the DAG scheduling contributes more than 60% to the final performance.

Sparse matrix computation. In solving a sparse matrix system, the SOR iterative methods could be used [1]. The left part of Fig. 7, shows an ITG based on the SOR method for a sparse matrix, assuming that the convergence test is not conducted in every iteration but every few hundreds of iterations instead. Some of the edges in this graph are marked distance 1. The unmarked edges have distance 0. The right part of Fig.7 is the simulated scheduling performance for sparse matrices in the Harwell-Boeing Test Suites [8]. We use matrix BCSSTK14 arising from structural analysis for the roof of Omni Coliseum at Atlanta and matrix BCSSTK15 for Module of an offshore platform. While parallelism in a sparse matrix computation is limited, this algorithm is able to explore a decent amount of parallelism.

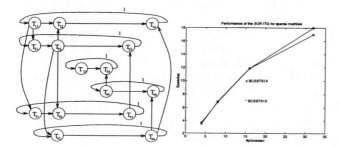

Figure 7: The left is a SOR ITG for a sparse matrix. The right is performance of the SOR ITG for BCSSTK14 and BCSSTK15.

7 Conclusions

Our experiments show that the automatic scheduling algorithm for ITGs delivers good performance on message-passing architectures. Our work is useful for assisting performance prediction in compilation optimization such as program partitioning [3, 17, 21]. Currently we are implementing code generation and runtime support for executing ITG schedules and investigating applications in sparse matrix computations.

Acknowledgement

This was supported in part by NSF RIA CCR-9409695 and by ARPA contract DABT-63-93-C-0064. We thank Pedro Diniz for his help in algorithm implementation, and the referees for their useful comments.

References

[1] L. Adams, and H. Jordan, Is SOR color-blind?, *SIAM J. Sci. Stat. Comp*, 7 (1986), pp 490-506.

[2] A. Aiken and A. Nicolau, Optimal Loop Parallelization, *SIGPLAN 88 Conf. on Programming Language Design and Implementation*. pp.308-317.

[3] L. Carter, J. Ferrante and S. F. Hummel, Efficient parallelism via hierarchical tiling, *Proc. of 7th SIAM Conf. on Parallel Processing for Scientific Computing*, Feb. 1995, 680-685.

[4] Y-C Chung and S. Ranka, Applications and performance analysis of a compile-time optimization application for list scheduling algorithms on distributed memory multiprocessors, *Supercomputing 92*, pp. 512-521.

[5] P. Chretienne, Cyclic scheduling with communication delays: a polynomial special case. Dec 1993. Tech Report, LITP.

[6] R. Cytron, Doacross: Beyond vectorization for multiprocessors, *Proc. of the 1986 Inter. Conf. on Parallel Processing*, St. Charles, Ill,1986, pp. 836-844.

[7] T. H. Dunigan, Performance of the INTEL iPSC/860 and nCUBE 6400 Hypercube, ORNL/TM-11790, Oak Ridge National Lab., TN, 1991.

[8] I. S. Duff, R. G. Grimes and J. G. Lewis, Users' Guide for the Harwell-Boeing Sparse Matrix Collection, TR-PA-92-86.

[9] H. El-Rewini, T. G. Lewis and H. H. Ali, Task Scheduling in Parallel and Distributed Systems, Prentice Hall, 1994.

[10] H. Gabow and R. Tarjan, Faster scaling algorithms for network problems, *SIAM J. Computing*, Oct 1989.

[11] F. Gasperoni and U. Schweigelshohn Scheduling Loops on Parallel Processors: A simple algorithm with close to optimum performance. *Proc. of CONPAR 92*, pp. 613-624.

[12] A. Gerasoulis, J. Jiao, and T. Yang, A multistage approach to scheduling task graphs. To appear in DIMACS Book Series on Parallel Processing of Discrete Optimization Problems. AMS publisher. Edited by P.M. Pardalos, K.G. Ramakrishnan, and M.G.C. Resende.

[13] A. Gerasoulis and T. Yang, On the Granularity and Clustering of Directed Acyclic Task Graphs, *IEEE Trans. on Parallel and Distributed Systems.*, Vol. 4, no. 6, June 1993, pp 686-701.

[14] J. J. Hwang, Y. C. Chow, F. D. Anger, and C. Y. Lee, Scheduling precedence graphs in systems with interprocessor communication times, *SIAM J. Comput.*, pp. 244-257, 1989.

[15] M. Lam, Software pipelining: an effective scheduling technique for VLIW machines, *ACM Conf. on Programming Language Design and Implementation*, 1988, 318-328.

[16] C. McCreary and H. Gill, Automatic Determination of Grain Size for Efficient Parallel Processing, *Communications of ACM*, Sept. 1989, 1073-1078.

[17] C. D. Polychronopoulos, *Parallel Programming and Compilers*, Kluwer Academic, 1988.

[18] K. K. Parhi and D. G. Messerschmitt, Static rate-optimal scheduling of iterative dataflow programs via optimum unfolding, *IEEE Trans. on Computers*, 40:2, 1991, pp. 178-195.

[19] R. Reiter, Scheduling parallel computations, *Journal of ACM*, Oct 1968, pp. 590-599.

[20] V. Sarkar, *Partitioning and Scheduling Parallel Programs for Execution on Multiprocessors*, MIT Press, 1989.

[21] V. Sarkar and R. Thekkath, A general framework for iteration-reordering loop transformations, *ACM Conf. on Programming Language Design and Implementation*, 1992, 175-187.

[22] V. H. Van Dongen, G. R. Gao and Q. Ning A polynomial time method for optimal software pipelining. *Proc. of CONPAR 92*, pp. 613-624.

[23] R. Wolski and J. Feo, Program Partitioning for NUMA Multiprocessor Computer Systems, *J. of Parallel and Distributed Computing*, 1993.

[24] M. Y. Wu and D. Gajski, Hypertool: A programming aid for message-passing systems, *IEEE Trans. on Parallel and Distributed Systems*, vol. 1, no. 3, pp.330-343, 1990.

[25] T. Yang and A. Gerasoulis. DSC: Scheduling parallel tasks on an unbounded number of processors, *IEEE Transactions on Parallel and Distributed Systems*, Vol. 5, No. 9, 951-967, 1994.

[26] T. Yang and A. Gerasoulis. List scheduling with and without communication. *Parallel Computing*, 19(1993), 1321-1344.

[27] T. Yang and A. Gerasoulis, PYRROS: Static Task Scheduling and Code Generation for Message-Passing Multiprocessors, *Proc. of 6th ACM Inter. Confer. on Supercomputing*, Washington D.C., 1992, pp. 428-437.

[28] T. Yang, C. Fu, A. Gerasoulis and V. Sarkar, Scheduling Iterative Task Graphs on Distributed Memory Machines, Technical Report TRCS95-07, UCSB, 1995.

A Submesh Allocation Scheme for Mesh-Connected Multiprocessor Systems

T. Liu   W-K. Huang   F. Lombardi   L. N. Bhuyan

Department of Computer Science
Texas A&M University
College Station, TX 77843-3112
E-mail: {tong,wkhuang,lombardi,bhuyan}@cs.tamu.edu

Abstract – *This paper proposes an efficient submesh allocation strategy based on a free list and compares the performance of the proposed strategy with existing strategies. The novel feature of the free list in the proposed scheme is its overlapped nature, i.e. free submeshes in the list may overlap each other. Allocation and deallocation algorithms are presented in this paper, both with a low complexity in $O(N_F^2)$, where N_F is the number of free submeshes in the free list. Extensive simulation results show that our approach is superior to previous strategies, yielding a lower waiting time and a substantial improvement in allocation time.*

1   INTRODUCTION

The two-dimensional (2D) mesh topology has become popular for multiprocessors because of its simplicity, regularity, and suitability for VLSI implementation. The mesh is a promising architecture for the future generation of supercomputers. In this paper, we develop a processor allocation scheme for 2D mesh architecture and show that the new scheme is superior to existing schemes [1]-[5].

In a 2D mesh connected system, incoming jobs are allocated to submeshes of different sizes. As the number of processors in mesh-based systems grows, designing efficient submesh allocation strategies becomes increasingly difficult, but interesting. A number of submesh allocation strategies have been proposed in the last few years [1]-[5]. Li and Cheng have proposed a Buddy based strategy [1] that is applicable only to square mesh systems whose side lengths must be a power of 2. Chuang and Tzeng have proposed a frame sliding (FS) strategy for meshes with arbitrary length and width [2]. The FS strategy totally eliminates internal fragmentation and also outperforms the Buddy strategy in terms of allocation

[0]This research supported by a grant from the Texas ATP.

time and processor utilizations. Zhu has proposed the First-Fit and the Best-Fit strategies which have better performance than the FS strategy [3]. Ding and Bhuyan have proposed an adaptive scan (AS) strategy which has still smaller external fragmentation and better performance than the FS strategy [4]. Das Sharma and Pradhan have proposed a strategy which fully uses the "deallocation list" and outperforms previous schemes [5].

The proposed approach is based on a free list. In our strategy a free list, similar to the one for the hypercube [6], is maintained. However for the mesh topology, creating the list is difficult, complex, and interesting. The free list used in the proposed strategy is an overlapped one, i.e. the free processors may belong to multiple free submeshes in the list. We show that the complexity of our strategy is not higher than existing strategies. Simulation results also show that our strategy performs better on allocation time and average waiting delay compared to the existing strategies.

This paper is organized as follows. In section 2, we introduce notations and definitions. In section 3, we give a brief discussion of the existing submesh allocation strategies. Our approach is presented in section 4. Section 5 provides complexity analysis and simulation results are given in section 6 to compare with other strategies. Finally, conclusions are given in the last section.

2   PRELIMINARIES

A two-dimensional mesh system $M(l, w)$ consists of $l \times w$ nodes structured as a $l \times w$ two-dimensional rectangular grid. A node or processor at column a and row b is addressed as (a, b). A submesh $S(l', w')$ in $M(l, w)$, denoted by $[x, y, l', w']$, is made of $l' \times w'$ nodes in the rectangular region defined by the lower left corner (x, y) to the upper right corner $(x + l' -$

$1, y + w' - 1)$.

Definition 1. A *free submesh* is a submesh in which all nodes are not occupied by allocated jobs. A *busy submesh* is a submesh in which all of its nodes are allocated to a job.

Definition 2. A *free list* is an ordered list of all free submeshes. The free list used in this paper is sorted by increasing order of shorter edge of submeshes. For those with the same shorter edges, it is ordered by increasing order of longer edge. A free node in $M(l, w)$ may belong to two or more free submeshes in the free list, i.e. submeshes in the free list may be overlapped. The length of the list is denoted as N_F.

Definition 3. A *busy list* is a list of all busy submeshes of the allocated jobs. The length of the list is denoted as N_B.

Definition 4. *The boundary value* of a free submesh is the number of nodes either neighboring with busy nodes of other jobs or on the border of the mesh system.

Definition 5. *Fragmentation* is said to occur when an incoming task can not be allocated to a free submesh even though the number of free processors in the system is more than that required by the incoming task.

3 PREVIOUS STRATEGIES

The *two-dimensional buddy* strategy (2DB) was proposed in [1]. However, this strategy requires the mesh as well as the incoming requests to be square with sides being a power of 2. The *Frame Sliding* (FS) strategy [2] was proposed to overcome the limitation of the 2DB strategy. It is applicable to any mesh system and causes no internal fragmentation. However, it fails to recognize free submeshes, because it searches for free submeshes in only certain horizontal and vertical strides.

In [3], two strategies were proposed as *First-Fit* (FF) and *Best-Fit* (BF). Both the FF and BF strategies are applicable to mesh systems of any size and can allocate a submesh of the exact size to an incoming task, thus eliminating the internal fragmentation experienced by the 2DB strategy.

The *Adaptive Scan* (AS) strategy [4] improved the FS strategy and eliminated the virtual fragmentation associated with the FS strategy. The AS uses scanning instead of sliding with fixed stride to search for the the candidate free submeshes. Rotating the orientations of incoming tasks was first used in the free submesh searching. However, the improvement is limited by its first-fit nature and thus it is not significant.

The *Busy-List* (BL) strategy was proposed [5]. A busy list is maintained in the system to keep records of tasks which are running in the system. By using the BL, the search of free submesh for a new incoming task is based on the busy list as opposed to all other strategies which are based on a two-dimensional array with each bit corresponding to a processor. This makes BL to perform more efficiently than the others because a set of processors' occupancy is obtained from each operation on the busy list while only one processor's usage status is know by each operation on the array.

4 PROPOSED APPROACH

Submesh allocation consists of finding a free submesh for the incoming job. The proposed strategy maintains a list of all free submeshes, not covered by any other submesh. There may be an overlap among free submeshes in the list. The *free list* is the main data structure maintained by our strategy. It is kept by increasing order of the shorter edge of submeshes and for those with same shorter edges, by an increasing order of longer edges. For same shorter and longer edges, the order is arbitrary. This ordering of the free list is for a faster search in the allocation process. By using overlapping, it is possible to find an existing free submesh very efficiently, however, some submeshes in the free list have to be adjusted after a free submesh is allocated. Because of this characteristic, we have developed efficient allocation and deallocation algorithms using an overlapped free submesh list.

The basic idea of our allocation approach is to search the free list for the first free submesh which is equal to or larger than the size required by the new job, and then to adjust the free list affected by this allocation. The proposed allocation algorithm implements a linear search starting from the head of the list to its tail until the first free submesh large enough to hold the new job, is found. Then, if the free submesh is larger than the new job, a decision should be made concerning where to put the new job in the free submesh. The allocation to one of the four corners will result in a smaller fragmentation to the system. The decision is then reduced to choosing one among eight positions: four corners, vertically and horizontally. The maximum of the boundary values is used for choosing the corner. Submeshes in the free list may be overlapping. Therefore, after allocating the new job, some free submeshes in the free list may be overlapped with the newly allocated job and thus

Algorithm Allocation
begin
  Step 1. /* *Locate a free submesh.* */
    for each submesh F_i in the free list
      if F_i can hold the new job then
        $F = F_i$;
        delete F_i from the free list;
        break;
    if no free submesh can fit the new job
      return Fail;
  Step 2. /* *Allocate the new job.* */
    for each job in the busy list and on the
    borders of the mesh system
      if it borders F **then**
        update the corner values;
      allocate the new job at the corner with the
      maximum boundary value;
      there are at most two free fragments in F
      after allocating the new job, put them
      (if exist) into the stack;
  Step 3. /* *Adjust the free list.* */
    for each submesh in the free list
      if it overlaps with the allocated new job
        delete it from the free list;
        put the fragments into the stack;
  Step 4. /* *Update the free list.* */
    for each submesh in the stack
      insert it into the free list;
end

Figure 1: The Allocation Algorithm

it needs an adjustment. The allocation algorithm is presented in Figure 1.

The proposed deallocation algorithm, which is shown in Figure 2, consists of maintaining a free list of all possible free submeshes. First, we set the departing job as a free submesh and expand it towards the four directions: up, down, left and right. Since the expansion may get different submeshes by starting from different directions, all four directions are used. This is shown in Figure 3. The rectangles with dotted lines are formed through the expansion of the departing job, FREE1 is from starting on top, FREE2 is from down which is the same as from right, FREE3 is from left. In general, this will result in at most four different free submeshes. If a free submesh is covered by any one of the newly formed free submeshes, then it should be deleted from the list, else if it was blocked from an expansion of one side only by the departing job, then it should be expanded and re-inserted into the free list. In Figure 3, the rectangles with solid

Algorithm Deallocation
begin
  Step 1. /* *Form new free submeshes* */
    for each submesh in the free list
      update west, east, north, south;
    for each submesh in the free list
      update west-up, west-down, east-up,
      east-down, north-left, north-right,
      south-left, south-right;
    Free_submesh1:
      [west:min(north-right, south-right)]
      [west-down:west-up];
    Free_submesh2, if $\neq$ Free_submesh1:
      [max(north-left, south-left):east]
      [east-down:east-up];
    Free_submesh3, if $\neq$ Free_submesh1,2;
      [south-left:south-right]
      [south:min(west-up, east-up)];
    Free_submesh4, if $\neq$ Free_submesh1,2,3;
      [north-left:north-right]
      [max(west-down, east-down):north];
    put the new free submeshes in the stack;
  Step 2. /* *Adjust the free list.* */
    for each new free submesh NF_i from Step 1.
      for each submesh F_i in the free list
        if F_i is covered by NF_i
          delete F_i from the free list;
        else if any edge of F_i is covered by
        or bordering with NF_i **then**
          extend F_i to the other end of NF_i;
          delete F_i from the free list and
          insert it into the stack;
  Step 3. /* *Update the free list.* */
    for each submesh in the stack
      insert it into the free list;
end

Figure 2: The Deallocation Algorithm

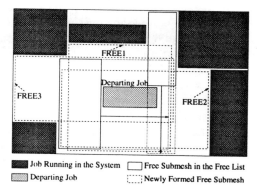

Figure 3: Description of the Deallocation Process.

Algorithm Insertion
begin
 let NF be the free submesh to be inserted;
 if the free list is empty **then**
  put NF in it;
  return;
 while not at the end of the free list
  let F_i be the current entry;
  if $NF's$ shorter edge $< F_i's$ shorter edge or
  ($NF's$ shorter edge $= F_i's$ shorter edge
  and $NF's$ longer edge $< F_i's$ longer edge)
  then
   for each submesh F_j of the remaining
   submeshes in the free list
    if NF is covered by F_j
     discard NF and **return**;
   put NF before F_i;
   return;
  else
   if F_i is covered by NF
    delete F_i from the free list;
  put NF at the end of the free list;
end

Figure 4: The Insertion Algorithm

lines are examples of such free submeshes in the free list.

The Insertion algorithm, which is presented in Figure 4, is used in the Allocation and Deallocation algorithms to insert a submesh into the free list. The insertion keeps the list by defined order. Also, it is discarded if it is covered by any submesh in the list. This algorithm also deletes all submeshes that are covered by the new submesh.

5 COMPLEXITY ANALYSIS

For the complexity analysis of the proposed algorithms, we first give a lemma showing that the lengths of the busy list (N_B) and the free list (N_F) are in the same order. The lemma makes the comparison at the same level.

Lemma 1. N_B and N_F are in the same order.

Proof. Let $s_1, s_2, s_3, ..., s_{N_B}$ be the N_B jobs currently in the system, and $f_1, f_2, f_3, ..., f_{N_F}$ be the N_F free submeshes in the free list. Let us consider a reverse situation, i.e. if currently $[i, j]$ is busy, we set it to free, while if $[i, j]$ is free, we set it to busy. The number of free submeshes after the reverse operation is equal to or less than N_B, because two or more previous busy submeshes may be merged into

one free submesh after the reverse operation. Similarly, by putting the overlapped free nodes into only one busy submesh in the reverse operation, the number of busy submeshes is equal to or less than N_F, because all nodes in some free submeshes may be overlapped and put into other submeshes by the reverse operation. Dynamically, a node changes from busy/free to free/busy, then N_B and N_F do not have fixed values, in some case N_B/N_F equals to N_F'/N_B' of some other case. So, we say they are in the same order. This conclusion is applicable to both the worst case and the average case. □

The time complexity of the Insertion Algorithm is therefore, $O(N_F)$. Consider now the complexity of the Allocation Algorithm. Each of Step 1, 2, 3 takes $O(N_F)$ in the worst case. In Step 4, each iteration executes an insertion and it takes $O(N_F)$. There are in the worst case $2(N_F + 1)$ free submeshes in the stack, thus Step 4 takes $O(N_F^2)$. Overall, the Allocation Algorithm takes $O(N_F^2)$ in the worst case.

Finally, the analysis for the Deallocation Algorithm is as follows. Step 1 takes $O(N_F)$. Step 2 takes $4N_F$. Step 3 takes $O(N_F^2)$ in the worst case. Overall, the time complexity of the Deallocation Algorithm is $O(N_F^2)$ in the worst case.

Thus, the time complexity of our approach, including allocation and deallocation, is $O(N_F^2)$ in the worst case. As the time complexity of the BL strategy is $O(N_B^3)$, and Lemma 1 has shown that $O(N_F)$ and $O(N_B)$ are in the same order, then our new approach has a lower complexity than BL. 2DB, FS, FF, BF, and AS strategies [1]-[4] have a time complexity of $O(N)$, where N is the number of processors in the mesh system. In the extreme case, each task utilizes one processor and there are N tasks running in the system, then $N_F = N$. However, in real applications each task requires a set of processors, otherwise, there is no need to use multiprocessors in the first case, and not all the processors are busy. Therefore, N_F is usually much smaller than N.

6 SIMULATION RESULTS

Simulations were performed for comparing the performance of the proposed approach with previous approaches. Since the BL strategy [5] has better performance than other previous strategies, most comparison here is with the BL. The size of the mesh system we use in our simulation varies from 8x8 to 128x128, the job interarrival time is 6 to 12 seconds, and the service time is fixed at 10 seconds per job. For comparison consistency, we use the same simulation model as in [4]-[5].

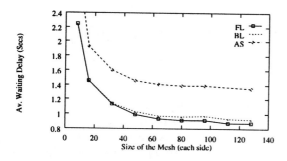

Figure 5: Average Waiting Delay for a Linear Distribution.

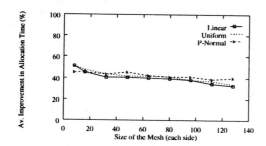

Figure 6: Improvement on Allocation Time vs Mesh Size.

First, we run the simulation by fixing the interarrival time rate at 7 seconds and service time rate at 10 seconds. The performance is measured in terms of the average waiting delay for a job before it is allocated a submesh. Figure 5 shows the average waiting delay of AS [4], BL, and and FL approaches under the Linear Distribution. We measure the real allocation time in our simulations, as shown in Figure 6. The allocation time by our approach is 32 to 51 percent better than that of BL. This is mostly due to the low complexity of our allocation algorithm. Next, we simulate by varying the interarrival time, but fixing the mesh size at 64x64, as well as all other parameters. The submesh sizes are generated under those three

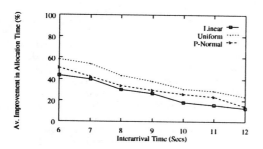

Figure 7: Improvement on Allocation Time vs Interarrival Time.

distributions for the given size of the mesh system. The improvement of FL over BL for the allocation time versus interarrival time is demonstrated in Figure 7 for the three distributions. The improvement is significant in all three cases, and higher the system load, larger is the improvement.

7 CONCLUSION

In this paper, we have proposed an efficient submesh allocation strategy suitable for 2D mesh multiprocessors. The proposed approach is based on an overlapped free list. We have developed a low complexity allocation algorithm $O(N_F^2)$ compared to $O(N_B^3)$ of the Busy-List strategy. Extensive simulation results show that the average waiting delay and the allocation time of the proposed strategy are always better than existing strategies. Overall, our approach is superior to the previous strategies.

References

[1] K. Li and K. H. Cheng, "A Two Dimensional Buddy System for Dynamic Resource Allocation in a Partitionable Mesh Connected System," In *Proc. ACM Computer Science Conf.*, pp. 22–28, 1990.

[2] P. J. Chuang and N. F. Tzeng, "An Efficient Submesh Allocation for Mesh Computer Systems," In *Proc. International Conference on Distributed Computing Systems*, pp. 256–263, May 1991.

[3] Y. Zhu, "Efficient Processor Allocation Strategies for Mesh-Connected Parallel Computers," *Journal of Parallel and Distributed Computing*, vol. 16, pp. 328–337, 1992.

[4] J. Ding and L. N. Bhuyan, "An Adaptive Submesh Allocation Strategy for Two-Dimensional Mesh Connected Systems," In *Proc. International Conference on Parallel Processing*, pp. 193–200, Aug. 1993.

[5] D. Das Sharma and D. K. Pradhan, "A Fast and Efficient Strategy for Submesh Allocation in Mesh-Connected Parallel Computers," In *Proc. IEEE Symposium on Parallel and Distributed Processing*, pp. 682–689, Dec. 1993.

[6] J. Kim, C. R. Das, and W. Lin, "A Top-Down Processor Allocation Scheme for Hypercube Computers," *IEEE Transactions on Parallel and Distributed Systems*, pp. 20–30, January 1991.

CONTIGUOUS AND NON-CONTIGUOUS PROCESSOR ALLOCATION ALGORITHMS FOR K-ARY N-CUBES *

Kurt Windisch and Virginia Lo,
Dept. of Computer and Information Science
University of Oregon
Eugene, OR 97403
Email: kurtw, lo@cs.uoregon.edu

Bella Bose
Dept. of Computer Science
Oregon State University
Corvallis, OR 97331
Email: bose@cs.orst.edu

Abstract

Efficient utilization of processing resources in a large, multi-user parallel computer depends on processor allocation algorithms that minimize system fragmentation. We propose three processor allocation algorithms for the k-ary n-cube class of parallel architectures, which includes the hypercube and multidimensional torus. The k-ary Partner strategy is a conventional contiguous processor allocation strategy that improves subcube recognition. The non-contiguous Multiple Buddy and Multiple Partner strategies lift the restriction of contiguity in order to address the problem of fragmentation associated with contiguous strategies. Simulations compare the performance of these three strategies with the performance of other k-ary n-cube allocation strategies, showing that non-contiguous allocation provides significantly increased system utilization by eliminating fragmentation.

1 Introduction

Our work addresses the problem of processor allocation in distributed memory multicomputers. **Allocation** involves assignment of a collection of processors to each incoming parallel computation, with the goal of maximizing throughput over a stream of jobs. Allocation policies can be categorized as **contiguous** or **non-contiguous** depending on whether the assigned processors are physically adjacent or not. Allocation algorithms used in commercially available multicomputers have been restricted to contiguous allocation schemes. In addition, many systems also require that the allocated processors form a subgraph of the original architecture (e.g. subcubes in hypercubes, and submeshes in meshes). Performance under contiguous allocation schemes suffers seriously from fragmentation. **Internal fragmentation** occurs when more processors are allocated to a job than it requests. **External fragmentation** exists when a sufficient number of processors are available to satisfy a request, but they cannot be allocated contiguously. Studies have shown that fragmentation severely limits the ability of all contiguous strategies to utilize processor resources.

Non-contiguous allocation techniques lift the contiguity constraint, thereby allowing a job's request to be satisfied by processors dispersed throughout the system. Under non-contiguous allocation, fragmentation is completely eliminated. However, performance is affected by the increased potential for message-passing contention due to the fact that processes from a given job may be required to send messages through communication links allocated to other jobs.

*This research is sponsored by NSF grant MIP-9108528 and the Oregon Advanced Computing Institute

We propose three new processor allocation strategies, both contiguous and non-contiguous, for the versatile k-ary n-cube class of architectures. k-ary n-cubes can be thought of as multidimensional tori, having a width of k processors in each of the n dimensions. This general class of architectures includes many of the topologies on which current commercially available distributed memory multicomputers are based: binary hypercubes, used in the N-Cube's machines and Intel's iPSC/860; and 3D tori, used in the Cray T3D. In addition, k-ary n-cubes having only two dimensions are also very similar to mesh-connected architectures, differing only in the additional wrap-around edges of the k-ary n-cube.

Section 2 formally describes the k-ary n-cube topology and related notation that will be used throughout the paper. Section 3 briefly describes related work in the area of processor allocation for k-ary n-cube topologies. Sections 4 and 5 introduce new contiguous and non-contiguous processor allocation strategies for the k-ary n-cube. Section 6 analyzes the performance of these strategies through simulation, and Section 7 summarizes our results and discusses future work.

2 Mathematical Preliminaries

A k-ary n-cube [4] [2], also denoted as Q_n^k, is a graph containing k^n nodes, each labeled with a distinct base-k, n-bit address, $(a_n - 1, a_n - 2, \ldots, a_i, \ldots, a_1, a_0)$, with $0 \leq a_i \leq k - 1$. Each base-k bit refers to an address in one of the n dimensions. $K_n(i)$ denotes the base-k representation of an integer i, using n bits. Each node is connected by an edge to $2n$ other nodes, two in each dimension. Specifically, for each bit, i, in $U = (a_n - 1, a_n - 2, \ldots, a_i, \ldots, a_1, a_0)$, U is connected to $V_1 = (a_n - 1, a_n - 2, \ldots, (a_i + 1) \bmod k, \ldots, a_1, a_0)$ and $V_2 = (a_n - 1, a_n - 2, \ldots, (a_i - 1) \bmod k, \ldots, a_1, a_0)$. The overall interconnection structure can be visualized as a multidimensional mesh with wrap-around edges. Note that the well-known hypercube is a 2-ary n-cube (Q_n^2).

From a complete k-ary n-cube, it is possible to identify subcubes, k-ary m-cubes, where $0 \leq m \leq n$. Each k-ary n-cube can be recursively partitioned into k distinct k-ary $(n-1)$-cubes. Subcubes having arity less than k, however, do not exist in a k-ary n-cube since such a cube would lack the wrap-around edges of a true k-ary n-cube. The resulting subgraph may, however, preserve the properties of a multidimensional mesh.

Subcubes may also be denoted by a distinct base-k, n-bit address, $(a_n - 1, a_n - 2, \ldots, a_1, a_0)$, with $0 \leq a_i \leq k - 1$. Each a_i may be enumerated or may contain a "don't care" bit, denoted by *. Each * in an address represents all of the k different enumerations of that bit from 0 to $k - 1$, and the number of *'s in an n-bit address is equal to the number of dimensions of the subcube. Subcubes also can be denoted by addresses having less than n bits, in which case, *'s for each of the missing bits are implied at the end of the address. Thus in a Q_4^2, for example, the address $K_2(1) = 01$ implicitly refers to the subcube having the full address $01**$.

Many hypercube and k-ary n-cube algorithms can be described using a tree representation of the cube and a set of its subcubes. The root (level 0) represents the complete k-ary n-cube topology, and intermediate nodes at level i of the tree represent $n-i$ dimensional subcubes. Leaves represent individual processors. The k edges from each node are labeled 0 through $k-1$ and a path from the root of the tree to any node is that node's address. An internal node in the tree is considered to be free if and only if all of its descendants are free, and a leaf node is free if and only if it has not already been allocated to a job.

3 Related Work

Extensive research has been conducted in the field of processor allocation for binary hypercube topologies. Since the hypercube is simply a k-ary n-cube where $k=2$, the most obvious allocation algorithms for k-ary n-cubes are generalizations of the common binary hypercube strategies.

Gautam and Chaudhary [5] extended the traditional hypercube algorithms, **Buddy** and **Gray Code** [3], to the k-ary n-cube topology. However, the Gray Code strategy is flawed, in that many of the recognized regions do not actually form subcubes, or even multidimensional submeshes. Both the Buddy and Gray Code strategies are limited in their subcube recognition capability. The Buddy strategy recognizes k^{n-m} subcubes of dimension m. Though the Gray Code strategy recognizes k^{n-m+1} possible allocations, only between k^{n-m} and $2*k^{n-m}$ (depending on whether k is even or odd) are actual subcubes.

Gautam and Chaudhary formulated these algorithms such that a subcube of any base (arity) could be searched for, but noted that they return incorrect results if the subcube base was not identical to the base of the topology. Our paper will assume that only requests for subcubes of the same base as the topology are allowed.

In addition to the above algorithms, Gautam and Chaudhary proposed a new algorithm called the **Sniffing** strategy [5]. They report that Sniffing is an improvement over the previous strategies, because it is able to satisfy requests for subcubes of bases, $j \neq k$, in a Q_n^k. The Sniffing strategy has recognition capability equivalent to the k-ary Buddy strategy when $j = k$, but is enhanced with the added capability of recognizing subcubes and meshes of different bases.

Qiao and Ni [9] developed a new processor allocation algorithm specifically for the 3D Tori (k-ary 3-cube) that is extendable to tori of any dimension. However, unlike the above strategies, the 3D Tori strategy allocates multidimensional submeshes, which are not necessarily strict subcubes of the k-ary n-cube, since they may lack the wrap-around edges.

4 New Contiguous Strategy

Contiguous processor allocation strategies, those that allocate processors comprising a single contiguous region or subgraph of the architecture (i.e., a subcube or multidimensional submesh), have been the primary focus of most processor allocation research. In some ways, they are preferable over non-contiguous strategies. Processors allocated contiguously are often able to communicate with each other through the interconnection network without interfering with the communications of other jobs in the system.

The biggest drawback of contiguous allocation is that system fragmentation typically runs very high. When servicing a dynamic workload under certain job scheduling policies, significant pockets of wasted processors, too small for most jobs, become scattered around the system and lead to impaired utilization of the system's processor resources. In binary hypercubes, Kreuger, Lai and Radiya [6] have shown that the maximum utilization attainable by contiguous allocation for uniform workloads under FCFS scheduling is 58%, and we have shown that the maximum for 2D mesh architectures is only 46% [8]. Significant improvement is achievable by the use of improved job scheduling policies. Despite this drawback, contiguous strategies remain the most commonly implemented processor allocation algorithms in real distributed memory parallel computers.

4.1 k-ary Partner Strategy

We propose a new contiguous strategy, the k-ary Partner strategy, by extending our earlier Partner strategy developed for the binary hypercube [1] to the k-ary n-cube. Although it is similar to the k-ary Buddy strategy, in that it also uses the buddy tree representation of subcubes, it provides significantly better performance. Like the k-ary gray code strategy, when an incoming task requests a Q_m^k subcube, level $n-m+1$ of the tree is searched from left to right for k distinct Q_{m-1}^k subcubes, which together form a single Q_m^k. However, the Partner strategy does not require these k subcubes to be adjacent in the tree, but still enforces adjacency in the topology, often allowing for recognition of many more subcubes than other tree-based strategies. We formally describe the Partner strategy below, beginning with a few definitions.

A node having the decimal node label z is denoted $K_m(z) = (a_{m-1}, a_{m-2}, \ldots, a_{\alpha+1}, a_\alpha, a_{\alpha-1}, \ldots, a_0)$, with $0 \leq \alpha \leq m-1$.

Definition 1 *If $a_\alpha = 0$, then the $k-1$ distinct α-th partners are each obtained by changing the α-th bit, a_α, of $K_m(z)$ to the unique bit i, where $1 \leq i \leq k-1$. This set of partners, denoted $K_m^{\alpha,i}$, are:*

$$K_m^{\alpha,1}(z) = (a_{m-1}, a_{m-2}, \ldots, a_{\alpha+1}, 1, a_{\alpha-1}, \ldots, a_0)$$
$$\vdots$$
$$K_m^{\alpha,i}(z) = (a_{m-1}, a_{m-2}, \ldots, a_{\alpha+1}, i, a_{\alpha-1}, \ldots, a_0)$$
$$\vdots$$
$$K_m^{\alpha,k-1}(z) = (a_{m-1}, a_{m-2}, \ldots, a_{\alpha+1}, k-1, a_{\alpha-1}, \ldots, a_0)$$

If $a_\alpha \neq 0$, then the set of α-th partners is undefined.

For example, the Q_4^4 node $K_3(8) = 020$ (also denoted $020*$), representing a 4-ary 3-cube, has three possible partner sets. The 0-th partner set consists of $\{K_3^{0,1}(8), K_3^{0,2}(8), K_3^{0,3}(8)\} = \{021, 022, 023\}$. The 1-st partner set is undefined, since $a_1 = 2 \neq 0$. The 2-nd partner set consists of $\{K_3^{2,1}(8), K_3^{2,2}(8), K_3^{2,3}(8)\} = \{120, 220, 320\}$.

Definition 2 *For an integer z, $0 \leq z \leq k^{n-m+1} - 1$, the node $K_{n-m+1}(z)$ is **free** if and only all of its descendant nodes are free.*

Allocation Algorithm: The k-ary Partner allocation algorithm attempts to allocate an m-dimensional subcube of a k-ary n-cube. The states, *busy* or *free*, of each of the k^n processors in the system are recorded in an array of allocation bits. First, the tree is searched for a free Q_{m-1}^k node, denoted $K_{n-m+1}(z)$, at level $n-m+1$ with the least z such that $0 \leq z \leq k^{n-m+1} - 1$. Then, partners of this node are found by searching for the least

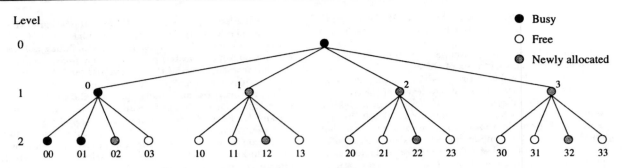

Figure 1: Allocation of a Q_1^4 in a 4-ary 2-cube with the k-ary Partner strategy. Black nodes are busy, grey nodes are newly allocated to the Q_1^4, and white nodes are free.

integer α ($0 \leq \alpha \leq n - m$), such that the $k - 1$ α-th partners of $K_{n-m+1}(z)$ (i.e., $K_{n-m+1}^{\alpha,1}(z), \ldots, K_{n-m+1}^{\alpha,k-1}(z)$) are defined and all free. If such a free partner set is found, the corresponding processors are allocated to the job, and their allocation bits are set to *busy*. Otherwise, the job cannot be allocated at this time.

Deallocation Algorithm: When a job is deallocated, the algorithm simply releases the k nodes in level $n - m + 1$ (i.e., $K_{n-m+1}(z), K_{n-m+1}^{\alpha,1}(z), \ldots, K_{n-m+1}^{\alpha,k-1}(z)$) that have been allocated to the job, and sets the allocation bits for their corresponding processors to *free*.

Figure 1 illustrates an example of the Partner allocation strategy in which a Q_1^4 is requested from a Q_2^4 containing two previously allocated processors. Step 2 of the algorithm searches level 2 of the tree for free nodes. The first free node encountered is 02, for which there are two possible partner sets. The 0-th partner set is undefined since $a_0 = 2 \neq 0$ for this node. However, the 1-st partner set contains the nodes 12, 22, and 32, all of which are free. Therefore, those four nodes, 02 and its 1-st partners, are allocated to the job.

Al-Dhelaan and Bose's subcube recognition proof [1] can be extended, without loss of generality, to show that the Partner strategy can recognize $(n - m + 1)k^{n-m}$ subcubes (Q_m^k's) from a Q_n^k, surpassing the recognition of the Gray Code strategy. Furthermore the allocation and deallocation time complexities are $O(k^{n+1}) = O(kP)$ and $O(k^n) = O(P)$, respectively, where P is the number of processors in the system. Fragmentation should be reduced due to the greater ability to successfully recognize subcubes and thus satisfy incoming requests.

4.2 Summary of Contiguous Strategies

The contiguous strategies studied all have the common characteristic of high levels of fragmentation and correspondingly low system utilization when the standard FCFS scheduling policy is used. This is due to the nature of the contiguity constraint and will be studied quantitatively in Section 6, where we will compare many allocation strategies by simulation. Improved performance requires exploration of other alternatives, including scheduling policies [6] and the approach we propose: non-contiguous allocation.

5 New Non-contiguous Strategies

Non-contiguous processor allocation algorithms overcome the fragmentation drawback of contiguous strategies by allowing a job's allocation to be dispersed between many non-adjacent regions of the topology when necessary. This reduces the system fragmentation by allowing the small pockets of unallocated processors to be utilized.

However, this increased utilization occurs at the price of increased contention within the interconnection network as dispersely allocated jobs route their internal communications through processors belonging to other jobs. Still, in large multiuser systems, especially those operating with high network bandwidth and with wormhole routing, noncontiguous allocation may have a positive net effect on performance [8].

We introduce three non-contiguous processor allocation strategies, all of which completely eliminate external fragmentation by allowing jobs to be allocated anywhere in the system, given enough free processors. Since these algorithms are not constrained to allocations of a specific shape, requests are specified as the number of processors needed, r, where $1 \leq r \leq k^n$.

5.1 Multiple Buddy

We now describe the first new strategy for allocating processors non-contiguously in k-ary n-cubes, the Multiple Buddy Strategy (MBS). MBS is an extension of our mesh-based MBS algorithm [8] and is based on the k-ary Buddy strategy and the mesh 2D Buddy strategy [7]. MBS eliminates the problems of fragmentation found in these contiguous strategies by allowing multiple contiguous blocks to be allocated to a job non-contiguously. This general approach allows the system to maintain contiguity within individual blocks, while still allowing it to disperse the allocation when necessary, in order to fully utilize free processors and overcome system fragmentation.

If there are enough free processors in the system, we know in advance that the allocation will succeed. MBS handles a request for r processors by first factoring the request into one or more k-ary subcube *blocks* by simply factoring the integer r into its base-k representation. In this form, each i-th digit ($0 \leq i \leq n$), represents the number of k-ary subcubes of dimension i that are needed. Next, the allocation algorithm attempts to allocate contiguous free processors to each of these factored blocks using a version of the k-ary Buddy system that uses free-lists to record maximal-sized subcubes of each dimension. If any individual i-dimensional block cannot be allocated contiguously, MBS then searches for a larger free subcube to split into i-cubes. If that fails, then MBS repeatedly splits the request into k blocks whose size is one dimension smaller until they all have been allocated. The total overhead for generating each block by splitting larger ones is $O(n)$ and at most k^n blocks will be allocated, giving $O(nk^n) = O(nP)$ time complexity for allocation, where P is the total number of processors.

Figure 2 shows how MBS would handle a request for 23 processors in a Q_3^4 that has many scattered processors already

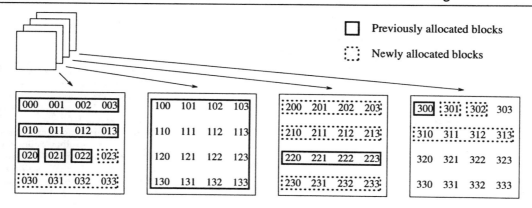

Figure 2: MBS example: a job requesting 23 processors is allocated in Q_3^4, which is represented by decomposing the three dimensional structure into its two dimensional planes. Due to scattered busy processors from previous allocations, the new job is broken into smaller, non-contiguous blocks.

busy. The request for 23 processors is factored into requests for one Q_2^4, one Q_1^4, and three Q_0^4s. However, since there are no available Q_2^4s in the system, that request is broken into four requests for Q_1^4s. Now, these blocks may all be allocated non-contiguously in the system.

When a job is deallocated, all of its blocks must be returned to the system, and when possible, those blocks must be merged with other free blocks into larger ones. Since the maximum number of buddy merges is $k^{n-1} + k^{n-2} + \cdots + 1 = \frac{k^n-1}{k-1} = O(k^{n-1})$, the overhead for deallocation in the worst case will not exceed $O(k^{n-1}) = O(\frac{P}{k})$, where P is the total number of processors.

5.2 Multiple Partner Strategy

Our second non-contiguous allocation strategy is based on the contiguous k-ary Partner strategy described in Section 4.1. The Multiple Partner strategy operates very similarly to the Multiple Buddy strategy. If there are enough processors in the system, we know in advance that the allocation will succeed. A request for r processors is factored (exactly as in Section 5.1) into a set of smaller request blocks. Then, the Multiple Partner strategy attempts to allocate each block using the k-ary Partner strategy described in Section 4.1. As in MBS, any i-dimensional request block, Q_i^k, that cannot be successfully allocated is repeatedly divided into k requests for Q_{i-1}^k's until they are all allocated. The deallocation step simply releases the allocation bits for all processors used by the job.

The total overhead for generating each block by splitting larger ones is $O(n)$ and at most $O(k^n)$ blocks will be allocated at a cost of $O(k^{n+1})$, giving $O(nk^{2n+1})$ time complexity for allocation. Deallocation simply costs $O(k^n)$ to reset the allocation bits of each allocated processor. An implementation of the k-ary Partner strategy using free lists as described in [1] and similar to those used in MBS, could be used to reduce this overhead.

5.3 Paging(i)

Another simple non-contiguous allocation strategy is Paging(i), which statically partitions the k-ary n-cube into k-ary i-cubes ($0 \le i \le n$). A request for r processors is satisfied by allocating the first $\lceil \frac{r}{k^i} \rceil$ free i-dimensional partitions found in a linear scan of the free partitions. Some degree of contiguity is maintained through the nature of the linear scan. There will be neither internal nor external fragmentation for $i = 0$, but

for $i \ge 1$, fragmentation may be introduced as a tradeoff for a greater degree of contiguity. The complexity of allocation is $O(k^n)$ and deallocation is $O(k^n) = O(P)$.

6 Performance Analysis

We conducted simulation experiments to analyze the performance of non-contiguous allocation strategies compared to contiguous ones, focusing on the effects of system fragmentation, both internal and external. Experiments were conducted using our discrete event simulator, ProcSimity [11].

The experiments model the arrival, service, and departure of a stream of 1000 jobs in k-ary n-cube systems using first-come, first-serve scheduling (FCFS). Jobs arrive, delay for an amount of time taken from an exponential distribution, and then depart. The overhead of allocation and deallocation is ignored in the simulation. The job request streams were modeled taking the job request sizes from the uniform distribution, and the exponential distribution, in which small jobs occur more frequently while very large jobs occur less frequently. The independent variable in these experiments was the *system load*, defined as the ratio of the mean service time to mean interarrival time of jobs. Higher system loads reflect the greater demands when jobs arrive faster than they can be processed.

Reported results represent the statistical mean after 10 simulation runs with identical parameters, and given 95% confidence level, mean results have less than 5% error.

6.1 Observations

Figure 3 shows the system utilization attained by contiguous and non-contiguous allocation strategies for three 1024-node k-ary n-cubes, and for exponentially distributed job sizes. We combine the data for the non-contiguous strategies after observing that all the noncontiguous strategies tested (Multiple Buddy, Multiple Partner, and Paging(0)), perform identically with respect to system fragmentation for all sizes of 1024-node k-ary n-cube. This is because they all completely eliminate system fragmentation, both internal and external, and thus are always able to allocate a job if there are enough free processors available anywhere in the system. Surprisingly, we also observed that the contiguous strategies tested (k-ary Buddy, Gray Code, and Partner) also performed very similarly, with the Partner strategy outperforming Buddy by only 0.04%. Therefore, Figure 3

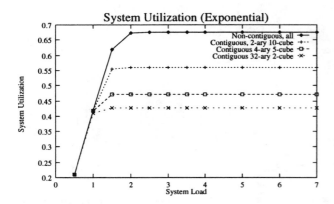

Figure 3: System utilization vs. system load for contiguous and non-contiguous allocation strategies with the exponential distribution of job sizes. Results are shown for k-ary n-cubes of different configurations with 1024 processors.

shows a single line for all the contiguous strategies for each size k-ary n-cube.

The system utilizations measured for exponentially distributed job sizes in Figure 3 show two important trends. The first important observation is that non-contiguous allocation performs much better than contiguous in all cases, achieving system utilizations increased by a ratio of 16-36%, depending on the topological dimensions.

Secondly, for k-ary n-cubes of equal numbers of processors, we observed that contiguous allocation performs best in higher dimensional topologies. As n decreases, ultimately yielding the 2D torus topology, system utilization falls off by a ratio of 24% from that of the binary hypercube.

Very similar trends were observed for the uniform distribution of job sizes.

Although non-contiguous allocation algorithms would appear to perform much better than contiguous ones, based on their levels of system utilization, special consideration must be given to the added network contention that will result from non-contiguously allocated jobs. Such allocations may interfere with each other's messages in the network, degrading overall performance. For mesh architectures, we have seen that the increased contention due to non-contiguous allocation is not as serious as the fragmentation effects of the contiguous allocation. Thus, we observed that non-contiguous allocation performed better overall, even when message-passing contention was considered [8]. We have run similar experiments, described in [10], to determine the feasibility of non-contiguity in k-ary n-cubes with respect to network contention.

7   Conclusion

We have presented several new contiguous and non-contiguous processor allocation algorithms, for k-ary n-cube architectures. Due to the great versatility and generality of the k-ary n-cube topology, these new algorithms are applicable on a wide range of parallel computers. Furthermore, we conducted preliminary experiments comparing these strategies by simulation on a variety of 1024-node k-ary n-cube configurations. We found that our contiguous Partner strategy has better subcube recognition ability than either Buddy or Gray Code; however, this improvement apparently has very little effect on performance. Further, we found that the non-contiguous strategies

dramatically outperform contiguous allocation strategies with respect to system fragmentation. As a result, system utilizations for non-contiguous strategies reach as high as 70%, compared to utilizations of 43-57% for contiguous schemes. It was also interesting to note that contiguous schemes performed better on the high-dimensional 1024-node k-ary n-cubes we tested, such as binary hypercubes, than on low-dimensional ones, such as 2D tori. Performance with respect to message-passing contention is further discussed and analyzed in [10].

Our ongoing work in this area includes examining the performance of a variety of job scheduling policies when used in conjunction with both contiguous and non-contiguous allocation schemes. Our goal is to develop efficient and practical resource management systems that fully utilize the computational capabilities of modern distributed memory multicomputers.

References

[1] A. Al-Dhelaan and B. Bose. A new strategy for processor allocation in an n-cube multiprocessor. In *Proceedings of the International Phoenix Conference on Computers and Communication*, pages 114–118, March 1989.

[2] B. Bose, B. Broeg, Y. Kwon, and Y. Ashir. Lee distance and topological properties of k-ary n-cubes. *IEEE Transactions on Computers*, 1995. To appear.

[3] M. Chen and K. G. Shin. Processor allocation in an n-cube multiprocessor using gray codes. *IEEE Transactions on Computers*, C-36(12):1396–1407, December 1987.

[4] W. J. Dally. Performance analysis of k-ary n-cube interconnection networks. *IEEE Transactions on Computers*, 39(6):775–784, June 1990.

[5] V. Gautam and V. Chaudhary. Subcube allocation strategies in a k-ary n-cube. In *Proceedings of the Sixth ISCA International Conference on Parallel and Distributed Computing Systems*, Louisville, KY, October 1993.

[6] Phillip Krueger, Ten-Hwang Lai, and Vibha A. Dixit-Radiya. Job scheduling is more important than processor allocation for hypercube computers. *IEEE Transactions on Parallel and Distributed Systems*, 5(5):488–497, May 1994.

[7] Keqin Li and Kam-Hoi Cheng. A two-dimensional buddy system for dynamic resource allocation in a partitionable mesh connected system. *Journal of Parallel and Distributed Computing*, 12:79–83, 1991.

[8] W. Liu, V. M. Lo, K. Windisch, and B. Nitzberg. Non-contiguous processor allocation algorithms for distributed memory multicomputers. In *Proceedings Supercomputing '94*, pages 227–236, 1994.

[9] W. Qiao and L. M. Ni. Efficient processor allocation for 3d tori. Technical report, Michigan State University, East Lansing, MI 48824-1027, 1994.

[10] K. Windisch, V. Lo, and B. Bose. Contiguous and non-contiguous processor allocation algorithms for k-ary n-cubes. Technical report, University of Oregon, 1995.

[11] K. Windisch, J. V. Miller, and V. Lo. Procsimity: an experimental tool for processor allocation and scheduling in highly parallel systems. In *Proceedings of the Fifth Symposium on the Frontiers of Massively Parallel Computation*, pages 414–421, February 1995.

Iteration Space Partitioning to Expose Reuse and Reduce Memory Traffic

Gary Elsesser, Viet N. Ngo
Cray Research, Inc.
655F Lone Oak Drive
Eagan, MN 55121
{gwe,vnn}@cray.com

Viet P. Ngo
Department of Mathematics
Cal State University Long Beach
Long Beach, CA 90840
viet@csulb.edu

Abstract: Standard scheduling techniques which partition the iteration space of parallel loops use barrier synchronization to resolve memory and cache coherency problems. Barrier synchronization, however, interferes with cache reuse and reduces processor utilization. We propose compile-time iteration space partitioning techniques which expose temporal reuse of memory. These loop restructuring methods group iterations so that memory reuse only occurs within groups. As as result each processor accesses a distinct set of array elements, thereby insuring memory coherency and reducing coherency traffic. In some cases barrier synchronization is also eliminated.

1 Introduction

Parallel processing of looping constructs written in high level languages, such as FORTRAN, are realized through a compiler transformation called DOALL conversion [10]. DOALL conversion is a compile-time technique for distributing the iterations of a parallelizable DO loop across available processors. Data dependencies and cache coherency between DOALL constructs is ensured by barrier synchronization.

Standard parallelization schemes such as contiguous or cyclic partitioning of the iteration space of DOALL loops will generate cache coherency traffic in the presence of memory reuse. In this paper we introduce a compile-time loop restructuring technique for minimizing cache coherency traffic between DOALL loops. Our restructuring methods group iterations so that the array references made by each processor are disjoint.

We consider the problem of transforming the DOALL loop systems shown in figure 1 for the purpose of memory/cache reuse, memory traffic reduction and barrier elimination. FORTRAN-90 array syntax statements often translate into type-II systems because, in the general case, B(a:b:c) = f(B(d:e:f)) is translated into the sequence A(:) = f(B(d:e:f)), B(a:b:c) = A(:). Therefore our domain of application includes a large body of FORTRAN 90 codes.

Since, in this paper, each instance of barrier synchronization is associated with the end of a DOALL

$$
\begin{array}{ll}
\textbf{doall}\ i = L_1, U_1 & \textbf{doall}\ i = L_1, U_1 \\
T_1:\ A(f_1(i)) = w_1(i) & T_1:\ A(f_{11}(i)) = \\
 & \quad w_1(i, B(f_{12}(i))) \\
\textbf{enddo}\ \otimes & \textbf{enddo}\ \otimes \\
\textbf{doall}\ i = L_2, U_2 & \textbf{doall}\ i = L_2, U_2 \\
T_2:\ w_2(i, A(f_2(i))) & T_2:\ B(f_{21}(i)) = \\
 & \quad w_2(i, A(f_{22}(i))) \\
\textbf{enddo}\ \otimes & \textbf{enddo}\ \otimes \\
\text{(type I)} & \text{(type II)}
\end{array}
$$

Figure 1: Loop Forms/Systems

loop, we indicate barrier synchronization, or the lack thereof, by annotating **enddo** statements. Loops ending in "**enddo** $\equiv$" require no barrier synchronization; while those ending with "**enddo** $\otimes$" require a barrier.

The following assumptions apply to the loop forms under consideration:

- work expressions, w_k, do not modify A or B.

- w_k only reads members of A and B explicity listed as arguments.

- w_k may read/write other variables, provided that this introduces no additional dependencies within the context of the loop pair.

- the indexing functions $(f_1, f_2, \ldots f_{22})$ have the form $ki + c$, where $k \neq 0$.

- The indexing functions and bounds are invariant within the context of the loop pair.

Let $[a : b : s]$ denote the bounded arithmetic sequence $\{a, a + s, a + 2s, \ldots, b\}$, where $[a : b]$ is a shorthand for $[a : b : 1]$. Set difference is denoted by $\setminus$; i.e., $A \setminus B = \{x \in A : x \notin B\}$. The iteration space of loop i is $\Gamma_i = [L_i : U_i]$. Each index map is an affine linear function, denoted by f with an identifying subscript.

$$
\begin{array}{ll}
[a : b : s] = [min(a,b), max(a,b)] \cap (a + s\boldsymbol{Z}) \\
\Gamma_j = [L_j, U_j] \cap \boldsymbol{Z} & f_s(x) = k_s x + c_s
\end{array}
$$

In [7] a method for partitioning parallel do-loops with non-uniform data dependencies is presented. Their method is suited to improving data locality on uniprocessors as well as parallelization on multiprocessor systems. Their method can not be easily extended to handle more than two data access functions because its complexity grows dramatically when the number of accesses increases to three or more. A scheduling technique that exploits reuse and eliminates cache thrashing due to non-uniformly generated dependences is presented in [2]. Our approach is similar to theirs in that we partition the iteration space into equivalence classes which access common array elements. The cache thrashing problem is also considered in [8], where an automatic loop transformation technique to reduce cache conflicts and a method for trading off this reduction against the possible increase in synchronization and scheduling overheads is presented. These approaches deal only with array accesses that are within a single nested loop construct.

Offset conflicts arising from uniformly-generated dependences have been investigated by [3] and [9]. Neither work addresses non-uniformly generated dependences. [1] develops methods for automatically partitioning loops with array accesses that have unit-coefficient. Their technique partitions the iteration space to satisify load balancing constraints and optimize cache-coherency traffic whereas we develop partitions that satisfy a zero communication constraint. Several other researchers have developed techniques for determining communication free partitions of index sets for subscript expressions having uniform dependencies [4, 6, 5].

Throughout this paper, the phrase *alignment on X* denotes a form of code motion whereby statements which access common elements of X are executed within a single iteration. For the loop forms defined in figure 1, alignment on X means that whenever $T_1(i)$ and $T_2(j)$ access a common element of X, they will be executed together. Alignment on X transforms all dependencies on X to loop-independent dependencies [10], thereby realizing maximum temporary reuse for X. When all inter-loop dependences are on object X, alignment on X also permits elimination of barrier synchronization.

Section 2 provides an algorithm for aligning type I systems on array A. In section 3 these techniques are expanded to support type II systems. Concluding remarks are offered in section 4.

2 TYPE I Systems

For type I systems, alignment on A provides maximum temporal reuse and permits removal of the synchronization barrier at the end of the first loop. When $\exists (i,j) \in \Gamma_1 \times \Gamma_2$ such that $f_1(i) = f_2(j)$, alignment of A requires that $T_1(i)$ and $T_2(j)$ be executed in the same iteration. $T_1(i)$ may be pushed down into the

$\begin{aligned}&\textbf{doall } i = L_1, U_1\\ &\quad \textbf{if } f_1(i) \notin f_2(\Gamma_2) \textbf{ then}\\ &T_1(i): \quad A(f_1(i)) = w_1(i)\\ &\quad \textbf{endif}\\ &\textbf{enddo } \equiv\\ &\textbf{doall } j = L_2, U_2\\ &\quad i = f_1^{-1}(f_2(j))\\ &\quad \textbf{if } i \in \Gamma_1 \textbf{ then}\\ &T_1(i): \quad A(f_2(j)) = w_1(i)\\ &\quad \textbf{endif}\\ &T_2(j): w_2(j, A(f_2(j)))\\ &\textbf{enddo } \otimes\end{aligned}$	$\begin{aligned}&\textbf{doall } i \in \beta_1\\ &\quad A(f_1(i)) = w_1(i)\\ &\textbf{enddo } \equiv\\ &\\ &\textbf{doall } j \in \beta_2\\ &\quad A(f_2(j)) =\\ &\quad w_1(f_1^{-1}(f_2(j)))\\ &\quad w_2(j, A(f_2(j)))\\ &\textbf{enddo } \equiv\\ &\textbf{doall } j \in \beta_3\\ &\quad w_2(j, A(f_2(j)))\\ &\textbf{enddo } \otimes\end{aligned}$
a) IF-form	b) β-form

Figure 2: TYPE I: conceptual intermediate forms

second nest (figure 2(a)) or $T_2(j)$ may be drawn into the first nest (algorithm A1 in section 3). The basic approach is illustrated in figure 2(a), which is labeled "IF-form" to emphasize the use of conditional statements. In figure 2(b) the tests are removed by the iteration space splitting technique described in section 2.1. Since the cost of determining the β sets is independent of the size of the loop iterations spaces, the β-form should be employed when $|\Gamma_1| + |\Gamma_2|$ is large.

In figure 2(a), $T_1(i)$ is transferred to the second DOALL loop iff $\exists j \in \Gamma_2$ such that $f_2(j) = f_1(i)$. Note that $A(f_1(i))$ is replaced by $A(f_2(j))$ in statement $T_1(i)$ of the second loop so that the reuse of A will be recognized. Alignment on A provides reuse of A and permits the barrier following the first DOALL loop to be removed.

When $\Gamma_1 = \Gamma_2$ and $f_1 = f_2$ alignment on A is equivalent to loop fusion. So for type I systems, our technique may be viewed as an extension of loop fusion.

2.1 Removal of Conditionals

Since the indexing functions, f_1 and f_2, are linear the conditional tests inside the loops may be removed via index set splitting. Equations 1 thru 4 define the iteration sets required by figure 2(b) in terms of the original index functions and iteration spaces. The notation $\pi_k(X)$ denotes the dimension k projection of the set X. For example: $\pi_2(\{(1,2),(2,3),(1,3)\}) = \{2,3\}$. The balance of this section provides a detailed approach for computing and exploiting the β sets.

$$\begin{aligned}
S &= \{(i,j) \in \Gamma_1 \times \Gamma_2 : f_1(i) = f_2(j)\} & (1)\\
\beta_1 &= \Gamma_1 \setminus \pi_1(S) & (2)\\
\beta_2 &= \pi_2(S) & (3)\\
\beta_3 &= \Gamma_2 \setminus \pi_2(S) & (4)
\end{aligned}$$

Key relationships are shown in figure 3. $R_A = f_1(\Gamma_1) \cap f_2(\Gamma_2)$, the elements of A written by T_1 and later read by T_2. By construction, $\beta_2 = f_2^{-1}(R_A)$, so all potential reuse of A is realized by the β_2 loop. Since β_2 also equals $f_1^{-1}(R_A)$, alignment on A could be achieved by moving selected T_2 to the first loop.

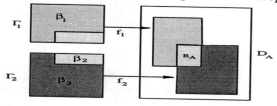

Figure 3: Type I: Domain Decomposition

To determine S we must solve the Diophantine equation $k_1 i + (-k_2)j = (c_2 - c_1)$. Since Diophantine equations are well understood, we proceed directly to the essential computations. The extented gcd algorithm determines d, a and b, which along with p, define h_1 and h_2. If d does not divide $(c_2 - c_1)$ there is no general solution; i.e., $S = \emptyset$.

$$
\begin{aligned}
d &= \gcd(|k_1|, |k_2|) = ak_1 - bk_2 \\
p &= (c_2 - c_1)/d \\
i &= h_1(t) = pa + \frac{k_2}{d}t \\
j &= h_2(t) = pb + \frac{k_1}{d}t
\end{aligned}
$$

Since $k_1 k_2 \neq 0$, the parametric solution functions, h_1 and h_2, and their respective inverses are well defined whenever the general solution is non-empty. By construction, $S \subseteq \{(h_1(t), h_2(t)) : t \in \mathbf{Z}\}$.

To determine S it is necessary to restrict t so that $(h_1(t), h_2(t)) \in \Gamma_1 \times \Gamma_2$. Equations 5 and 6 define $[t_1 : t_2]$ such that $(h_1(t), h_2(t)) \in \Gamma_1 \times \Gamma_2$ iff $t \in [t_1 : t_2]$. If $\Gamma_1 = \emptyset \lor \Gamma_2 = \emptyset$, then $t_1 > t_2$, so boundary cases are handled gracefully.

$$
\begin{aligned}
t_1 &= \max(\lceil h_1^{-1}(\text{if } k_2 > 0 \text{ then } L_1 \text{ else } U_1 \text{ fi})\rceil, \\
&\quad \lceil h_2^{-1}(\text{if } k_1 > 0 \text{ then } L_2 \text{ else } U_2 \text{ fi})\rceil) \quad (5)
\end{aligned}
$$

$$
\begin{aligned}
t_2 &= \min(\lfloor h_1^{-1}(\text{if } k_2 > 0 \text{ then } U_1 \text{ else } L_1 \text{ fi})\rfloor, \\
&\quad \lfloor h_2^{-1}(\text{if } k_1 > 0 \text{ then } U_2 \text{ else } L_2 \text{ fi})\rfloor) \quad (6)
\end{aligned}
$$

Thus far we have determined that:

$$
\beta_1 = \Gamma_1 \setminus h_1([t_1 : t_2]) \qquad \beta_2 = h_2([t_1 : t_2])
$$
$$
\beta_3 = \Gamma_2 \setminus h_2([t_1 : t_2])
$$

For convenience in loop construction, we would like a positive stride through each iteration space. By construction, (k_2/d) divides $(h_1(t_2) - h_1(t_1))$, so when $k_2 < 0$, a positive stride is obtained by exchanging interval bounds. The key intermediate sets, $[\mathcal{L}_1 : \mathcal{U}_1 : \mathcal{S}_1]$ and $[\mathcal{L}_2 : \mathcal{U}_2 : \mathcal{S}_2]$, are constructed as follows (where $j = 3 - i$):

$$
[\mathcal{L}_i : \mathcal{U}_i : \mathcal{S}_i] = \begin{cases} \left[h_i(t_1) : h_i(t_2) : \frac{k_j}{d}\right], & \text{if } k_j > 0 \\ \left[h_i(t_2) : h_i(t_1) : \frac{-k_j}{d}\right], & \text{if } k_j < 0 \end{cases}
$$

Let $h(t) = (h_1(t), h_2(t))$. Then, by construction, $S = h([t_1 : t_2])$. If $t_1 > t_2$ then S is empty. In equations 7 thru 9 the stride in each $[x : y : s]$ set is positive. It follows that $[\mathcal{L}_i : \mathcal{U}_i : \mathcal{S}_i] \subseteq \Gamma_i$. Therefore the β sets, as defined by equations 7 thru 9, are well suited to loop construction.

$$
\begin{aligned}
\beta_1 &= [L_1 : U_1] \setminus [\mathcal{L}_1 : \mathcal{U}_1 : \mathcal{S}_1] &(7) \\
\beta_2 &= [\mathcal{L}_2 : \mathcal{U}_2 : \mathcal{S}_2] &(8) \\
\beta_3 &= [L_2 : U_2] \setminus [\mathcal{L}_2 : \mathcal{U}_2 : \mathcal{S}_2] &(9)
\end{aligned}
$$

While mapping **doall** $j \in \beta_2$ to a standard loop form is trivial, conversion of the β_1 and β_3 forms is less obvious. As shown above, the β_1 and β_3 sets may be written in the form $[L : U] \setminus [a : b : s]$, where $s > 0$. Figure 4 shows how iterations sets of this later form may be constructed.

(a) set difference	(b) nested loops
doall $i \in [L : U] \setminus$ $[a : b : s]$ iteration(i) **enddo** $\otimes$	**doall** $i = L$, $a - 1$ iteration(i) **enddo** $\equiv$ **doall** $j = a$, $b - s$, s **doall** $i = j + 1$, $j + s - 1$ iteration(i) **enddo** $\equiv$ **enddo** $\equiv$ **doall** $i = b + 1$, U iteration(i) **enddo** $\otimes$

Figure 4: Nest forms for β_1 and β_3

2.2 Example

(a) original	(b) transformed
doall $i = 1$, 1000 A(i+1) = X(i) **enddo** $\otimes$ **doall** $j = 5$, 1005 Y(j) = A(2*j + 3) **enddo** $\otimes$	**doall** $i = 1$, 11 A(i+1) = X(i) **enddo** $\equiv$ **doall** $i = 13$, 999, 2 A(i+1) = X(i) **enddo** $\equiv$ **doall** $j = 5$, 499 A(2*j+3) = X(2*j+2) Y(j) = A(2*j+3) **enddo** $\equiv$ **doall** $j = 500$, 1005 Y(j) = A(2*j+3) **enddo** $\otimes$

Figure 5: Example of Type I Conversion

We conclude our discuss of type-I with an example and a brief discussion of reuse measurement. Figure 5 shows a candidate system and the corresponding transformed system. For clarity, zero and one trip loops have been removed. For this example:

$$d = 1,\ a = 1,\ b = 0,\ p = 2$$
$$h_1(t) = 2 + 2t,\ h_2(t) = t,\ [t_1 : t_2] = [5 : 499]$$
$$\beta_1 = [1 : 11] \setminus [12 : 1000 : 2]$$
$$\beta_2 = [5 : 499]$$
$$\beta_3 = [500 : 1005] \setminus [5 : 499]$$

In a type-I loop system, alignment achieves all available temporal reuse of A. Equation 10 defines a measure of reuse potential, $R \in [0,1]$. $R = 0$ indicates no potential reuse, while $R = 1$ indicates ideal opportunity.

$$R = \frac{|f_1(\Gamma_1) \cap f_2(\Gamma_2)|}{\min(|\Gamma_1|, |\Gamma_2|)} = \frac{|\beta_2|}{\min(|\Gamma_1|, |\Gamma_2|)} \quad (10)$$

Since potential for reuse, R, is influenced by the size of β_2, it is useful to establish bounds for $|\beta_2|$. By inspection of h_1 and h_2, we may observe that $|\beta_2|$ becomes smaller as $|k_1|$ or $|k_2|$ grow. More precisely,

$$|\beta_2| \leq \min\left(\frac{|\Gamma_1|}{|k_2|}, \frac{|\Gamma_2|}{|k_1|}\right) d + 1$$

Therefore reuse potential is greatest when indexing functions have small strides. For our example $R = 0.495$, a value bounded by $1/k_2$.

3 TYPE II Systems

While type-I systems have a single flow dependence (on A), type-II systems have both flow (on A) and anti-flow (on B) dependence. In general, anti-flow dependences can be eliminated by storage splitting, so anti-flow dependences are storage related. But the reuse benefit of alignment on A can easily be overshadowed by the cost of storage splitting. Therefore, to ensure a net benefit from storage alignment, our analysis of type-II systems addresses both dependences without storage splitting. Having examined the basic mechanics of alignment in our analysis of type-I systems, treatment of type-II systems will focus on the interplay between the flow and anti-flow dependences.

In figure 1 we defined type-II systems in terms of four index maps (f_{11}, f_{12}, f_{21} and f_{22}) and two iteration spaces (Γ_1 and Γ_2). Flow dependence on A is captured by the dependence relationship defined in equation 11. As with S in section 2, $(i,j) \in S^A$ if and only if $T_1(i)$ writes an element of A which is later read by $T_2(j)$. Likewise S^B, as defined by equation 12, has the property that $(i,j) \in S^B$ if and only if $T_1(i)$ reads an element of B which is later written by $T_2(j)$. The algorithm for solving bounded diophantine systems presented in section 2 is applicable to these systems, and by extension, to the set intersection which follow. As with S in section 2, let $S_k^X = \pi_k(S^X)$; so $S_1^A = \pi_1(S^A)$, $S_2^B = \pi_2(S^B)$, etc. Equation 13 defines

as S_1 and S_2, so that S_k contains the members of Γ_k which have dependences on both A and B.

$$S^A = \{(i,j) \in \Gamma_1 \times \Gamma_2 : f_{11}(i) = f_{22}(j)\} \quad (11)$$
$$S^B = \{(i,j) \in \Gamma_1 \times \Gamma_2 : f_{12}(i) = f_{21}(j)\} \quad (12)$$
$$S_1 = S_1^A \cap S_1^B\ ;\ S_2 = S_2^A \cap S_2^B \quad (13)$$

The relationships between the foundational maps and sets employed in our analysis of type-II systems are show in figure 6. $D(A)$ and $D(B)$ denote the index domains for the arrays A and B. $C(A) = f_{11}(\Gamma_1) \cap f_{22}(\Gamma_2)$, the part of A written in the first loop and read in the second. Similarly, $C(B) = f_{12}(\Gamma_1) \cap f_{21}(\Gamma_2)$, the part of B read in the first loop and written in the second. By construction, the map $f_{11} : S_1^A \rightarrow C(A)$ is one-to-one and onto. Likewise, f_{12}, f_{21} and f_{22} are each one-to-one and onto. For sake of convenience, all functions extend to $\mathbf{R}$ in the obvious way. Restrictions on domain will be made explicit, as appropriate.

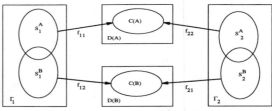

Figure 6: Type II: Maps and Working Sets

The analysis that follows rests on three functions: h_A, h_B and g. h_A describes flow dependence on A; so when $(i,j) \in S^A$, $h_A(i) = j$. Likewise, h_B describes anti-flow dependence on B. h_A and h_B are defined by composition of index maps.

$$h_A(x) = f_{22}^{-1}(f_{11}(x)) \qquad h_B(x) = f_{21}^{-1}(f_{12}(x))$$

Equation 14 defines g, a map that provides a conservative approximation of the order dependences which storage alignment on A induces on Γ_1. Subsequent occurrences of m and b refer to equation 14.

$$g(x) = f_{12}^{-1}(f_{21}(f_{22}^{-1}(f_{11}(x))))$$
$$= h_B^{-1}(h_A(x)) = mx + b \quad (14)$$

In type-II systems storage alignment may induce a dependence chain. When the statements $T_2(j)$, where $j \in S_2^A$, are moved from second loop to the first, an execution order requirement may be induced on Γ_1. In particular, storage alignment on A forces iteration k to proceed iteration i when:

$$T_1(i) \xrightarrow{A} T_2(j) \xleftarrow{B} T_1(k)$$

Theorem 1

if $i, k \in \Gamma_1 \wedge h_A(i) = j = h_B(k) \in \Gamma_2$ then

(i) $i \in S_1^A$, (ii) $j \in S_2$, (iii) $k = g(i) \in S_1^B$

Theorem 1 follows directly from the definitions of h_A, h_B and g. So when storage alignment on A requires iteration k to precede iteration i, $k = g(i)$. Therefore g provides a conservative predictor for alignment induced ordering of Γ_1.

Since g is a conservative estimate of alignment induced ordering, a dependence chain predicted by g may include few, if any, actual dependences. Consider $f_{11}(x) = f_{21}(x) = x$, $f_{12}(x) = 3x$, $f_{22}(x) = 6x$ and $\Gamma_1 = \Gamma_2 = [1 : 100]$. Then $g(x) = x/2$, so the following forms a g-chain in Γ_1: 64, 32, 16, 8, 4, 2, 1. But none of these chain members is aligned with a member of Γ_2 because $h_A(x) = x/3$. Thus, an estimate of chain length derived from g, even when the domain of g is constrained to Γ_1, may substantially overestimate the length of induced dependence chains.

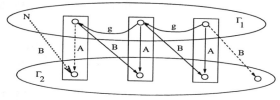

Figure 7: Alignment Induced Ordering of Γ_1

Figure 7 illustrates the key properties of a dependence chain. Since the A links are defined by h_A, the B links by h_B, and h_A and h_B are each one-to-one, the dependence graph is the union of two bipartite graphs. Therefore, it is a collection of disjoint cycles and chains. Each link of a chain (or cycle) is associated with a member of S_2 and each order dependence is predicted by g^{-1}. These observations undergird the taxonomy presented in figure 8.

When $m \neq 1$, $x_0 = -b/(m-1)$ is the fixed point of g. Let Γ_1^* denote the non-fixed-points of Γ_1; so when $m \neq 1$ then $\Gamma_1^* = \Gamma_1 \setminus \{x_0\}$, otherwise $\Gamma_1^* = \Gamma_1$.

Figure 8 shows our algorithm selection criterion. Each algorithm, along with its precondition, is described below. Algorithms A1 thru A4 are very similar in form; their pseudo-code forms are presented in figure 9. Algorithms A5, A6.1 and A6.2 are more subtle since they handle dependence chains or cycles induced by alignment on A.

3.1 Algorithm A1: $S^B \subseteq S^A$

Clearly, $S^B = \emptyset \Rightarrow S^B \subseteq S^A$. If $g(x) \equiv x$ then $h_A = h_B$, so $S^B \subseteq S^A$. Since alignment on A ensures that the data dependencies on A are satisfied and $S^B \subseteq S^A$, alignment on A ensures satisfaction of the dependencies on B. Algorithm A1 is essentially the same as the type I algorithm. The β-form for type I, as

```
if  S^B = ∅ ∨ g(x) ≡ x then
   A1: Align on A, no barrier. (TYPE I)
elsif  S^A = ∅ then
   A2: align on B, no barrier.
elsif  S_1 = ∅ then
   A3: align on A, need barrier, move T_1 to 2nd loop.
elsif  S_2 = ∅ then
   A4: align on A, need barrier, move T_2 to 1st loop.
elsif  m = -1 then
   A5: 2-cycle merge, need barrier
else
   A6.1: chain by post/wait  (or)
   A6.2: chain decomposition.
end if
```

Figure 8: Algorithm Taxonomy for **TYPE II**

presented in section 2, moves statements from the first loop to the second, while A1 moves statements from the second loop to the first. In general, either choice is valid, though for a given system one approach may perform better then the other.

If both S^A and S^B are empty then there are no data dependences between the loops, and therefore no reuse opportunity. As a type-I systems, $\beta_1 = \Gamma_1$, $\beta_2 = \emptyset$ and $\beta_3 = \Gamma_2$; which is equivalent to simple barrier removal.

3.2 Algorithm A2: No dependence on A

Since S^A is empty, there is no reuse opportunity. Therefore A2 alignments on B in order to eliminate barrier synchronization. When barrier synchronization overhead is significant, elimination of the barrier may be worthwhile – even when there is no memory reuse opportunity.

3.3 A3: move $T_1(S_1^A)$ to second loop

Because there are dependences on both A and B, and A's dependencies do not cover B's, the inter-loop barrier must be retained. S_1 is empty, so the longest dependence chain possible has the form

$$T_1(i) \xrightarrow{A} T_2(j) \xleftarrow{B} T_1(k)$$

By moving $T_1(i)$ to the second loop, rather than $T_2(j)$ to the first, we avoid the need for additional synchronization in either loop. Barrier synchronization between the parallel loops ensures that all data dependencies on B are respected. If S_2 is also empty then statements can be moved either way; the dependencies on A and B would then be completely disjoint.

3.4 A4: move $T_2(S_2^A)$ to first loop

Algorithm A4 is the dual of A3. As with A3, dependence chains are extremely limited. When S_2 is empty, the longest dependence chain has the form

$$T_2(j) \xleftarrow{A} T_1(i) \xrightarrow{B} T_2(k)$$

doall $i \in S_1^A$	doall $i \in S_1^B$
$T_1(i)$	$T_1(i)$
$T_2(h_A(i))$	$T_2(h_B(i))$
enddo $\equiv$	enddo $\equiv$
doall $i \in \Gamma_1 \setminus S_1^A$	doall $i \in \Gamma_1 \setminus S_1^B$
$T_1(i)$	$T_1(i)$
enddo $\equiv$	enddo $\equiv$
doall $j \in \Gamma_2 \setminus S_2^A$	doall $j \in \Gamma_2 \setminus S_2^B$
$T_2(j)$	$T_2(j)$
enddo $\otimes$	enddo $\otimes$
Algorithm A1	Algorithm A2
doall $i \in \Gamma_1 \setminus S_1^A$	doall $i \in S_1^A$
$T_1(i)$	$T_1(i)$
enddo $\otimes$	$T_2(h_A(i))$
doall $j \in S_2^A$	enddo $\equiv$
$T_1(h_A^{-1}(j))$	doall $i \in \Gamma_1 \setminus S_1^A$
$T_2(j)$	$T_1(i)$
enddo $\equiv$	enddo $\otimes$
doall $j \in \Gamma_2 \setminus S_2^A$	doall $j \in \Gamma_2 \setminus S_2^A$
$T_2(j)$	$T_2(j)$
enddo $\otimes$	enddo $\otimes$
Algorithm A3	Algorithm A4

Figure 9: Algorithms A1 - A4

By moving $T_2(j)$ immediately after $T_1(i)$ in the body of the first loop, the need for additional synchronization is avoided. As with A3, the inter-loop barrier enforces the dependence on B. Since $S_1 \neq \emptyset$, there is a least one $T_1(i)$ that must be moved between loops in order to align on A.

3.5 A5: 2-cycle merge method

Before exploring the implications of $m = -1$, we show that alignment on A can induce cycles when $m = -1$. Any chain or cycle that may occur due to storage alignment on A, must be present in g. Since g has no cycles (excluding fixed points) when $m \neq -1$, it follows that non-trivial cycles may occur only when $m = -1$.

g predicts a cycle only if $\exists x \in \Gamma_1$ and $k > 1$ such that $g^{(k)}(x) = x$. If $m = 1$ then $g^{(k)}(x) = x + kb$, so only trivial cycles are possible. When $g(x) \equiv x$ all points are fixed points; otherwise $b \neq 0$ and there are no cycles. If $m \neq 1$, we find that

$$g^{(k)}(x) - x = 0 \;\; \Leftrightarrow (x - x_0)\, m^k + x_0 - x = 0$$
$$\Leftrightarrow (m^k - 1)\, (x - x_0) = 0$$
$$\Leftrightarrow m^k = 1 \vee x = x_0$$

Since $m \in Q$ and $m \neq 1$, if follows that non-trivial cycles can occur in only when $m = -1$. Hence, induced dependence cycles are handled exclusively by algorithm A5.

For the balance of this section assume that $m = -1$, so $g(x) = -x + b$. If $\exists x, g(x) \in \Gamma_1$ then $b = x + g(x) \in$

Z. We observe that if $b \notin Z$ then $g(\Gamma_1) \cap \Gamma_1 = \emptyset$, which implies that $S_2 = \emptyset$, so algorithm A4 would be selected. Therefore, algorithm A5 may assume that $b \in Z$ and $x_0 = b/2 \in Z/2$.

$g(x) = -x + b$ decomposes Q into $\{x_0\}$ and $\{\{x, g(x)\} : x \in \Gamma_1^*\}$. Since $\Gamma_1 \neq Z$, g partitions Γ_1 into three types of points: I_0, a fixed point (when $x_0 \in \Gamma_1$); $I_2 \cup g(I_2)$, 2-cycles constained in Γ_1; and I_1, members of Γ_1 whose partner falls outside. Formally, $I_0, I_1, I_2 \subseteq \Gamma_1$ are:

$$I_0 = \{\, x \in \Gamma_1 : x = x_0 \,\}$$
$$I_1 = \{\, x \in \Gamma_1 : g(x) \notin \Gamma_1 \,\}$$
$$I_2 = \{\, x \in \Gamma_1 : x < g(x) \in \Gamma_1 \,\}$$

By construction, $I_0, I_1, I_2, g(I_2) \vdash \Gamma_1$. If $x_0 \notin [L_1, U_1]$ then $I_0 = I_2 = \emptyset$ and $I_1 = \Gamma_1$. But then $\Gamma_1 \setminus g(\Gamma_1) = \emptyset$, which implies $S_2 = \emptyset$, so algorithm A4 would have been selected. Therefore, algorithm A5 may assume that $x_0 \in [L_1, U_1]$. The computation of I_0 follows directly from its definition, while I_1 and I_2, which are less obvious, are determined by these less obvious formulae:

$$I_1 = [\, L_1 : b - U_1 - 1\,] \cup [\, b - L_1 + 1 : U_1\,]$$
$$I_2 = [\, \max(b - U_1, L_1) : \lfloor x_0 \rfloor - |I_0|\,]$$

It is useful to observe that if $x_0 > (L_1 + U_1)/2$ then $I_1 = [L_1 : b - U_1 - 1]$, while if $x_0 < (L_1 + U_1)/2$ then $I_1 = [b - L_1 + 1 : U_1]$. When $x_0 = (L_1 + U_1)$, $I_1 = \emptyset$ because Γ_1^* decomposes entirely into two-cycles.

Algorithm A5, as presented in figure 10, uses I_2 to identify iteration pairs that may induce an ordering constraint on Γ_1. Since alignment on A is performed by moving $T_2(S_2^A)$ to the first loop (lines 3, 7 and 8), special care must be taken with $T_2(S_2)$. Dependence constraints for other iterations of Γ_2 are satisfied through alignment (when $j \in S_2^A \setminus S_2^B$) or as a consequence of the barrier on line 9 (when $j \in S_2^B \setminus S_2^A$). If $j \in S_2$ and $i = h_A^{-1}(j)$ then $h_B^{-1}(j) = g(i)$. Therefore, all potential dependence order problems in Γ_1 lie in $I_2 \cup g(I_2)$.

We may partion I_2 into four subsets:

$P1 = I_2 \cap S_1$	$(i, g(i) \;\in S_1^A)$
$P2 = (I_2 \cap S_1^A) \setminus S_1^B$	$(i \;\in S_1^A)$
$P3 = (I_2 \cap S_1^B) \setminus S_1^A$	$(g(i) \;\in S_1^A)$
$P4 = I_2 \setminus (S_1^A \cap S_1^B)$	

If the iteration space is sufficiently large, it may be appropriate to split the iteration space of the loop 5 to remove tests on lines 7 and 8. When $I_2 \subseteq S_1$ the tests on lines 7 and 8 may be removed (since $P1 = I_2$). When $I_2 \cap S_1^A = I_2 \cap S_1^B$, P2 and P3 are empty so the tests on lines 7 and 8 may be merged; i.e., rewritten as "if $i \in S_1^A$ then $T_2(h_A(i)); T_2(h_A(g(i)))$ fi". In general, its difficult to remove the inner tests.

```
 1:  doall i ∈ I_0 ∪ I_1
 2:      T_1(i)
 3:      if i ∈ S_1^A then T_2(h_A(i)) fi
 4:  enddo ≡
 5:  doall i ∈ I_2
 6:      T_1(i);  T_1(g(i))
 7:      if i ∈ S_1^A then T_2(h_A(i)) fi
 8:      if g(i) ∈ S_1^A then T_2(h_A(g(i))) fi
 9:  enddo ⊗
10:  doall j ∈ Γ_2 \ S_2^A
11:      T_2(j)
12:  enddo ⊗
```

Figure 10: A5: 2-cycle merge algorithm

```
 1:  doall i ∈ Γ_1
 2:      T_1(i)
 3:      if i ∈ S_1^B then post(i) fi
 4:      if i ∈ S_1^A then
 5:          if h_A(i) ∈ S_2^B then wait(g(i)) fi
 6:          T_2(h_A(i))
 7:      end if
 8:  enddo ≡
 9:  doall j ∈ Γ_2 \ S_2^A
10:      if j ∈ S_2^B then wait(h_B^{-1}(j)) fi
11:      T_2(j)
12:  enddo ⊗
```

Figure 11: A6.1: Post/Wait

The statements on lines 6, 7 and 8 are ordered to ensure satisfaction of the following potential dependencies:

WR on A:
$T_1(i) \rightarrow T_2(h_A(i))$
$T_1(g(i)) \rightarrow T_2(h_A(g(i)))$

RW on B:
$T_1(i) \rightarrow T_2(h_A(g(i)))$
$T_1(g(i)) \rightarrow T_2(h_A(i))$

Any time both statements of a potential dependence pair exist, the indicated dependence constraint applies. The statements on line 6 may be performed concurrently. Likewise, the statements on lines 7 and 8 may be executed concurrently.

The first loop services unconstrained iterations of Γ_1 along with corresponding alignment candidates from Γ_2. As with algorithms A1 thru A4, the test on line 3 may be removed via iteration set splitting. The third loop (line 10) services iterations of Γ_2 which are not pulled into the first loops via alignment on A. The barrier at line 9 is necessary to ensure that $T_2(j)$ is not executed too early when $j \in S_2^B \setminus S_2^A$. When $j \in S_2^A \setminus S_2^B$, $T_2(j)$ is executed on line 3. But when $j \in S_2$, $T_2(j)$ is executed on line 7 or 8.

3.6 A6: handling induced Chains

Algorithm A6 handles induced dependence chains, but not cycles; cycles are addressed by algorithm A5. We present two approaches for achieving reuse in the context of alignment induced dependence chains: a post/wait algorithm (section 3.6.1) and a direct decomposition approach (section 3.6.2). The strengths and weaknesses of each approach are discussed.

3.6.1 A6.1: post/wait

The post/wait algorithm described in figure 11 ensures satisfaction of A's flow dependencies through storage alignment and satisfaction of B's anti-flow dependencies by post/wait synchronization [5]. When $|S^B| \ll |\Gamma_1|$ the simplicity of the iteration space of the post/wait algorithm (line 1) compensates for its synchronization overhead (lines 3, 5, 10).

Tests within loop bodies can be removed using the iteration space splitting techniques outlined in section 2. Exact branch frequencies, which may be computed in advance, are:

Line	prob. True	Line	prob. True
3	$|S^B|/|\Gamma_1|$	5	$|S_2|/|S^A|$
4	$|S^A|/|\Gamma_1|$	10	$|S^B|/(|\Gamma_2| - |S_A|)$

3.6.2 A6.2: direct decomposition

Figure 12 presents a direct chain decomposition algorithm. Each iteration of the first loop processes an induced dependence chain, while the second loop performs instances of T_2 which exhibit no reuse opportunity. Justification for the key conditions, lines 1 and 6, follow from the structure of the dependence chains (figure 7). Let $H = \Gamma_1 \setminus h_A^{-1}(S_2)$ and $F = h_B^{-1}(S_2)$. If $i \in H$ then i is a chain head because there is no $k \in \Gamma_1$ such that $h_B(k) \in \Gamma_2$ and $h_A^{-1}(h_B(k)) = i$. Conversely, if $i \in \Gamma_1 \setminus H$ then $h_A(i) \in S_2$. But $h_A(i) \in S_2$ implies that $h_B^{-1}(h_A(i)) = g(i) \in \Gamma_1$, so $T_1(g(i))$ must precede $T_2(h_A(i))$; i.e., iteration i is not a chain head. By similar argument, F is the set of iterations which have an induced successor. On line 7, g^{-1} advances to the next member of an induced dependence chain.

Direct chain decomposition (A6.2) is a general purpose approach which may be used in place of all algorithms except A5. When the precondition for A1 holds, the chain exit condition (line 6) is always true because when $S^B \subseteq S^A$ it follows that $h_B^{-1}(S_2) = S_1^B \subseteq S_1^A = h_A^{-1}(S_2)$. When A2 is applicable, algorithm A6.2 has no effect on execution order, synchronization or reuse; it does not align on B, as is done by algorithm A2. When A3 is applicable, movement of $T_2(S_2^A)$ to the first loop induces short (length 2) dependence chains. When A4 is applicable, S_2 is empty, so A6.2 is identical to A4 (modulo control overhead). Code size may be reduced, in exchange for control overhead, by a three-fold approach:

```
1:    doall i ∈ Γ₁ \ h_A⁻¹(S₂)
2:       k = i
3:       loop
4:          T₁(k)
5:          if k ∈ S₁ᴬ then T₂(h_A(k)) fi
6:          exit loop when  k ∉ h_B⁻¹(S₂)
7:          k = g⁻¹(k)
8:       end loop
9:    enddo ⊗
10:   doall j ∈ Γ₂ \ S₂ᴬ
11:      T₂(j)
12:   enddo ⊗
```

Figure 12: A6.2: Direct Chain Decomposition

if $m = 0$ **then** process loops separately
elsif $m = -1 \wedge |S_1| + |S_2| > 0$ **then** use A5
else use A6.2

In principle, algorithm A5 may be replaced by A6.1 because the post precedes the wait, thereby ensuring deadlock-free synchronization. Unfortunately, in practice, the post/wait approach is subject to the possibility of deadlock if the number of independent threads of control, P, is less than $|I_2|$. When $P < |I_2|$ it is possible for all P threads to be waiting at line 5 (figure 11). So algorithm A6.1 should not be adopted as a general purpose replacement for A5 unless the scheduling mechanism can preempt waiting threads.

Algorithm A6.2 is not a suitable replacement for A5 because it is not able to handle induced dependence cycles in Γ_1. The direct decomposition approach enforces the following execution sequence:

$$T_1(i);\ \ T_2(h_A(i));\ \ T_1(g^{-1}(i));\ \ T_2(h_A(g^{-1}(i)))$$

When the preconditions of algorithm A5 are satisfied, $g = g^{-1}$. So when $i \in I_2$, $T_1(g^{-1}(i))$ reads a member of B which has been prematurely updated by $T_2(h_A(i))$. If $H \cap I_2 = \emptyset$ then $T_1(I_2)$ and $T_1(g(I_2))$ are never executed. Hence, A6.2 may not replace A5; though it may safely replace any of A1 thru A4.

4   Concluding Remarks

Since the speed of processors has been increasing at a greater rate than memory bandwidth, and this trend is likely to continue, memory reuse techniques will become critical to overall systems performance. The loop restructuring techniques presented here are designed to provide maximum memory reuse in exchange for additional loop startup cost and a potential reduction in parallelism. Though at present the overhead of these techniques appears to be prohibitive, as the ratio of processor speed to memory bandwidth grows these technique will become more attractive.

This paper provides the theoretical foundation for a family of loop restructuring techniques. Additional research is required to determine if these methods can be extented to other loop systems and reference patterns and empirical studies are needed to determine the conditions under which these algorithms are appropriate.

References

[1] S.G. Abraham and D.E. Hudak, "Compile-time partitioning of iterative parallel loops to reduce cache coherency traffic", *IEEE Transactions on Parallel and Distributed Systems*, (July 1991), pp. 318–328.

[2] J. Fang and M. Lu, "An iteration partition approach for cache or local memory thrashing on parallel processing", *Proc. of the 4th Workshop on Languages and Compilers for Parallel Computing*, 1991, pp. 313–327.

[3] D. Gannon, W. Jalby and K. Gallivan, "Strategies for cache and local memory management by global program transformation", *Journal of Parallel and Distributed Computing*, (Oct, 1988), pp. 587–616.

[4] E. D'Hollander, "Partitioning and labeling of index sets in do loops with constant dependence vectors", *International Conference on Parallel Processing*, 1989, pp. 139–144.

[5] S.P. Midkiff and D.A. Padua, "Compiler algorithms for synchronization", *IEEE Transactions on Computers*, (Dec, 1987), pp. 1485–1495.

[6] J. Peir and R. Cytron, "Minimum distance: A method for partitioning recurrences for multiprocessors", *IEEE Transactions on Computers*, (Aug, 1988), pp. 1203–1211.

[7] K.A. Tomko and S.G. Abraham, "Iteration partitioning for resolving stride conflicts on cach-coherent multiprocessors", *International Conference on Parallel Processing*, 1993, pp. II95–102.

[8] S. Venugopal and W. Eventoff, "Aotomatic transformation of FORTRAN loops to reduce cache conflicts", *1991 International Conference on Supercomputing*, Cologne, Germany, 1991, pp. 183–193.

[9] M., Wolfe and M. Lam, "A data locality optimizing algorithm", *SIGPLAN '91 Conference on Program Language Design and Implementation*, 1991, pp. 30–44.

[10] H. Zima and B. Chapman, *Supercompilers for Parallel and Vector Computers*, ACM Press Frontier Series, Addison-Wesley, Menlo Park, CA, (1991), 376 pp.

THE CDP^2 ALGORITHM
A COMBINED DATA AND PROGRAM PARTITIONING ALGORITHM ON THE DATA PARTITIONING GRAPH

Tsuneo Nakanishi[†], Kazuki Joe[†], Hideki Saito[‡], Akira Fukuda[†], Keijiro Araki[†]

† Graduate School of Information Science,
Nara Institute of Science and Technology
8916-5 Takayama-cho, Ikoma, Nara, 630-01 JAPAN
‡ Center for Supercomputing Research and Development,
University of Illinois at Urbana-Champaign
1308 W. Main St., Urbana, IL, 61801 U.S.A.
nakasu-para@is.aist-nara.ac.jp

Abstract — *In this paper we propose the CDP^2 algorithm to perform data partitioning, data distribution and program partitioning simultaneously on the Data Partitioning Graph (DPG), an intermediate representation for parallelizing compilers. So far, program partitioning and data partitioning have been treated as separate problems. However, we cannot obtain communication costs without a data partitioning and location decision. Likewise, we cannot decide data partitioning without a program partitioning decision. Therefore both problems should be dealt with simultaneously. The CDP^2 algorithm handles both problems in an integrated manner and offers better data and program partitioning solutions.*

1 INTRODUCTION

The introduction of large-scale distributed shared memory multiprocessors has disclosed the problem of data partitioning. Since the cost of a data access on such a machine largely depends on where the data is located, appropriate partitioning and distribution of a program and its data are very important to reduce the number of high-cost remote memory accesses.

The program partitioning problem to obtain tasks with optimal granularity and the processor assignment (scheduling) problem to assign the tasks to the processors have been discussed in earlier papers[1,2]. However, the problem of data partitioning has not been taken seriously partly because those previous works are based on small-scale machines or centralized shared memory.

So far, parallelizing compilers have often employed the dependence graph to perform scheduling, program partitioning, restructuring, etc. The dependence graph has been convenient for these issues because of its simplicity. However, the dependence graph is not powerful enough in the distributed shared memory environment, since the dependence graph does not carry the information on the data referred to by the tasks. This information is essential for data partitioning and data distribution. Therefore, the dependence graph will not be able to deal simultaneously with the data partitioning and distribution problems and the program partitioning problem.

Based on the observation discussed above, we have proposed the Data Partitioning Graph (DPG). The DPG is an intermediate representation which can handle program partitioning, processor assignment, data partitioning and data distribution in an integrated manner. The DPG has nodes representing tasks as the dependence graph does, however, it also has nodes representing the data referred to by the tasks.

In this paper we propose the CDP^2 algorithm, which works on the DPG, to perform program and data partitioning simultaneously. Chapter 2 gives a brief introduction to the Data Partitioning Graph. Chapter 3 describes our parallel execution model and discusses the program partitioning (grain packing) and data-program partitioning. Chapter 4 summarizes Girkar's program partitioning algorithm. In Chapter 5, the overview of the CDP^2 algorithm is followed by the fundamentals and the detail of the algorithm. Chapter 6 discusses the advantages of the CDP^2 algorithm against Girkar's algorithm. Finally, Chapter 7 gives the conclusion of the paper.

2 DATA PARTITIONING GRAPH

Dependence Graph

Parallelizing compilers employ various kinds of task graphs. The simplest task graph will be the dependence graph which is often used for scheduling. Nodes of the dependence graph express tasks. An edge $e = (u, v)$ of the dependence graph expresses that the task v cannot begin its execution before the completion of the task u. Therefore, edges of the dependence graph represents dependencies between tasks. The dependence graph is formally denoted by $G = (V, E)$, where V and E is a set of nodes and edges respectively. We may refer to *task* and *node* interchangeably since they have a one-to-one correspondence and likewise for *dependence* and *edge*.

The dependence graph can hold costs as the attributes of its nodes and edges. When we use the dependence graph for scheduling (and program partitioning), we assign the execution time of a task to the corresponding node and the communication time on a dependence to the corresponding edge. We denote the cost of a node v by $\omega(v)$ and the cost of an edge e by $\omega(e)$. We define the length of the path on the dependence graph as the total cost of the nodes and the edges on the path. We call the maximum length path the *critical path*. The critical path of a dependence graph gives the optimal parallel execution time of the program.

Data Partitioning Graph

The DPG has not only nodes for tasks like the dependence graph but also other nodes for sets of variables or array elements referred to by the tasks. We call the former *C-nodes* and the latter *D-nodes*. An edge between C-nodes represents a control dependence or a data dependence while an edge between a C-node and a D-node expresses that the C-node access a set of variables represented by the D-node. In short, the DPG has four kinds of edges: control dependence edge, data dependence edge, read access edge and write access edge. Read access edges and write access edges are collectively called *data access edges*. We assume DPGs to be acyclic in terms of data dependence edges in the rest of this paper. (It is not necessary to be acyclic for data access edges.)

The DPG holds the communication time of a data access as an attribute of the corresponding data access edge and the execution time of a task as an attribute of the corresponding C-node. We denote the communication time of a data access edge de by $\omega(de)$, and the execution time of a C-node cv by $\omega(cv)$. A data dependence edge can hold the communication time on the corresponding dependence, which can be computed from the communication times of data access edges. We denote the communication time of a data dependence edge ce_d by $\omega(ce_d)$.

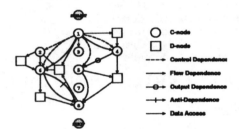

Figure 1: DPG: Data Partitioning Graph

Table 1: a classification of variables

	0	1	2
A	(R=8,W=5)	(R=x,W=5)	(R=x,W=5)
B	(R=3,W=x)	(R=3,W=4)	(R=3,W=4)
C	(R=x,W=x)	(R=2,W=x)	(R=x,W=2)
D	(R=8,W=5)	(R=x,W=5)	(R=x,W=5)

In order to introduce D-nodes, we classify all variables of a given program according to their access patterns. We provide an explanation of the variable classification by showing a simple example. See [3] for detail. In Table 1, we have three tasks $\{0, 1, 2\}$ and four variables $\{A, B, C, D\}$. R and W indicate costs for read accesses and write accesses respectively. x indicates the lack of accesses of that type. Since the variables A and D have the same access pattern, we have three classes of variables $\{A, D\}$, $\{B\}$ and $\{C\}$. A D-node is generated for each class. For instance, D-node Δ is generated for class $\{A, D\}$, and then four data access edges $(0, \Delta)$, $(\Delta, 0)$, $(1, \Delta)$ and $(2, \Delta)$ are linked.

The DPG DG is formally denoted by $DG = (\{CV, DV\}, \{CE_c, CE_d, DE_r, DE_w\})$, where CV, DV, CE_c, CE_d, DE_r and DE_w is the set of C-nodes, D-nodes, control dependence edges, data dependence edges, read access edges and write access edges respectively. Note that we may refer to *task* and *C-node*, and *data* and *D-node* interchangeably.

3 DATA-PROGRAM PARTITIONING

Model

__Target Architecture__ In this paper we assume the following architecture on which parallel programs are executed. i) The number of processors is infinite. ii) Each processor owns its local memory module which can be locally accessed with zero latency. iii) Each processor can access remote memory modules with some latency. iv) The latency required for accessing remote memory modules does not depend on their locations.

__Task Execution__ We assume that all tasks can be assigned to processors with a static scheduling scheme and that task execution is non-preemptive. Each processor executes a task in the following manner. i) Read values of variables located on remote memory modules. ii) Execute its computation. iii) Write values to variables located on remote memory modules. iv) Wait for the succeeding task to be executable. v) Go to i).

Variables located on remote memory modules are read and written collectively at steps i) and iii), respectively. Local memory accesses are available during the task computation.

Grain Packing

For efficient parallel execution, we have to partition a given program to tasks of appropriate granularities according to the overhead of parallel execution and the amount of available parallelism. Grain packing is a method to obtain a program partition of better granularity by constructing a new coarser task through the merging of multiple tasks. The number of tasks of a given program decreases and the communication between merged tasks becomes local. In other words, grain packing reduces the overhead as well as the parallelism.

Grain packing is performed on the dependence graph. When two tasks u and v are merged into one new task n, all edges between u and v are removed and all edges connecting to u or v are reconnected to n. Multiple edges may be generated between the node n and another node m in the case where there are edges e_1 and e_2 between u and m and between v and m, respectively. These multiple edges are unified into one new edge e between n and m. The cost $\omega(n)$ is a total of $\omega(u)$ and $\omega(v)$. Similarly, the cost $\omega(e)$ is a total of $\omega(e_1)$ and $\omega(e_2)$. Figure 2(a) shows an example of node merging.

If any merged task is made of the set of nodes in a weakly connected subgraph of the original dependence graph, the program partition is said to be *connected*. Girkar proved Theorem 1 on the optimal program partition[2]. This partition gives the minimum critical path length.

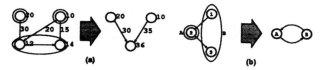

Figure 2: node merging

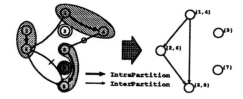

Figure 3: a connected program partition

Theorem 1 *There exists an optimal program partition to be connected. (See [2] for the proof.)* ◇

When the tasks are executed in the model described in the previous section, naive grain packing can introduce a deadlock. For example, the program partition shown in Figure 2(b) incurs a deadlock since task A cannot begin its computation before task B completes, and vice versa. To avoid deadlocks, we perform grain packing under the *convex constraint*. For any two nodes u and v which are merged into node n, we say the program partition satisfies the convex constraint if all nodes on any path from u to v are merged into n. The program partition which satisfies the convex constraint is said to be *convex*.

Data-Program Partitioning

We perform data partitioning, data distribution, and program partitioning simultaneously (data-program partitioning) on the DPG by merging C-nodes and D-nodes. When a C-node cv and a D-node dv are merged, the data represented by dv is placed on the local memory module of the processor on which the task cv runs. In data-program partitioning, C-nodes are also merged in the same way as program partitioning to obtain tasks of appropriate granularities.

4 GIRKAR'S ALGORITHM

Our CDP^2 algorithm is an extension of Girkar's program partitioning algorithm[2]. Girkar's algorithm accepts the dependence graph $G_0 = (V, E)$ as its input and generates the optimal convex program partition through grain packing. V and E are the set of nodes and edges of G_0, respectively. The granularities of the tasks of G_0 should be fine enough. According to Theorem 1, we can find the optimal program partition out of connected program partitions. Therefore, Girkar's algorithm searches the set of connected partitions of the given dependence graph for the optimal program partition by using a branch-and-bound method.

To define a connected program partition, Girkar's algorithm classifies all edges of G_0 into two sets *IntraPartition* and *InterPartition*. *IntraPartition* is a set of edges connecting the nodes to be merged. *InterPartition* is a set of remaining edges, the edges connecting the nodes not to be merged. The edges of *Intra-Partition* define a connected program partition in the

manner shown in Figure 3. A weakly connected component of graph $G' = (V, IntraPartition)$ becomes one merged task. Girkar's algorithm employs a branch-and-bound method for the edge classification. The partial solution of the branch-and-bound method used in Girkar's algorithm is defined as $\langle g, A, a \rangle$, where g is the dependence graph in the middle of merging, a is the critical path length of g, and A is a set of edges of the original dependence graph G_0 which are classified as neither *IntraPartition* nor *InterPartition*. a is used for the bounding operation. When the classification is accomplished, namely, when $A = \phi$, g is the optimal program partition. We show Girkar's algorithm in Algorithm 1.

Algorithm 1 (Girkar's algorithm)

1. *Put the initial solution $\langle G_0, E, a_0 \rangle$ into list lis.*
2. *Pick a partial solution $\langle g, A, a \rangle$ with the smallest a out of list lis.*
3. *Terminate the algorithm if $A = \phi$. g gives the optimal program partition.*
4. *Choose an edge $e = (u, v) \in A$.*
5. *Derive a new partial solution $\langle g', A', a' \rangle$ by classifying e into InterPartition and put this to the list lis. Here, $g' = g$, $A' = A - \{e\}$, and a' is the length of the longest path in g'*
6. *Let B be a set of nodes of g on all paths from u to v. These nodes should be merged to preserve the convex constraint.*
7. *Go to 2 if there is an edge $(a, b) \in E - A$ such that $\{a, b\} \subseteq B$, or the convex constraint will be violated.*
8. *Derive a new partial solution $\langle g'', A'', a'' \rangle$ by classifying e as IntraPartition and put this into the list lis. $g'' = (V'', E'')$ is a graph obtained by merging all nodes in B into one new node n. At this time $V'' \leftarrow (V - B) \cup \{n\}$ and $\omega(n) = \sum_{b \in B} \omega(b)$. a'' is the length of the longest path in g''.*
9. *Go to 2.* ◇

5 THE CDP^2 ALGORITHM

Our CDP^2 algorithm performs data partitioning and program partitioning simultaneously by merging C-nodes and D-nodes of the DPG (data-program partitioning). The granularities of the tasks of the input DPG must be fine enough.

Three steps of operations are to be performed before applying the CDP^2 algorithm. First, control dependence is converted to data dependence. This is because the CDP^2 algorithm assumes that all the C-nodes, or tasks, are executed (as assumed in Girkar's algorithm) while the DPG has control dependence edges. Second, all transitive data dependences are eliminated. Finally, for each data access edge (cv, dv) (or (dv, cv)), the communication cost of the remote memory accesses from the task cv to the data dv is assigned.

Overview of the algorithm

The algorithm iterates until a partial solution becomes the optimal solution. At each iteration, the best partial solution is picked up, the edge classification is performed and the new partial solutions are computed. The algorithm uses a branch-and-bound method on the classification of the edges.

The first step of the edge classification is to choose an unclassified data dependence edge in the partial solution. Then, a data dependence edge is classified into the sets *IntraPartition* and *InterPartition*. At the same time, the data access edges derived from the data dependence edge are classified into the sets *IntraPartitionD* and *InterPartitionD*. An edge in *IntraPartitionD* connects a C-node and a D-node to be merged. An edge in *InterPartitionD* connects a C-node and a D-node not to be merged. There can be a conflict between the data dependence edge classification and the classification of data access edges, however, those cases can easily be excluded. Each nonconflicting classification later produces a new partial solution (or a branch).

After the edge classification, the costs of data dependence edges as well as costs of nodes are computed according to the costs of data access edges and the classification decision of data access edges. The computed costs of data dependence edges are used to determine the optimal parallel execution time. We can obtain the corresponding dependence graph by removing D-nodes and data access edges from the DPG. The critical path length of the dependence graph gives the optimal parallel execution time with the given data-program partitioning.

The partial solution of the branch-and-bound method in this algorithm is defined as $\langle dg, I, A, DI, DA, a \rangle$. dg is the current DPG for data-program partitioning. I and DI are sets of data dependence edges classified as *IntraPartition* and *IntraPartitionD*, respectively. A and DA are sets of data dependence edges and data access edges whose classification are not decided, respectively. a is the critical path length of the dependence graph generated from dg by removing D-nodes and data access edges.

Fundamentals

The following are the properties of the DPG $DG = (\{CV, DV\}, \{CE_c, CE_d, DE_r, DE_w\})$ and the definition of the communication cost used in the CDP^2 algorithm.

Property 1 *For any data dependence edge $ce_d = (cv_a, cv_b) \in CE_d$ of the given DPG,*

- *if ce_d is a flow dependence edge,*
  $\exists dv \in DV : (cv_a, dv) \in DE_w, (dv, cv_b) \in DE_r$
- *if ce_d is an output dependence edge,*
  $\exists dv \in DV : (cv_a, dv) \in DE_w, (cv_b, dv) \in DE_w$
- *if ce_d is an anti-dependence edge,*
  $\exists dv \in DV : (dv, cv_a) \in DE_r, (cv_b, dv) \in DE_w$ ◇

Let us denote by $DV(ce_d)$ the set of D-nodes derived from Property 1 for any data dependence edge ce_d. Property 2 assures the consistent classification of data dependence edges and data access edges.

Property 2 *dv is any element of $DV(ce_d)$ specified for any data dependence edge $ce_d = (cv_a, cv_b)$. Let de_a be the data access edge between cv_a and dv. Let de_b be the data access edge between cv_b and dv. (See Figure 4).*

- *if $ce_d \in InterPartition$,*
  i) $de_a \in IntraPartitionD$ and $de_b \in InterPartitionD$, ii) $de_a \in InterPartitionD$ and $de_b \in IntraPartitionD$, iii) $de_a \in InterPartitionD$ and $de_b \in InterPartitionD$
- *if $ce_d \in IntraPartition$, iv) $de_a \in IntraPartitionD$ and $de_b \in IntraPartitionD$, v) $de_a \in InterPartitionD$ and $de_b \in InterPartitionD$* ◇

Figure 4: edge classification

In the DPG communication costs of data dependence edges are unknown since the DPG has communication costs in form of attributes of data access edges. Here we introduce the following definition for communication costs of data dependence edges.

Definition 1 *The communication cost $\omega(ce_d)$ of any data dependence edge $ce_d = (cv_a, cv_b)$ is defined by the following formula.*

$$\omega(ce_d) = \sum_{dv \in DV(ce_d)} (\delta(de_a)\omega(de_a) + \delta(de_b)\omega(de_b))$$

where de_a and de_b are data access edges; the former is between cv_a and dv, the latter is between cv_b and dv. Furthermore $\delta(de)$ is a function which returns 1 if data access edge de is in $InterPartitionD$, otherwise it returns 0. ◇

Algorithm

The following is the flow of the CDP^2 algorithm. The CDP^2 algorithm accepts DPG DG_0 as its input and generates a dependence graph which gives the optimal data-program partition. This partition gives the minimum critical path length. Note that the nodes of the dependence graph consist of C-nodes and D-nodes, namely tasks and data.

Algorithm 2 (The CDP^2 algorithm)

1. *Insert the initial solution $\langle DG_0, \phi, CE_d, \phi, DE_r \cup DE_w, a_0 \rangle$ to list lis.*
2. *Pick a partial solution $\langle dg, I, A, DI, DA, a \rangle$ with the smallest a out of list lis.*
3. *Terminate the algorithm if $A = \phi$. The dependence graph constructed by dg gives the optimal data-program partition.*
4. *Choose an edge $ce_d = (cv_a, cv_b) \in A$.*
5. *For each $dv \in DV(ce_d)$,*
   i. *Let de_a be the data access edge between cv_a and dv. Let de_b be the data access edge between cv_b and dv.*
   ii. *$DA' \leftarrow DA - \{de_a, de_b\}, DI'_1 \leftarrow \phi, DI'_2 \leftarrow \phi, DI'_3 \leftarrow \phi$.*
   iii. *If $de_a \in DA$ and $de_b \in DA$, then $DI'_1 \leftarrow DI \cup \{de_a\}, DI'_2 \leftarrow DI \cup \{de_b\}, DI'_3 \leftarrow DI$.*

   iv. If $de_a \notin DA$ and $de_b \in DA$,

     a. If $de_a \in DI$, then $DI'_1 \leftarrow DI$.

     b. If $de_a \notin DI$, then $DI'_2 \leftarrow DI \cup \{de_b\}$, $DI'_3 \leftarrow DI$.

   v. If $de_a \in DA$ and $de_b \notin DA$,

     a. If $de_b \notin DI$, then $DI'_1 \leftarrow DI \cup \{de_a\}$, $DI'_3 \leftarrow DI$.

     b. If $de_b \in DI$, then $DI'_2 \leftarrow DI$.

   vi. If there is no data access edge between dv and cv_a or between dv and cv_b in DA', then $A' \leftarrow A - \{ce_d\}$ and $I' \leftarrow I$.

   vii. Put new partial solutions $\langle dg', I', A', DI'_i, DA', a' \rangle$ into list lis if $DI_i \neq \phi$, $i = 1, 2, 3$. dg' is a new DPG obtained by merging and a' is the critical path length of the dependence graph constructed by dg'.

6. *Let B be a set of data dependence edges on all paths consisting of only data dependence edges from cv_a to cv_b. These edges should be classified as IntraPartition to guarantee the convex constraint if cv_a and cv_b are merged.*

7. *Go to 2 if there is a data dependence edge e such that $e \in B$ and $e \in CE_d - A - I$, or the convex constraint will be violated.*

8. *For each $ce' = (cv'_a, cv'_b) \in B$,*

   i. For each $dv \in DV(ce')$,

     a. Let de_a be the data access edge between cv_a and dv. Let de_b be the data access edge between cv_b and dv.

     b. $DA'' \rightarrow DA - \{de_a, de_b\}$, $DI''_4 \leftarrow \phi$, $DI''_5 \leftarrow \phi$.

     c. If $de_a \in DA$ and $de_b \in DA$, then $DI''_4 \leftarrow DI \cup \{de_a, de_b\}$, $DI''_5 \leftarrow DI$.

     d. If $de_a \notin DA$ and $de_b \in DA$,

       - If $de_a \in DI$, then $DI''_4 \leftarrow DI \cup \{de_b\}$.

       - If $de_a \notin DI$, then $DI''_5 \leftarrow DI$.

     e. If $de_a \in DA$ and $de_b \in DI$,

       - If $de_b \in DI$, then $DI''_4 \leftarrow DI \cup \{de_a\}$.

       - If $de_b \notin DI$, then $DI''_5 \leftarrow DI$.

     f. If there is no data access edge between dv and cv_a or between dv and cv_b in DA'', then $A'' \leftarrow A - \{ce_d\}$ and $I'' \leftarrow I \cup \{ce_d\}$.

     g. Put new partial solutions $\langle dg'', I'', A'', DI''_i, DA'', a'' \rangle$ into list lis if $DI_i \neq \phi$, $i = 4, 5$. dg'' is a new DPG obtained by merging and a'' is the critical path length of the dependence graph constructed by dg''.

9. *Go to 2.◇*

The data dependence edge chosen at step 4 is classified as *InterPartition* at step 5, and as *IntraPartition* at step 8.

6 DISCUSSION

In this chapter we discuss the improvements from Girkar's algorithm to the CDP^2 algorithm.

Girkar's algorithm gives the optimal program partitioning by grain packing which employs a branch-and-bound method. Girkar's algorithm refers to the costs of task execution and inter-task communication,

which are given in advance, in grain packing. However, the inter-task communication cost depends on the location of the referred data in the distributed shared memory environment. The CDP^2 algorithm decides the location of data first and then computes inter-task communication costs. In brief, the CDP^2 algorithm applys more precise inter-task communication costs for grain packing and performs data partitioning, data distribution and program partitioning simultaneously.

In return for the advantages described above, the complexity of the CDP^2 algorithm is higher than that of Girkar's algorithm. The complexity of Girkar's algorithm is exponential in the worst case. However, the worst case is not in common in real programs[2]. An additional work of the CDP^2 algorithm against Girkar's algorithm is the classification of data access edges. The complexity of that job depends on the number of data access edges, and in the worst case it becomes exponential. Nevertheless, we don't think it is a serious problem because of locality of data references. Moreover, sophisticated selection of data dependence edges and data access edges on classifying will contribute to the efficiency of the algorithm.

7 CONCLUSION

In this paper, we proposed the CDP^2 algorithm, an algorithm to partition the program and its data simultaneously. It is an extension of Girkar's program partitioning algorithm. The CDP^2 algorithm employs a branch-and-bound method to find a grouping decision of DPG C-nodes and D-nodes which gives the optimal partition of the program and its data.

Performance evaluation of the CDP^2 algorithm using real application programs and the development of static scheduling algorithms for the result of the CDP^2 algorithm will be in our future work.

ACKNOWLEDGMENT

This work is greatly indebted to Dr. Constantine D. Polychronopoulos for his advice and encouragement.

REFERENCES

[1] V.Sarkar, *Partitioning and Scheduling Parallel Programs for Multiprocessors*, The MIT Press, Cambridge, (1988).

[2] M. B. Girkar and C. D. Polychronopoulos, "Partitioning Programs for Parallel Execution", *Proc. 1988 Int'l Conf. on Supercomputing*, (1988), pp.216–229.

[3] T. Nakanishi, K. Joe, H. Saito, C. D. Polychronopoulos, A. Fukuda and K. Araki, "The Data Partitioning Graph: Extending Data and Control Dependencies for Data Partitioning", *Proc. 7th Annual Workshop on Language and Compilers for Parallel Computing*, (1994), pp.170–185.

MEMORY EFFICIENT FULLY PARALLEL NESTED LOOP PIPELINING*

Nelson L. Passos and Edwin H.-M. Sha

Dept. of Computer Science & Eng.
University of Notre Dame
Notre Dame, IN 46556

Liang-Fang Chao

Dept. of Electrical & Computer Eng.
Iowa State University
Ames, Iowa 50011

Abstract—— *Scientific computing applications usually contain significant portions of code consisting of iterative algorithms. Such algorithms are commonly implemented as nested loops. Superscalar, VLIW, and other parallel architectures are highly dependent on the execution sequence of the instructions. In this paper, nested loops are modeled as cyclic multi-dimensional data flow graphs (MDFGs), and a new technique, called multidimensional interleaving, is applied to the loops in order to obtain full parallelism among its instructions, such that the additional memory accesses are problem size independent. The resulting execution time optimization combined with the low overhead on the memory access is significantly superior to those obtained by other known methods.*

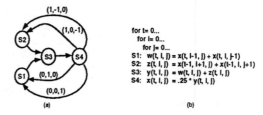

```
for t= 0...
  for i= 0...
    for j= 0...
S1:  w(t, i, j) = x(t, i-1, j) + x(t, i, j-1)
S2:  z(t, i, j) = x(t-1, i+1, j) + x(t-1, i, j+1)
S3:  y(t, i, j) = w(t, i, j) + z(t, i, j)
S4:  x(t, i, j) = .25 * y(t, i, j)
```

Fig. 1: (a) MDFG for the temperature distribution problem (b) section of the code - loop body

INTRODUCTION

Advances in computer design produced different categories of parallel systems, whose performance is highly dependent on the sequence that the instructions are executed. If operations within a loop are independent of each other, the scheduling mechanisms of those systems are able to achieve a high utilization of the parallel resources. In this paper, we propose a loop transformation technique, applicable to uniform nested loops, in order to obtain a fully parallel configuration of the operations in the loop body, while minimizing any additional memory access, producing the required conditions for the desired high performance.

Previous research on nested loop optimization has focused on the distribution of iterations among multiple processors. Some of these studies include the Doacross technique [2], and the unimodular transformations [8]. These methods do not change the internal structure of each iteration and, consequently, are not able to obtain the required fine-grain parallelism. Other methods that work at the operation level, such as software pipelining [4], and loop quantization [6], are dependent on unfolding (or unrolling) operations which increase the complexity of the problem being investigated. Most recent studies, derived from the

traditional wavefront method, have shown that the target parallelism can be achieved by changing the execution order of the iterations, usually known as the schedule vector. Among these new techniques are the index-shift method [5], the affine-by-statement scheduling [3], and the chained multi-dimensional retiming [7]. However, the change in the schedule vector, usually introduces long delays, between the production and consumption of some data value. This characteristic reduces the possibility of maintaining those values in register files and/or fast access memories, requiring external memory accesses that will negatively affect the final execution performance. In our study, the loop being optimized is modeled as a multi-dimensional data flow graph (MDFG) [1] and the fully parallel solution is achieved while keeping the new production/consumption delays independent on the problem size, and maintaining the original execution order.

Figure 1 shows a problem representing the computation of the steady state temperature distribution in a two-dimensional closed environment. The instructions are labeled $S1$ to $S4$ in order to provide the labels for the nodes of the MDFG. The edges are labeled according to the multi-dimensional (MD) distance between the iteration where the data is produced and the one where it is required. For example, data produced at statement $S4$ will be used at statement $S1$. This implies the existence of two dependencies between $S4$ and $S1$, labeled $(0, 0, 1)$ and $(0, 1, 0)$. Statements $S1$ and $S2$ could be executed in parallel, however,

*This work was supported in part by ORAU Faculty Enhancement Award under Grant No. 42265, by the NSF Research Initiation Award MIP-9410080, and by the William D. Mensch, Jr. Fellowship.

instruction $S3$ depends on the results from $S1$ and $S2$, and $S4$ depends on $S3$ on the same iteration, which implies a sequential execution of these statements. The proposed transformation, called *multi-dimensional interleaving*, focuses on modifying the graph in order to allow a final MD retiming that will produce a fully parallel solution. Additional information on MD retiming can be found in [7].

EXPANSION AND MIGRATION

In order to improve a program performance, all instructions should be executed in parallel. This simultaneous execution requires the results of each operation in one iteration to be stored until they are referenced in a future iteration. We call this waiting time the *storage delay*. It is intuitive that if all instructions in one iteration depend on data previously stored in some register or memory, then they can be fired simultaneously. However, our example has the delay $(0, 0, 1)$ in the cycle $S1 \rightarrow S3 \rightarrow S4$ that restricts the possibility of further retiming and does not allow us to achieve the fully parallel solution, while maintaining the original execution sequence (schedule vector).

In [7], it has been proven that a solution for this problem can be found by using an MD retiming, associated with a new schedule vector. However, this new schedule vector may require longer storage delays. The consequence of these long delays is that local memory, such as register files would not be able to accommodate all data values awaiting to be referenced. In order to avoid this situation, we imposed as objective of our method that the original execution sequence should remain the same. To solve this new problem, we propose a new loop transformation, consisting of an expansion of the iteration space, followed by a rearrangement of the iterations.

The *expansion* of the iteration space is obtained by the application of an *expansion vector transformation* τ_e to any iteration $I = (i_1, i_2, \ldots, i_n)$, changing the indices of I to $I^\tau = \tau_e \times I$. The new iteration space contains iterations equivalent to the original problem and a new set of integral points not related to the original iterations. We say that these new iterations are *empty* or *inactive*, while the original ones are considered *valid* or *active*. After the expansion of the iteration space, the storage delays between iterations have an apparent larger value.

A consequence of these new delay values is that the required retiming is no more constrained by the number of delays in a cycle since we can increase such number of delays as much as necessary to obtain a legal retiming. However, a real fully parallel solution is still far away. Considering that our final objective is to allow the usage of an MD retiming, specifically multiples of $(0, 0, \ldots, 0, 1)$, in order to obtain the parallelism among all operations in the loop body, we assume the expansion transformation as a neces-

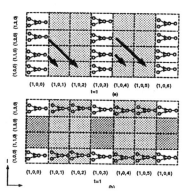

Fig. 2: (a) expanded iteration space and iterations being moved (b) iteration space after migration

sary stage to get the final solution. A row-wise expansion is chosen because it follows the row-wise sequence of execution introducing small storage delays, not dependent on the problem size, and it is sufficient to achieve the target fully parallel solution.

For an expansion vector $\tau_e = (1, 1, \ldots, 1, \tau)$, an *expanded loop body* (ELB), is defined as the set of τ consecutive iterations in a same row of the expanded iteration space, such that the first iteration of the set is one of the original iterations in G and the next $\tau - 1$ are empty iterations. We call this first iteration *head* of the expanded loop body. The theorem below presents the way to compute the expansion vector in order to support the MD retiming required by a fully parallel solution.

Theorem 1 *Given an MDFG $G = (V, E, d, t)$ submitted to an expansion vector $\tau_e = (1, \ldots, 1, \tau)$, resulting in an MDFG $G_e = (V, E, d_e, t)$, there exists a MD retiming function such that $d_e(e) \neq (0, 0, \ldots, 0), \forall e \in E$, if for any cycle l, consisting of k edges, $\tau = max \left\{ \left\lceil \frac{k}{d_n} \right\rceil \right\}$, $\forall d(e) = (0, 0, \ldots, 0, d_n)$, $e \in l$, and $d_n > 0$.*

The solely expansion of the iteration space is not sufficient to achieve the final parallel solution. Therefore, we apply an *iteration migration* after the row-wise expansion, as defined below.

Definition 1 *Given an MDFG G_e, resulting from a row-wise expansion by $\tau_e = (1, 1, \ldots, 1, \tau)$, the migration operation moves valid iterations from rows $h + 1$ to $h + \tau - 1$ to the inactive iterations found in the ELBs of row h. We say that row h is a target row.*

Intuitively, we know that the iteration space contains several target rows. Without loss of generality, we assume that the lower bounds of the loop control indices are zero. This assumption implies that the first target row is the row zero, and the remaining target rows have index $\alpha \times \tau$, for $\alpha > 0$, integer. Figure 2 presents a migration operation applied to an expanded iteration space. In this example, an original iteration $(1, 1, 1)$ is being moved to the new iteration $(1, 0, 4)$. Due to this movement, some of the original dependencies in the vertical direction became dependencies in the row-direction and may restrict the application of the MD retiming. Another problem is the possibility of new dependencies conflicting with the execution sequence. In order to avoid such problems, we will migrate iterations in such a way that no dependencies that conflict with the row-wise execution are produced, imposing an uniform solution. We will map iterations from row $h+1$ to the first iteration after the head of an ELB at the target row h, and in a more general form, iterations from row $h + \delta$ will be mapped to the δth position after the head, for $0 < \delta < \tau$. To identify such ideal iterations, we partition the expanded iteration space in subspaces containing τ rows, being the first one a target row. On each partition, we move all valid iterations to the target row by using a *migration vector*. Since our objective is to migrate iterations in the column direction, no changes in any other dimension should occur during the migration. Therefore, a migration vector with the form $c = (0, \ldots, 0, -1, g)$ is sufficient to perform such mapping. Consequently, an iteration $P = (p_1, p_2, \ldots, p_n)$, is expanded and moved to a target row in position $\tau_e \times P + mod(p_{n-1}, \tau) * c$. The next theorem shows what happens to the dependencies after the expansion and migration.

Theorem 2 *Given an MDFG $G = (V, E, d, t)$, submitted to an expansion vector $\tau_e = (1, \ldots, 1, \tau)$, and a migration vector $c = (0, \ldots, 0, -1, g)$, producing $G_m = (V, E, d_m, t)$, the dependence vectors obtained from a dependence $d = (d_1, \ldots, d_{n-2}, d_{n-1}, d_n)$ are:*
(a) if $gcd(d_{n-1}, \tau) = \tau$, $d_m = (d_1, \ldots, d_{n-1}, \tau \times d_n)$.
(b) if $gcd(d_{n-1}, \tau) \neq \tau$ two new edges replace the original one s.t. d_m has the form $(d_1, \ldots, d'_{n-1}, d'_n)$ and:

1) $d'_n = \tau * d_n + mod(d_{n-1}, \tau) * g$, $d'_{n-1} = \left\lfloor \frac{d_{n-1}}{\tau} \right\rfloor * \tau$,

2) $d'_n = \tau * d_n + (mod(d_{n-1}, \tau) - \tau) * g$, $d'_{n-1} = \left\lceil \frac{d_{n-1}}{\tau} \right\rceil * \tau$.

We notice that after migration there is a non-regularity on the dependence vectors at every τ iterations in the row direction. To model this non-regularity, we introduce a second edge between the affected nodes, representing the additional dependence vector. We compute the migration vector according to the theorem below.

Theorem 3 *Given an MDFG $G = (V, E, d, t)$ submitted to an expansion vector $\tau_e = (1, \ldots, 1, \tau)$, there exists an MD retiming function such that $d_n(e) \neq (0, \ldots, 0)$, $\forall e \in E$, if we apply a migration vector*

$c = (0, \ldots, 0, -1, g)$, *such that for any cycle l, consisting of k edges, $g = \min\{1, \gamma * \tau - \tau * p + 1\}$ with $\gamma = \max\left\{0, \left\lceil \frac{k - \tau * d_n - d_{n-1} + d_{n-1} * \tau * \mu}{d_{n-1} * \tau} \right\rceil\right\}$ and $\mu = \min\left\{0, \left\lfloor \frac{d_n}{d_{n-1}} \right\rfloor\right\}$, $\forall e \in E$ s.t. $d(e) = (0, \ldots, 0, d_{n-1}, d_n)$, $0 < d_{n-1} < \tau$.*

Changes in the loop controls

As consequence of this migration procedure, a start-up sequence will be required at the first row to achieve the full pipelined execution. The length of such sequence is obtained from the ELB that contains the iteration $(i_1, \ldots, i_n)$, such that $i_n = (\tau - 1) * g$. We also notice that after migration there are $\tau - 1$ rows without any valid iteration in between the target rows. Since these rows are not useful for the desired computation, we remove them by changing the respective loop control index to execute at every τ rows. At the same time, due to the expansion in the row direction, the innermost loop has its upper bound multiplied by τ. The lower bound will be adjusted according to the start-up sequence and the MD retiming.

Considering such changes, the utilization of the loop indices inside the loop body also need some adjustment. For simplicity, in this paper, we will indicate the transformation of the original indices by using implicit functions shown in comments inside the loop body.

MULTI-DIMENSIONAL INTERLEAVING

We reduce our problem to an algorithm based on a modified topological sort of the MDFG, combined with the formulation developed along the previous sections. This algorithm is called *M-DIP* for *Multi-Dimensional Interleaving Parallelization*, and it is shown below:

$$M\text{-}DIP(G)$$
$$LEV \leftarrow toposort(G)$$
$$\forall u \xrightarrow{e} v \in E \text{ s.t. } d(e) = (0, 0, \ldots, 0, d_n) \neq (0, 0, \ldots, 0)$$
$$\tau \leftarrow \max\left\{1, \left\lceil \frac{LEV(v) - LEV(u) + 1}{d_n} \right\rceil\right\}$$
$$\forall d(e) = (0, 0, \ldots, 0, d_{n-1}, d_n), \; 0 < d_{n-1} < \tau, \; e \in E$$
$$\mu \leftarrow \min\left\{0, \left\lfloor \frac{d_n}{d_{n-1}} \right\rfloor\right\}$$
$$\forall u \xrightarrow{e} v \in E \text{ s.t. } d(e) = (0, 0, \ldots, 0, d_{n-1}, d_n), \; 0 < d_{n-1} < \tau$$
$$\gamma \leftarrow \max\left\{0, \left\lceil \frac{LEV(v) - LEV(u) + 1 - \tau * d_n - d_{n-1} + d_{n-1} * \tau * \mu}{d_{n-1} * \tau} \right\rceil\right\}$$
$$g \leftarrow \gamma * \tau - \tau * \mu + 1$$
/* compute the retiming function for each statement*/
$$\forall v \in V, \text{ compute } r(v) = LEV(v) \times (0, \ldots, 0, 1)$$

Applying this algorithm to our example on figure 1, the topological sort procedure computes $LEV(S1) = LEV(S2) = 0$, $LEV(S3) = -1$, and $LEV(S4) = -2$. From this results, we can compute τ based on the edge with delay $(0, 0, 1)$, obtaining $\tau = 3$. In the next step, μ is computed from the dependencies $(1, -1, 0)$ and $(0, 1, 0)$

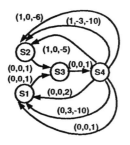

Fig. 3: Transformed MDFG

```
for t= 0...
  for i= 0...step 3
/* -- start-up sequence -- */
/* -- multi-dimensional retiming prologue -- */
    for j= 8...3*J
/*    F(i,j) = i + mod(j,3)
      H(j) = (j - mod (j,3)*4)/3            */
S1:  w(t, F(i,j), H(j)) = x(t, F(i-1,j), H(j)) + x(t, F(i,j), H(j-1))
S2:  z(t, F(i,j), H(j)) = x(t-1, F(i+1,j), H(j)) + x(t-1, F(i,j), H(j+1))
S3:  y(t, F(i,j-1), H(j-1)) = w(t, F(i,j-1), H(j-1)) + z(t, F(i,j-1), H(j-1))
S4:  x(t, F(i,j-2), H(j-2)) = .25 * y(t, F(i,j-2), H(j-2))
```

Fig. 4: Transformed fully parallel loop body

resulting $\mu = 0$. Examining dependence $(0, 1, 0)$, between $S4$ and $S1$, we determine $\gamma = 1$, therefore, $g = 3 * 1 - 3 * 0 + 1 = 4$. The new dependencies are: from $(0, 0, 1)$ we get $(0, 0, 3)$; from $(1, 0, -1)$ we obtain $(1, 0, -3)$; from $(0, 1, 0)$, two new dependencies are produced $(0, 0, 4)$ and $(0, 3, -8)$. while $(1, -1, 0)$ produces $(1, -3, -8)$ and $(1, 0, -4)$. This new dependencies are not directly used on the transformation but they satisfy the MD retiming computed in the last step of the algorithm, i.e., $r(S4) = (0, 0, -2)$ and $r(S3) = (0, 0, -1)$. Figure 3 shows the final MDFG and figure 4 shows the code for this example.

EXPERIMENTS

Four practical problems are reported in this paper to show the efficiency of our algorithm. The simulation of a two-dimensional Infinite Response Filter, *2D IIR*. A loop used to simulate a differential pulse-code modulation (*DPCM*), commonly used in image data compression. The optimization of the heat transfer problem, *HEAT*, and finally, the Livermoore Loop 23 representing a 2-D implicit hydrodynamics code fragment.

Table 1 summarizes the comparison between our results and other methods in an ideal VLIW system with sufficient logical units to execute all operations in parallel and a register file of 256 storage positions. We assumed that original dependencies requiring memory access were not optimized and the target goal of the test was to improve the exe-

| Method | original | | M-DIP | | ISM | | uni. | | chained | |
Test	t	m	t	m	t	m	t	m	t	m
2D IIR	5	0	1	0	1	16	5	2	1	2
DPCM	6	0	1	0	1	8	6	1	1	1
HEAT	3	0	1	0	1	7	3	1	1	1
LL 23	6	0	1	0	1	11	6	1	1	1

Table 1: Comparison of results of different methods

cution time while keeping any additional memory accesses restricted to the existent register file. The column t represents the achieved execution time per iteration computed as the number of clock cycles required, assumed that any operation requires one clock cycle. The column m shows the additional memory accesses required for every iteration, after optimization. The results include the original values for the non-optimized code, those obtained from our method (M-DIP), those from the index shift method (ISM) [3, 5], from the unimodular transformation, when applying loop skewing, and those from the chained multi-dimensional retiming [7]. We notice that when the fully parallel solution was achieved, the number of new memory accesses required by the optimized code has increased, except on the application of the M-DIP algorithm. Therefore we can conclude that the M-DIP algorithm is able to always achieve a fully-parallel solution, with a minimum increase in the number of stored variables. The algorithm, by construction, keeps such a number independent of the problem size, which guarantees the best utilization of fast storage devices such as register files.

REFERENCES

[1] L.-F. Chao and E. H.-M. Sha, " Static Schedulings of Uniform Nested Loops," *Proceedings of 7th Int. Parallel Processing Symposium*, Newport Beach, CA, pp. 1421-1424, 1993.

[2] R. Cytron," Doacross: Beyond Vectorization for Multiprocessors". *Proc. Int. Conf. on Parallel Processing*, pp. 836-844, 1986.

[3] A. Darte and Y. Robert, " Constructive Methods for Scheduling Uniform Loop Nests," *IEEE Transactions on Parallel and Distributed Systems*, 1994, Vol. 5, no. 8, pp. 814-822.

[4] M. Lam, " Software Pipelining: An Effective Scheduling for VLIW Machines".*ACM SIGPLAN Conf. on Prog. Lang. Design and Implementation*, 1988, pp. 318-328.

[5] L.-S. Liu, C.-W. Ho and J.-P. Sheu, " On the Parallelism of Nested For-Loops Using Index Shift Method," *Proc. of the Int. Conf. on Parallel Processing*, 1990, Vol. II, pp. 119-123.

[6] A. Nicolau, " Loop Quantization or Unwinding Done Right," *Proc. of the 1987 ACM Int. Conf. on Supercomputing*, Springer Verlag Lecture Notes on Computer Science 289, 1987, pp. 294-308.

[7] N. L. Passos and E. H.-M. Sha " Full Parallelism in Uniform Nested Loops using Multi-Dimensional Retiming". *Proceedings of 23rd Int. Conf. on Parallel Processing*, 1994, vol. II, pp. 130-133.

[8] M. Wolf and M. Lam, " A Loop Transformation Theory and an Algorithm to Maximize Parallelism". *IEEE Trans. on Parallel and Distributed Systems*, vol. 2, n. 4, pp. 452-471, 1991.

Path-Based Task Replication for Scheduling with Communication Costs [†]

Jayesh Siddhiwala

Liang-Fang Chao
lfc@iastate.edu

Department of Electrical and Computer Engineering
Iowa State University, Ames, Iowa 50011, USA

ABSTRACT

Task replication is an effective mechanism to trade off computational power for reduction in communication delays. We propose efficient algorithms that obtain optimal schedules with a minimal amount of task replications under a minor assumption on an unlimited number of processors. Tasks are replicated path by path judiciously on critical parts of the task graph. The experimental results show that a significant reduction in the amount of replication is achieved with the optimal schedule length.

1 Introduction

The purpose of task scheduling and allocation on a set of interconnected processors is to reduce the overall task execution time by exploiting parallelism hidden in the program. Distributed memory multiprocessors offer significant advantages over the shared memory multiprocessors in terms of cost, memory access time and scalability. Extensive studies have shown that inter-processor communication (IPC) cost is one of the major factors responsible for performance degradation. This paper proposes a static scheduling algorithm that reduces inter-processor communication bottleneck with minimal task replication.

An application program is first partitioned into tasks along their natural boundaries such that the amount of data transfers between tasks is minimized. Data transfers take place in the beginning and the end of a task and control constructs are restricted within the tasks. A partitioned program is usually represented as a directed acyclic task graph, where each node is a task and each edge represents a precedence relationship labeled with estimated IPC cost.

It is observed that schedules with communication costs can be improved using *task replication*, also known as *task duplication*. However, it has been shown that scheduling with task replication and even constant communication cost is NP-complete on an un-

limited number of processors [1]. A few heuristic algorithms have been proposed to reduce the schedule length using task replication when necessary [2, 3, 4]; however, optimal schedule length cannot be guaranteed even for special cases. The algorithms in [2, 3] suffers from the difficulties that some replications performed earlier would restrict future replications because no space is available in a chosen processor. Thus, insufficient and improper replications might result in a suboptimal schedule. The algorithm proposed in [4] tries to overcome these problems by allocating unused processors when necessary, while obtaining an optimal schedule, but it generates many unnecessary replications.

In order to reduce the time complexity of replication algorithms, polynomial-time algorithms are proposed under the following assumption [5]:

> **Assumption A:** at most one immediate predecessor of a task is allocated to the same processor as the task.

Darbha and Agrawal [5] proposed a polynomial-time algorithm under an assumption equivalent to Assumption A. This algorithm obtains shortest schedules under the assumption, but a large amount of unnecessary replication is performed.

In this paper, we propose a polynomial-time optimal algorithm under Assumption A which uses an efficient path-based replication scheme that replicates a portion of a path at each step. Since task replication is carefully applied only to those tasks in the task graph that are critical to the entire schedule length, only a small number of tasks are replicated. Our processor allocation scheme ensures that enough space on processors are always available for task replication. Hence, the difficulties encountered in previous works [2, 3] are resolved. Our algorithm not only obtains the shortest schedule length but also uses significantly less amount of task replication,. It is assumed that an unlimited number of processors are available, as previously mentioned works.

[†] This work was supported in part by NSF Research Initiation Award MIP-9410080.

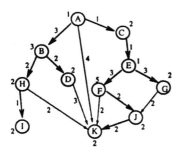

Figure 1: A task graph

2 Foundations

A partitioned parallel program is represented by a directed acyclic graph (DAG) $G = (V, E, t, c)$, where V is a set of nodes representing tasks to be executed; E is a set of edges representing precedence relations (data dependencies); $t : V \longrightarrow N$ denotes the computation time of tasks in V; $c : E \longrightarrow N$ denotes the communication cost to transfer data across the interconnection network. For example, in Figure 1 the number attached to a node (resp. an edge) is its computation time (resp. communication cost). The set of predecessors and the set of successors of a task v are denoted by $pred(v)$ and $succ(v)$, respectively.

For a task u, $S(u)$ denotes the starting time of u; $P(u)$ denotes the processor where the task u is allocated to. Under a processor allocation P, the time taken to transfer data from task u to task v is represented as $C^P(u, v)$. That is, $C^P(u, v) = 0$ if $P(v) = P(u)$; $c(u, v)$ otherwise. Thus, the task v cannot start execution until the data from the predecessor u arrives at time $S(u) + t(u) + C^P(u, v)$.

For a task v, the *earliest starting time* (EST) and the *latest starting time* (LST) under Assumption A are denoted as $EST(v)$ and $LST(v)$, respectively. Let us assume that initially none of the predecessors of task v are scheduled on the same processor as the task. We define $Y_1 = \max\{EST(w) + t(w) + c(w, v) \mid w \in pred(v)\}$. If predecessor u gives this maximum then $Y_1 = EST(u) + t(u) + c(u, v)$. The predecessor u is called an *EST_parent* of v and is denoted as $EST\_par(v)$. In the case of a tie the selection can be made arbitrarily. Thus, $EST(v) = \max\{EST(u) + t(u), \max\{EST(w) + t(w) + c(w, v) \mid w \in pred(u) \text{ and } w \neq u\}\}$. For a task with no predecessors, the EST is 0. In Figure 1 each darker edge connects a task with its EST_parent. The LST values can be derived analogously.

A schedule S (with arbitrary amount of replication) is optimal if its schedule length is equal to the lower bound under assumption A, which is $L_{opt} = \max\{EST(v) + t(v) \mid v \in V\}$.

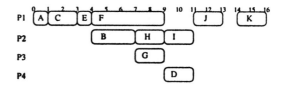

Figure 2: The priority-based initial schedule

One can always construct a schedule with arbitrary replication to achieve the optimal schedule length L_{opt}. Although an algorithm is proposed in [5], we briefly describe a much simpler linear-time algorithm, based on breadth-first search (BFS). A more detailed description can be found in [6]. During the BFS traversal of the task graph, starting from the leaf tasks, a task v is replicated into k copies, where k equals to the number of its successors of which task v is *the* EST-parent. As a result of the BFS traversal a replication graph is obtained such that each task in the graph has exactly one EST-parent and is the EST-parent of exactly one task. Now each task can be assigned to the same processor as its EST-parent and an optimal schedule with length L_{opt} can be achieved.

From the above construction, we know that an optimal schedule can be obtained so that every task is scheduled to the same processor as one replica of its EST-parent. The information about EST-parents will be used to guide our minimal replication scheme towards optimal schedules such that only important replications are performed.

3 Scheduling with Minimal Replication

In this section, we propose a scheduling algorithm that uses a task replication scheme to achieve the optimal schedule length, with the minimal amount of replication. The algorithm has two phases: priority-based initial scheduling (Sec. 3.1) and path-based replication (Sec. 3.2). The basic replication scheme is further refined to reduce the amount of replication to a minimal while maintaining optimal schedule length (Sec. 3.3).

3.1 Priority-based Scheduling

The tasks are ordered in the task pool according to the lexicographical order of (EST, LST) with EST in ascending order and LST in ascending order.

In each step a task v with the highest priority is removed from the task pool and an attempt is made to schedule it on the same processor as its EST-parent, if possible, otherwise, an unused processor is used to schedule it. Such processor allocation scheme ensures that there is always enough space on the processors to

Path-based Replication Algorithm
Compute $EST(v)$ and $EST\_par(v)$ for every task v;
Compute the optimal schedule length L_{opt};
$L \longleftarrow$ the current length of schedule S;
while $L \neq L_{opt}$ **do begin**
    Form a set Q of critical paths with at least
      one potential and important node;
    Order these critical paths in ascending order
      of idle time and in descending order
      of communication cost;
    while Q is not empty and $L \neq L_{opt}$ **do begin**
      Pop a critical path p from Q;
      $tgt \longleftarrow$ an important and potential node in p;
      $u \longleftarrow EST\_par(tgt)$;
      $(S, P) \longleftarrow$ `Minimal_Replication`(u, tgt, S, P);
      Update the schedule of affected tasks ;
    end
end

Figure 3: Path-based target selection and replication

accommodate replicas, generated in the second phase, such that an optimal schedule can always be achieved. Due to the processor allocation strategy, tasks assigned to the same processor form a path such that each task is the EST-parent of the task scheduled immediately following it on the same processor. The task graph in Figure 1 will be used as an example throughout this section. Its initial schedle is shown in Figure 2.

3.2 Path-based Replication Scheme

In each replication step, the algorithm, shown in Figure 3, identifies a task whose schedule will be improved if a path in its ancestor trees is replicated to the same processor as this task.

The starting time of a task v, $S(v)$, is the maximum of $S(w) + t(w) + C^P(w, v)$ over all $w \in pred(v)$. A predecessor u that gives this maximum is called *a very important parent (VIP)*. The schedule of task v cannot be improved unless $S(u)$ or $C^P(u, v)$ of every VIP u are improved.

A task v is a *potential* task if the EST-parent u of v is not scheduled on the same processor as v. i.e. $EST(v) < S(v)$ and $P(EST\_par(v)) \neq P(v)$. So, the schedule of v has potential to be improved. A task v is *important* if it is a leaf task or it is a VIP of one of its successors, i.e. an improvement of the schedule of v would possibly improve the entire schedule length. Replication will only be applied to the tasks which are both potential and important tasks, termed here as *target tasks*.

For a path $p = v_1 \longrightarrow v_2 \longrightarrow \ldots \longrightarrow v_k$, the *computation time* of the path is $\sum_{1 \leq i \leq k} t(v_i)$. The *com-*

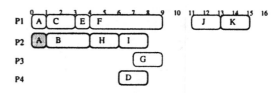

(a) After the first replication targeted on Node B

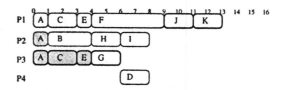

(b) After the second replication targeted on Node G

Figure 4: The final schedule using complete path replication

munication cost is $\sum_{1 \leq i < k} C^P(v_i, v_{i+1})$. The *execution time* of the path p, denoted by $exec(p)$, is $S(v_k) + t(v_k) - S(v_1)$; the *effective execution time*, denoted by $eff(p)$, is the summation of computation time and communication cost. We call the difference $exec(p) - eff(p)$ to be the *idle time* of path p.

A path is called *critical* if its execution time equals the current schedule length. A critical path with large idle time is not likely to be shortened unless other paths are shortened first. Hence, a critical path with larger communication cost is given higher priority for replications.

The selected critical path is scanned to find a target task. The algorithm replicates a path leading to the EST-parent of the target task to improve the schedule of the target task.

The replication process continues until no target node can be identified. The number of processors used remains the same as the initial schedule. It can be shown that the optimal schedule length L_{opt} can be achieved in time polynomial in the number of original tasks and replicas.

Theorem 3.1 *Let $G = (V, E, t, c)$ be the task graph, and R the number of replicas in the final schedule. The replication algorithm in Figure 3 can achieve the optimal schedule length L_{opt} under Assumption A. The time complexity of the algorithm is $O(R^2 + R \cdot |V| + R^2 \log R)$.*

Proof: (sketch) When the algorithm terminates, there is no target node on any critical path. That is, the schedule of each node must equal to its EST or it does not affect the overall schedule. Since every node

Procedure Minimal_Replication(u, tgt, S, P)
    Find the path q of nodes scheduled on processor
        $P(u)$ upto node u;
    if $q = u$ then begin
        Replicate node u;   return; end ;
    $\delta(tgt) \longleftarrow \min\{L - L_{opt}, S(tgt) - EST(tgt),$
        $\min_{(tgt,w) \in E}(S(w) - EST(w))\};$
    if $\delta(tgt) = 0$ then   return; /* No replication */
    $x_{m+1} \longleftarrow u;$ $l \longleftarrow m + 2;$
    $slack \longleftarrow c(u, tgt) - \delta(tgt);$
    repeat $l \longleftarrow l - 1;$
    until $(l = 1)$ or $(c(x_{l-1}, x_l) \leq slack);$
    Replicate nodes from x_l to u to processor $P(tgt)$;
    if $l > 1$ then   Adjust schedule times of these
        replicas by the comm. cost $c(x_{l-1}, x_l)$;

Figure 5: Minimal Replication Procedure

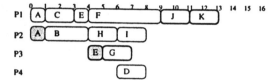

Figure 6: Final schedule using Procedure Minimal_Replication

becomes a target node at most once, the total number of critical paths ever identified is linear in R. Detailed proof can be found in [6]. $\square$

In the initial schedule of Figure 2, Nodes B, G, and D are potential and important. The set of critical paths Q with potential and important nodes is {ABDK(idle time = 0), ACEGJK (idle time = 1)}. The path ABDK is selected first. The target node would be B and EST-par(B) = A. The schedule length is reduced from 16 to 15 with replication of Node A, as shown in Figure 4-(a). Next, the path ACEGJK is selected. The target node is Node G and EST-par(G) = E. The entire path leading to E, ACE, is replicated. The final schedule of length 13 is shown in Figure 4-(b), which is the optimal.

3.3 Minimal Replication

For a target node v whose EST-parent is u, instead of replicating the entire path from the root to the task u, the replication of a portion of a path (subpath) might be sufficient in some cases. When the entire path $q = x_1 \longrightarrow x_2 \ldots \longrightarrow x_m \longrightarrow u$ is replicated, the communication cost $c(u, v)$ from u to v is saved. The minimal replication procedure is shown in Figure 5.

The amount of reduction in $S(v)$ sufficient to achieve the optimal schedule, denoted by $\delta(v)$, can be computed as shown in Figure 5 ($\delta(tgt)$). Although the replication of q can achieve a maximum of $c(u, v)$ reduction in communication cost, only $\delta(v)$ amount of reduction is sufficient. Therefore, we can replicate only a subpath of q as long as the amount of reduction is $\geq \delta(v)$. We define $slack = c(u, v) - \delta(v)$. If $\delta(v)$ is 0, any improvement in the schedule of task v will not affect the entire schedule; thus, replication will not be performed. If the subpath $q' = x_l \longrightarrow \ldots \longrightarrow x_m \longrightarrow u$ is replicated, the communication cost from x_{l-1} to x_l will be introduced. Thus the amount of reduction is $c(u, v) - c(x_{l-1}, x_l)$. We would like to choose a subpath q' such that $c(u, v) - c(x_{l-1}, x_l) \geq \delta(v)$, i.e. $c(x_{l-1}, x_l) \leq c(u, v) - \delta(v)$. Since the time complexity of the Minimal_Replication procedure is still linear in the number of replicated nodes. Therefore, Theorem 3.1 holds for the Minimal_Replication procedure.

Let's re-examine the replication procedure on target node G. When the Minimal_Replication procedure is called, $\delta(G)$ and $c(E, G)$ are derived to be 2 and 3, respectively. The $slack$ is one, and the edge (C,E) satisfies that $c(C, E) = 1 \leq slack$. Therefore, only node E is replicated in the new schedule. In the final schedule, shown in Figure 6, the number of replicated nodes are further reduced by 2 for the optimal schedule length.

4 Experimental Results

In the paper [4], the task graph shown in Figure 7 is used. The schedule obtained in [4] has the optimal schedule length 246 with 16 task replications. The schedule obtained by our minimal replication algorithm, shown in Figure 7, achieves the same schedule length with 9 task replications only. The amount of reduction in total task computation time is 23%. Due to such reduction, a processor compaction process can reduce the number of processors by 2.

Experimental results on six graph structures with randomly generated data are performed, shown in Table 1. Two of the six graphs are tree-structured (In1, In2). Significant reduction in the amount of replication is achieved while the optimal schedule length is guaranteed. Our results are compared with the results obtained by our implementation of the SDBS algorithm [5]. Since both algorithms achieve the optimal schedule length L_{opt}, the schedule length is not shown. The column #Nd represents the number of tasks in the original graphs. For each graph structure, three groups of experiments with different ranges of

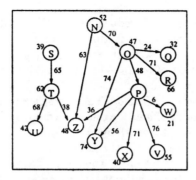

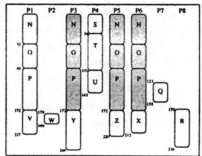

Figure 7: Our schedule for graph in [4]

Table 1: Average percentage reduction

Cmp Time		10-50		10-50		10-50	
Cmm. Cost		10-50		50-100		100-500	
	#Nd	CT	CC	CT	CC	CT	CC
G2	10	45.00	−0.4	49.0	9.5	48.8	−26.8
G3	24	35.8	−93.5	37.3	−155.6	37.6	−285.7
G4	17	32.0	−76.5	29.8	−168.1	32.0	−148.3
G5	13	48.3	12.7	48.3	35.3	50.0	−21.7
In1	25	66.2	32.4	81.9	56.3	70.6	27.9
In2	40	75.7	54.4	75.6	50.9	76.1	47.4

computation times and communication costs are performed. In each group of experiments, five sets of random values are generated. The figures in the table show the percentage reduction averaged over the five experiments. Columns CT and CC represent percentage reduction in the total computation time and the total communication cost, respectively. Our algorithm provides a significant amount of reduction in task replication in all cases. Intuitively, a schedule with less amount of replication requires more communications. However, our algorithm even reduces the communication cost in half of the cases.

5 Conclusion

We have proposed a scheduling and allocation algorithm which achieves the optimal schedule length with minimal task replication under the assumption that no

two predecessors or successors of a task are scheduled to the same processor as the task. Although the number of processors is unlimited in our algorithm, a processor compaction can be performed on the optimal schedule to reduce the processor space. The proposed algorithms are efficient, achieves optimality, and uses the minimal amount of replication.

6 REFERENCES

References

[1] Papadimitriou, C. and Yannakakis, M., " Towards an architecture-independent analysis of parallel algorithms", *SIAM J. of Comp.*, Vol 19, pp. 322-328, 1990.

[2] Kruatrachue, B. and Lewis, T. G., "Grain Size Determination for Parallel Processing", *IEEE software*, pp. 23-32, 1988.

[3] Chung, Y. C. and Ranka, S., "Application and Performance Analysis of a Compiler Time Optimization Approach for List Scheduling Algorithms on Distributed Memory Multiprocessors", *Intl. Conf. on Supercomputing*, pp. 512-521, 1992.

[4] Ahmad, I. and Kwok, Y.-K., "A New Approach to Scheduling Parallel Programs using Task Duplication", *Intl. Conf. on Parallel Processing*, pp. II 47-51, Aug. 1994.

[5] Darbha, S. and Agrawal, D. P., "SDBS: A Task Duplication based Optimal Scheduling Algorithm", *Scalable High performance Computing Conference*, May 1994.

[6] Jayesh Siddhiwala and Liang-Fang Chao, "Path-Based Task Replication for Scheduling with Communication Costs," Technical Report, Dept. of Elec. and Computer Engr., Iowa State University, April 1995.

Mapping Arbitrary Non-Uniform Task Graphs onto Arbitrary Non-Uniform System Graphs

Song Chen, Mary M. Eshaghian and Ying-Chieh Wu

Department of Computer and Information Science

New Jersey Institute of Technology

Newark, NJ 07102

Abstract

In this paper, a generic technique for clustering and mapping arbitrary task graphs onto arbitrary system graphs is presented. The task and system graphs studied in this paper have non-uniform computation and communication weights associated with the nodes and edges. Using two clustering algorithms, a multi-level clustered graph called Spec graph can be obtained from a given task graph and a multi-level clustered graph called Rep graph can be obtained from a given system graph. We present a mapping algorithm which produces a sub-optimal matching of a given Spec graph containing M task modules, onto a Rep graph of N processors, in $O(MP)$ time, where $P = \max(M, N)$. Our experimental results indicate that our mapping results are better than or similar to those of other leading techniques which work only for restricted task or system graphs.

Keywords: mapping, task scheduling, task allocation, clustering, portable parallel programming.

1 Introduction

A program task can be represented by a task graph, with each node representing a task module and each edge representing data communication between two modules. Each node is associated with a weight representing the computation amount of the corresponding task module, while the weight of an edge represents the communication amount. Similarly, a parallel computer system can be modeled as an weighted undirected system graph, whose weights represent processor speeds and transmission rates of communication links. If the task graphs and the system graphs are known before program execution, then mapping of the task graphs onto the system graphs is called static mapping. In static mapping, the assignments of the nodes of the task graphs onto the system graphs are determined prior to the execution and are not changed until the end of the execution. Static mapping can be classified in two general ways. The first classification is based on the topology of task and or system graphs [3]. Based on this, the mappings can be classified into four groups: (1) mapping of specialized tasks onto specialized systems, (2) mapping of specialized tasks onto arbitrary systems, (3) mapping of arbitrary tasks onto specialized systems and (4) mapping of arbitrary tasks onto arbitrary systems. The second classification can be based on the uniformity of the weights of the nodes and the edges of the task and or the system graphs. Based on this, the mappings can be categorized into the following four groups: (1) mapping of uniform tasks onto uniform systems [2, 7, 1, 6, 3], (2) mapping of uniform tasks onto non-uniform systems, (3) mapping of non-uniform tasks onto uniform systems [9, 10, 12, 5, 13] and (4) mapping of non-uniform tasks onto non-uniform systems [11, 8].

In this paper, we concentrate on static mapping of arbitrary non-uniform task graphs onto arbitrary non-uniform system graphs. The existing mapping techniques in this group include Lo's max flow min cut mapping heuristic [8] and Shen and Tsai's A* searching heuristic [11]. These algorithms have $O(M^2N|E_p|\log M)$ and $O(N^M)$ time complexity respectively, where M is the number of nodes in the task graph, $|E_p|$ is the number of edges in the task graph and N is the number of nodes in the system graphs. In this paper, we present an algorithm which can map arbitrary non-uniform architecturally independent task graphs onto arbitrary non-uniform system graphs in $O(MP)$ time, where $P = \max(M, N)$. This technique is based on Cluster-M programming tool and consists of two clustering algorithms and a mapping algorithm, which are extensions to those presented in [3]. Our experimental results indicate that our mapping results are better than or similar to those of other leading techniques which work only for restricted task or system graphs.

The rest of the paper is organized as follows. In

section two, we present the clustering algorithms. The mapping algorithm is given in section three. We show our experimental results in comparing our algorithm with several other existing mapping algorithms in section four. Brief conclusions are given in section five.

2 Clustering Arbitrary Task and System Graphs

In this section, we present two algorithms for clustering non-uniform directed task graphs, and for clustering non-uniform undirected system graphs. The clustering is done only once for a given task graph (system graph) independent of any system graphs (task graphs). It is a machine-independent (application-independent) clustering, therefore, it is not repeated for different mappings.

2.1 Clustering directed task graphs

The clustering algorithm given in this section is an extension to the clustering-directed-graphs algorithm presented in [3]. The purpose of this algorithm is to produce a multi-level clustered graph called Spec graph from a given task graph. Each clustering level in the Spec graph represents a set of concurrent computations, called Spec clusters. There are quadruple parameters $(\sigma_S, \delta_S, \Pi_S, \pi_S)$ associated with every Spec cluster. Each of these parameters is described as follows. The size of a Spec cluster is denoted by σ_S, and represents the maximum number of nodes in this Spec cluster that can be computed in parallel. The number of levels in a Spec cluster represents the maximum sequential computation length of each node in the Spec cluster, and is denoted by δ_S. The total amount of communication among all clustering levels is denoted by Π_S and the average communication amount at current (top) level is denoted by π_S. There are two types of operations performed on the Spec clusters which effect the values of their quadruples: embedding and merging [4]. Embedding is when two or more sequential Spec clusters are combined into one Spec cluster. Merging is when a number of concurrently executable sub-clusters are grouped to form a new Spec cluster.

The clustering is done step by step. Each clustering step corresponds to a computation step. At every step, we cluster the nodes (clusters) as follows. If a node is a Merge-node, we first embed it onto one of its parent nodes (the one to which it is connected by an edge having the highest communication weight), then merge all the parent nodes

into a larger Spec cluster, similar to [3]. If a node is a Broadcast-node, we will embed one of its child nodes (the one to which it is connected by an edge having the highest communication weight) to this Broadcast-node and then merge the rest of the child nodes with the Broadcast-node into a larger Spec cluster. Note that since our task graphs are independent of the system graphs (unlike [10, 12, 13]), they do not contain the information about computation time and communication delay. Therefore, we can only embed one child node to the parent node in both the merge and broadcast cases as shown above. However, the embedding of multiple child nodes can be done as part of the mapping. This is explained in the next section.

The time complexity of the clustering-non-uniform-directed-graphs algorithm is bounded by the number of edges in the task graph, which is $O(M^2)$, where M is the number of nodes.

2.2 Clustering undirected system graphs

The algorithm for clustering the system graph is used to generate a multi-level clustered graph called Rep graph. At every clustering level of the Rep graph, there are a number of clusters called Rep clusters. Each Rep cluster represents a set of processors with a certain degree of connectivity. Similar to a Spec cluster, each Rep cluster is associated with quadruple parameters $(\sigma_R, \delta_R, \Pi_R, \pi_R)$ which represents the number of processors contained in the cluster, average computation speed of the processors in the cluster, total communication bandwidth and the average communication bandwidth at the current (top) clustering level. To construct a clustered graph (Rep graph) from an undirected system graph, initially, every node with computation speed of S_i forms a cluster with parameters $(1, S_i, 0, 0)$. Then clusters which are completely connected are merged to form a new cluster and the parameters of the new cluster are calculated correspondingly. This is continued until no more merging is possible. Two clusters x and y are connected if x contains a node p_x and y contains a node p_y, such that node p_x and p_y are connected by a direct communication link. The algorithm for clustering non-uniform undirected graphs is an extension to the clustering-undirected-graphs algorithm in [3]. The time complexity of the new clustering algorithm is $O(N^3 \log N)$ [4].

3 Non-Uniform Mapping

Given a Spec graph and a Rep graph, we present an efficient mapping algorithm which produces a sub-optimal matching of the two graphs in $O(MP)$ time, where $P = \max(M, N)$.

The non-uniform mapping algorithm is an extension to the Cluster-M uniform mapping algorithm presented in [3]. In the following, we give an overview of the algorithm. Before the mapping begins, we need to compute a reduction factor denoted by f, which is essential for mapping of task graphs having more nodes than the system graphs. The reduction factor f is the ratio of the total size of the Rep clusters over the total size of the Spec clusters. It is used to estimate how many computation nodes will share a processor. The mapping is done recursively at each clustering level, where we find the best matching between Spec clusters and Rep clusters. For matching Spec clusters to Rep clusters, the Spec and Rep clusters are first sorted in descending order with respect to the four parameters $(\sigma, \delta, \Pi, \pi)$. For example, Spec clusters with larger sizes are sorted before those with smaller sizes and for Spec clusters of the same size, those with larger number of levels are sorted first.

Second, we map each of the Spec clusters (denoted by κ_{S_i}) as follows. We first search for the Rep cluster (denoted by κ_{R_j}) with the best matched size, i.e., closest to $f \times \sigma_{S_i}$. Therefore, we try to minimize the function in Equation (1). If multiple Rep clusters with the matching size are found, we select the one with the minimum estimated execution time. The estimated execution time of mapping Spec cluster κ_{Si} onto Rep cluster κ_{Rj}, $\tau(\kappa_{Si}, \kappa_{Rj})$, is equal to the number of clustering levels of κ_{Si} times the average computation and communication time at each level, as formulated in Equation (2). If no Rep cluster with a matching size can be found for a Spec cluster, we either merge or split (unmerge) Rep clusters until a matching Rep cluster is found.

$$|f_m| = \sum_i |f \times \sigma_{S_i} - \sigma_{R_{f_m(\kappa_{S_i})}}| \qquad (1)$$

$$\tau(\kappa_{Si}, \kappa_{Rj}) = \delta_{Si}(\frac{1}{\delta_{Rj}} \times \frac{1}{f} + \frac{\Pi_{Si} \times \delta_{Rj}}{\Pi_{Rj} \times \delta_{Si}}) \qquad (2)$$

Thirdly, for every matched pair of the Spec and Rep clusters, we do the following to embed communication intensive nodes together. (This is similar to the clustering process in [10, 12, 13]. However, in this paper, we only do it in the mapping step so that the clustering of the task graph is kept independent

of the system graph, as described in the previous section.) If a Spec cluster has multiple sub-clusters and the average communication time between these sub-clusters is greater than the possible computation time of a sub-cluster as formulated in Condition (3), we then embed the sub-clusters onto a sub-cluster having the largest size, and calculate the parameter quadruple for the new cluster. We then insert it in the proper position in the sorted list of Spec clusters for mapping and repeat the matching as described above by Equation (1) and (2) for the remaining Spec clusters in the list. If no embedding is necessary, then the mapping of this Spec cluster onto a Rep cluster is done for this level, and therefore this Spec cluster is removed from the list.

$$\frac{\pi_{Si}}{\pi_{Rj}} > \frac{\min(\sigma_{sub-cluster} \times \delta_{sub-cluster})}{\delta_{Rj}} \times \frac{1}{f} \qquad (3)$$

In the above mapping algorithm, the worst case of a mapping at a level i happens either when (case 1) for each Spec cluster, all the remaining Rep clusters have the matching size, therefore Equation (2) is used to select the best Rep cluster; or in (case 2) for each Spec cluster, no Rep cluster of matching size is found. Therefore Rep clusters are merged/split recursively until a Rep cluster of matching size is obtained. Suppose the number of Spec clusters at level i is K_i. In both cases described above, or in any combination of the two cases, it takes $O(K_i N)$ time to find the best match for all K_i Spec clusters, as the total number of clusters in the Rep graph is $O(N)$, where N is the number of processors. For each pair of matching Spec and Rep clusters, if Condition (3) is satisfied, the extra time taken in embedding will be $O(M)$. Since the total number of Spec clusters is $O(M)$, i.e., $\sum_i K_i = O(M)$, where M is the number of nodes in original task graph. Therefore, the total time complexity of this mapping algorithm is $\sum_i (K_i N + M) = O(MN) + O(M^2) = O(MP)$, where $P = \max(M, N)$.

Figure 1 shows a mapping example of a Gaussian elimination task. Suppose it takes 1 unit of time to add or subtract, and it takes 2 units of time to multiply or divide two real numbers. We also assume that the communication amount of sending/receiving each real number to be 1. A task graph for computing the Gaussian elimination of a 5×5 matrix is shown in figure 1(a). In each task module T_j^k, column j is modified by using column k. Suppose that the system running this task contains only two workstations p_1 and p_2. p_1 and p_2 have speed of 2 and 1.6 respectively and are connected with a link of bandwidth 1. The mapping result

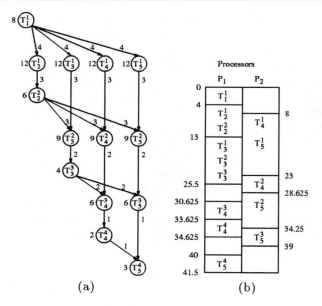

(a)   (b)

Figure 1: (a) The task graph and (b) the mapping result of the Gaussian elimination on a 5 × 5 matrix.

using our technique is illustrated in figure 1(b).

4   Comparison Results

In this section, we present a set of experimental results which have been obtained in comparing our algorithm with other leading techniques. The examples selected here are not designed by us, rather, they are those presented and studied by the authors of papers which report the leading techniques. The following three criteria are used for evaluating the performance of the algorithms examined: (1) the total time complexity of executing the mapping algorithm T_r; (2) the total execution time of the generated mappings, T_m; and (3) the number of processors used, N_m. From (2) and (3), we can obtain the speedup $S_m = \frac{T_s}{T_m}$ and efficiency $\eta = \frac{S_m}{N_m}$, where T_s is the sequential execution time of the task.

Since there is no existing mapping technique which maps a machine-independent arbitrary non-uniform task onto an arbitrary non-uniform system, it is not easy to choose candidates for the comparison study. Therefore, we focus on comparing the leading mapping techniques designed for arbitrary non-uniform tasks but specialized systems only. The comparison results are shown in Figure 2, 3 and 4.

5   Conclusions

In this paper, we have presented the first generic algorithm for mapping non-uniform arbitrary task

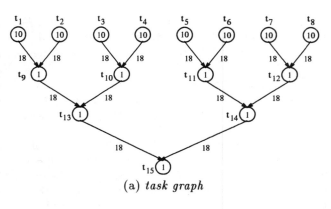

(a) *task graph*

	T_r	T_m	N_m
Clan	$O(M^3)$	59	4
Cluster-M	$O(MP)$	59	4

(b) *mapping results*

Figure 2: Comparison example with Clan.

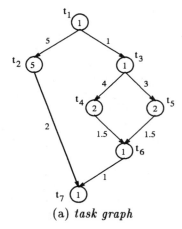

(a) *task graph*

	T_r	T_m	N_m
MCP	$O(M^2 \log M)$	10.5	3
Sarkar	$O(|E_t|(M + |E_t|))$	10	3
DSC	$O((M + |E_t|) \log M)$	9	2
Cluster-M	$O(MP)$	10	3

(b) *mapping results*

Figure 3: Comparison example with MCP, Sarkar and DSC.

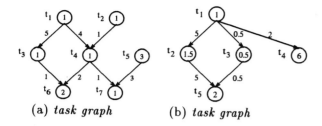

(a) *task graph*   (b) *task graph*

	T_r	T_m	N_m
DSC	$O((M + |E_t|) \log M)$	5	3
Cluster-M	$O(MP)$	5	3

(c) *mapping results of task graph (a)*

	T_r	T_m	N_m
DSC	$O((M + |E_t|) \log M)$	9.5	2
Cluster-M	$O(MP)$	9	2

(d) *mapping results of task graph (b)*

Figure 4: More comparison examples with DSC.

graphs onto non-uniform arbitrary system graphs. Given a task graph and system graph, we showed efficient techniques for producing two clustered graphs called Spec graph and Rep graph, which are the input to the mapping algorithm. The clustering is done only once for a given task graph (system graph) independent of any system graphs (task graphs). This is a machine-independent (application-independent) clustering and is not repeated for different mappings. The complexity of the mapping algorithm is $O(MP)$, where M is the number of task modules, N is the number of processors and $P = \max(M, N)$. We presented our experimental results in comparing the performance of our generic algorithm with others that lead but are more restricted. We showed that we can obtain similar results in less time. Furthermore, we use machine-independent task graphs. Therefore, the presented mapping algorithm can be intergrated as part of portable parallel programming tools.

References

[1] F. Berman and L. Snyder. On mapping parallel algorithms into parallel architectures. *Journal of Parallel and Distributed Computing*, pages 439–458, April 1987.

[2] S. H. Bokhari. On the mapping problem. *IEEE Trans. on Computers*, c-30(3):207–214, March 1981.

[3] S. Chen and M. M. Eshaghian. Tools for design and mapping of portable parallel programs. In *Proc. Workshop on Challenges for Parallel Processing, International Conference on Parallel Processing*, August 1995.

[4] S. Chen, M. M. Eshaghian, and Y. Wu. Mapping arbitrary non-uniform task graphs onto arbitrary non-uniform system graphs. Technical report, Dept. of Computer and Information Science, New Jersey Institute of Technology, 1995.

[5] H. El-Rewini and T. G. Lewis. Scheduling parallel program tasks onto arbitrary target machines. *Journal of Parallel and Distributed Computing*, pages 138–153, 9 1990.

[6] F. Ercal, J. Ramanujam, and P. Sadayappan. Task allocation onto a hypercube by recursive mincut bipartitioning. *Journal of Parallel and Distributed Computing*, pages 33–44, 1990.

[7] S. Lee and J. K. Aggarwal. A mapping strategy for parallel processing. *IEEE Trans. on Computers*, 36:433–442, April 1987.

[8] V. M. Lo. Heuristic algorithms for task assignment in distributed systems. *IEEE Trans. on Computers*, 37(11):1384–1397, November 1988.

[9] C. McCreary and H. Gill. Automatic determination of grain size for efficient parallel processing. *Communications of ACM*, 32(9):1073–1078, September 1989.

[10] V. Sarkar. *Partitioning and Scheduling Parallel Programs for Execution on Multiprocessors*. MIT Press, 1989.

[11] C. Shen and W. Tsai. A graph matching approach to optimal task assignment in distributed computing systems using a minmax criterion. *IEEE Trans. on Computers*, c-34(3):197–203, March 1985.

[12] M. Y. Wu and D. Gajski. Hypertool: A programming aid for message-passing systems. *IEEE Trans. on Parallel and Distributed Systems*, 1(3):101–119, 1990.

[13] T. Yang and A. Gerasoulis. DSC: Scheduling parallel tasks on an unbounded number of processors. *IEEE Trans. on Parallel and Distributed Systems*, 5(9):951–967, September 1994.

Portable Runtime Support for Asynchronous Simulation *

Chih-Po Wen and Katherine Yelick

Computer Science Division, University of California

Berkeley, CA 94720

cpwen@cs.berkeley.edu yelick@cs.berkeley.edu

Abstract

We present library and runtime support for portable asynchronous applications, using event-driven simulation as an example. Although event-driven simulation has a natural source of parallelism between the simulated entities, real speedups have been hard to obtain because of the fine-grained, unpredictable communication patterns. Language and systems software support is also lacking for asynchronous problems. Our runtime supports makes the applications portable, eases performance tuning, and allows code re-use between applications. Our goal is to support a range of platforms with varying performance characteristics, from special-purpose multiprocessors through networks of workstations. We discuss the performance issues in the runtime support, describe a distributed event graph data structure, and present performance numbers from a parallelized timing simulator called SWEC.

1 Introduction

Parallel applications can be classified according to their degree of irregularity. Fox identifies three classes: *synchronous* algorithms have little or no data-dependent behavior and fit into an SIMD execution model; *loosely synchronous* problems are spatially irregular but temporally regular and can be expressed in SPMD languages with alternating communication and computation phases; *asynchronous* problems are irregular in time and space and are the most difficult to parallelize [Fox91]. Data parallel languages like HPF [Hig93] and NESL [Ble93] are well-suited to synchronous problems, and with sufficient compiler [BCF+93, BC90] and run-time support (e.g., the PARTI system [SBW91]), can be used for loosely synchronous problems. In this paper we describe library and runtime support for asynchronous applications.

The goal of our research is to provide software systems to make asynchronous applications easier to parallelize, using extensive runtime support and a distributed data structure library. Portability is a primary goal. The programs developed using our system run and exhibit good performance on a range of machines. In this paper, we use a conservative parallel version of the SWEC simulator, an asynchronous event-driven simulator, as a running example to study programmability, performance, and portability issues of our system. The main distributed data structure in this problem is an *event-graph*, a component of our Multipol library [CDI+95]. In previous work, we described a parallel implementation of the SWEC simulator that used optimistic concurrency [WY93]. The implementation performed well on the CM5 multiprocessor, but contained machine-specific techniques and was not organized to allow for code re-use in other simulators. This work addresses those concerns.

Our research targets distributed memory architectures such as the CM5, Paragon, SP1/SP2, and networks of workstations. A common characteristic of distributed memory machines is that the overhead and latency of accessing remote data is much larger than accessing local data, and the bandwidth of remote access improves with the size of the data. Solving irregular problems on these machines is particularly challenging, because communication is fine-grained and unpredictable, leaving few opportunities for compile-time optimization or runtime preprocessing.

Our runtime system employs two techniques to

*This work was supported in part by the Advanced Research Projects Agency of the Department of Defense monitored by the Office of Naval Research under contract DABT63-92-C-0026, by the Department of Energy grant DE-FG03-94ER25206, and by the National Science Foundation (number CCR-9210260 and number CDA-8722788). The information presented here does not necessarily reflect the position or the policy of the Government and no official endorsement should be inferred.

reduce communication costs: *split-phase operations* built from atomic threads, which hide the latency of remote operations, and *message aggregation*, which reduces the overhead of communication. On top of this, the Multipol data structures provide reusable abstractions for irregular structures. One of these, an *event graph*, provides order-preserving, flow-controlled message delivery between a set of logical processes. It also takes advantage of the process graph structure to optimize for communication efficiency.

The rest of the paper is organized as follows. Section 2 gives an overview of the application, which is a timing simulator called SWEC. Section 3 describes our runtime support, drawing examples from SWEC to motivate the design. Section 4 describes the interface and semantics of the event graph and sketches the algorithm used in the parallelized SWEC. Section 5 gives the performance results and uses statistics from our experiments to evaluate the effectiveness of our runtime system. Section 6 describes previous work on portable software support and discrete event simulation. Section 7 concludes the paper.

2   Overview of SWEC

The SWEC program is mainly used to perform timing simulation for digital MOS circuits [LKMS91]. It partitions a circuit into loosely coupled subcircuits (as shown in Figure 1), each of which can be simulated independently within a time step. At the end of a time step, if the subcircuit's new state cannot be extrapolated linearly from the old state within some error margin, the new state is propagated to the fanout subcircuits. This communication is referred to as an event. Events occur at unpredicatable times and vary in frequency depending on the non-linearity of the voltage. The time step size used by the subcircuits depend on their states. SWEC uses smaller timesteps for subcircuits with more activity to improve accuracy, and larger timesteps for subcircuits with little activity to save computation. There are no global synchronization points, since each subcircuit is simulated with a variable number of timesteps that cannot be predicted in advanced.

In the sequential implementation, the subcircuits are always scheduled in strict ascending time order, so that all relevant events are processed before any affected subcircuit computation takes place. Figure 2 sketches the sequential algorithm. A subcircuit

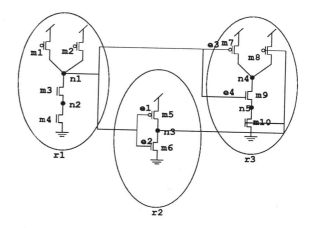

Figure 1: Partitioning the circuit into subcircuits. The subcircuits (denoted by **r**) contains MOS transistors (denoted by **m**) and voltage points (denoted by **n**). The transistors are connected to their fanin voltage points via the gates (denoted by **e**).

s has two attributes: **s.time**, the current simulation time of the subcircuit and **s.step** the current duration of a time step. Subcircuits are placed in a priority queue, **p**, ordered by the estimated next time point (**s.time + s.step**).

The priority queue in SWEC offers little room for parallelization, because it imposes a total order on the scheduling of subcircuits. Our parallel implementation distributes the subcircuits among the processors, and the simulation proceeds in a dataflow manner. The parallel algorithm is a typical case of distributed asynchronous simulation, where each subcircuit corresponds to a logical process, and each event corresponds to a collection of messages sent to the fanout subcircuits. We adopt the conservative approach to parallel asynchronous simulation [KC81], scheduling a subcircuit only when all input events have been processed. Previously, we showed that conservative parallel simulation is primarily useful for combinational circuits (circuits without feedback paths); optimistic scheduling is much more effective for sequential circuits[WY93]. We are developing an optimistic analog to the conservative Parallel SWEC, but in this paper we focus on the conservative example.

Parallel SWEC is irregular in computation and communication. The number of time steps as well as their costs depend on the state of simulation, which is not known in advance. The number of messages is also data dependent, so their storage cannot be preallocated. Furthermore, the messages, usually small,

```
For all subcircuits s,
  initialize s.time and s.step
  insert s into p using key (s.time + s.step)

While min key of p < end of simulation,
  delete s with min key from p
  simulate s from s.time up to (s.time + s.step) and
      update its state
  for each voltage point v of s,
    if new state of v cannot be linearly extrapolated
        from previous state
      for each fanout subcircuit s' of s at v
        update the state of the mosfet driven by v in s'
        reduce s'.step if necessary
        delete s' from p
        insert s' in p using key (s'.time + s'.step)
  s.time = s.time + s.step
  s.step = projected time step based on new state of s
  insert s into p using key (s.time + s.step)
```

Figure 2: The sequential SWEC algorithm.

are sent at unpredictable times. To resolve load imbalance caused by computation irregularities, we use simple static load balancing heuristics for distributing the subcircuits across processors. Our heuristics work well in practice and avoid the communication overhead of dynamic load balancing. In the remainder of this paper, we focus on communication irregularities and how small asynchronous messages are handle in a portable manner.

3 Runtime Support

To achieve good performance on distributed memory machines, three communication costs must be addressed: overhead, latency, and bandwidth. In this section, we explore these issues and present the solutions used in our runtime layer.

3.1 Overhead Reduction

Communication overhead refers to the processor time spent in setting up the communication and injecting data into the network. It can be decomposed into the fixed start-up overhead for setting up the communication, such as allocating storage or performing kernel calls, and the overhead per byte for injecting the message. Even for machines with like the CM5 with small hardware messages and lightweight communication such as Active Messages [vECGS92], the startup cost observed by irregular applications may be significant. Active messages avoid storage allocation by using fixed hardware packet sizes and by requiring that the user provide an address range in memory for the message data to be written. Structuring the program based on a fixed packet size limits portability, and for asynchronous programs, the cost of buffer management may resurface in the users' code.

Our runtime system accumulates small, asynchronous messages in a continuous buffer until its size exceeds a certain threshold, or until no other computational threads are eligible for execution. The aggregated messages in the buffer are then sent as a single message using a bulk communication primitive. Message aggregation, a technique that is well-known for bulk-synchronous algorithms and libraries [BSS91, KB94], reduces communication overhead by amortizing the start-up overhead over many messages.

Although message aggregation reduces the amortized communication start-up, it incurs the overhead for data copying, and thus increases the overhead per byte. It may also increases the observed latency of remote operations, since messages may be delayed in the aggregation buffer. The exact tradeoff depends on the application, the machine architecture, as well as the input. Therefore, we expose the degree of aggregation to the programmer for performance tuning, but ensure that all message will eventually be delivered.

3.2 Latency Hiding

Irregular applications typically have dynamic data structures such as graphs, trees, and tables. Rather than simple fetch and store operations, as one would have on arrays or static graphs, dynamic structures have operations to add or delete nodes from linked structures as well as non-trivial operations to observe the state of the structure. For such data structures, there can be many sources of latency in an operation: the delay due to message aggregation, the network latency, the delay due to scheduling at the remote processor, the computation time at the remote processor, and the latency of sending back the result. Since the total latency can be long, support for multithreading is required to overlap the latency with other computation.

Our runtime system supports split-phase operations implemented as user-level *atomic threads* for latency hiding. A atomic thread is a finite computation that is guaranteed to run atomically. A split-phase operation is built from two or more such threads. High-level synchronization constructs such as suspension are implemented on top of atomic threads using continuations. Atomicity eases programming by eliminating locking for simple read-modify-write operations. Unlike system level threads, the context of our threads are explicitly managed by the programmer, so only the required variables are passed from one thread to its continuation. The threads synchronize using counters, similar to those used in Split-C [CDG+93]. Every split-phase operation takes a counter as input argument, which is incremented when the operation completes. A separate continuation thread can be created to wait on the value of the counter. The counter automatically enables the continuation thread for execution when the operation completes, without requiring polling by the programmer.

The runtime system also provides the programmer with mechanisms for building customized schedulers. For example, a scheduler which has knowledge of the simulation time is used to schedule the subcircuit threads in parallel SWEC. The user schedulers are fairly invoked by the system, so the scheduling policy of one data structure can be fine-tuned for performance without introducing unexpected deadlocks.

3.3 Bandwidth Optimization

Using message aggregation, the overhead per byte, or the communication bandwidth, becomes the determining factor of communication efficiency. Communication bandwidth can be improved by implementing bulk communication with nonblocking primitives. The runtime system interfaces with all machine architectures using the *nonblocking store* primitive, which transfers a block of data from a local address to a remote address, increments a local counter when the local buffer can be reused, and invokes a remote thread when the transfer is complete. The nonblocking store primitive saves the copying cost for queuing messages at the sending processor, and may improve network utilization by interleaving the injection of packets from multiple messages that are to be sent to different processors.

3.4 Summary of Runtime Support

The runtime system supports a small set of primitives for managing atomic threads and three forms of split-phase communication: *put*, which transfers a block of data from a local address to a remote address, *get*, which transfers a block of data from a remote address to a local address, and *remote thread*

invocation, which creates a thread on a remote processor. The remote threads are like any local threads and may perform arbitrary computation. They are different from the network handlers, which the runtime system uses in a restricted manner to drain data from the network as fast as possible. The put and get operations are used for bulk, regular communication, while remote threads are usually used for irregular communication such as sending events in SWEC.

4 A Distributed Event Graph Data Structure

The dominant communication in asynchronous simulation is the propagation of events. Storage must be allocated for the event messages before they are sent, and the messages must arrive in their time stamp order for the conservative method to work. In this section, we introduce a data structure called an event graph, which supports in-order, flow-controlled message delivery in a static network of nodes. The event graph also takes advantage of the graph structure to optimize for locality.

4.1 Interface and Semantics

The event graph can be thought of as a graph with fixed-sized FIFO buffers attached to the edges, with which the nodes send and receive events. We refer to the size of these FIFO buffers as the *edge capacity*. There are five operations on the event graph, all of them split-phase:

- **MakeEventGraph**: create a distributed event graph on all processors based on an input graph. The programmer can write custom partitioners to assign nodes to processors. The programmer also specifies the edge capacity. **MakeEventGraph** completes when the event graph is ready for use on all processors.

- **SendEvent**: propagate an event from a node to all its fanout nodes. **SendEvent** completes when storage for the messages has been allocated in all fanout edges and the event has been copied. Its completion does not guarantee that the event messages are immediately available, although it is guaranteed to arrive in finite time.

- **ReceiveEvent**: read and/or remove an event message from an incoming edge of a node. The removal frees up space for more incoming messages. The programmer has the option to wait for new messages when the edge contains no message. The programmer can also query the status of each edge, and uses non-split-phase versions of `ReceiveEvent` for better performance.

- `WaitForEvent`: wait until some event message is available for a node to receive.

- `Freeze/UnFreeze`: freeze and unfreeze the state of the event graph for taking *distributed snapshots*. `Freeze` suspends all split-phase mutators, such as `SendEvent`. When `Freeze` completes, the events are available using `ReceiveEvent`. No event can be "stuck" in the communication layer or the network. `Unfreeze` re-enables all mutators. The `Freeze/Unfreeze` operations are useful for detecting global properties in applications using distributed data structures. For example, they are used in parallel SWEC to detect deadlocks. They are also used to compute the global time in an optimistic parallelization of SWEC (currently under development).

Two operations on a data structure are said to *interfere* if their behavior under concurrent execution is undefined. For the event graph, `Freeze` interferes with all mutator operations that can temporarily leave the data structure in an inconsistent state. Some of the `ReceiveEvent` operations that modify the edge buffers are not split-phase, and as a result they interfere with `Freeze`. The programmer must insert sufficient synchronization in the program to ensure these operations are not issued when `Freeze` is in progress. Also, `SendEvent` interferes with `SendEvent` on the same node, since the meaning of concurrent enqueues to the same FIFO is not clear.

4.2 Implementation Techniques

The implementation of event graph uses the following techniques to improve performance:

- Lazy evaluation of `SendEvent`. We weaken the semantics of `SendEvent` so that its completion guarantees the event will arrive in finite time, but does not guarantee it is immediately available. The semantics requires no acknowledgement from the remote processor. The programmer can use the `Freeze` operation to ensure that all events are globally visible.

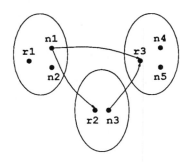

Figure 3: Mapping of circuit to event graph. The circuit shown in the previous example is mapped to an event graph. The nodes derived from the same subcircuit are allocated on the same processor, so that the simulation steps can be performed using local data.

- Caching event messages. The edges are allocated on the receiving processors, and all `ReceiveEvent` operations are handled locally. To further reduce communication, if an event is sent to two different node on the same processor, its value is cached on the remote processor. The caching is static because the graph structure remains fixed, so the storage for caching can be pre-allocated.

- Replicating the buffer control state. Sending an event requires that space be available on all fanout edges. We replicate the flow-control state of the edges (such as the number of elements) on the sending processor, and update the replica lazily so that the flow-control overhead can be amortized over many operations. When the edge capacity is large, most `SendEvent` operations can usually proceed without requiring communication for buffer space allocation.

4.3 Parallelizing SWEC using Event Graph

We parallelize SWEC using the event graph as follows. One distributed event graph is created for the entire circuit to handle all messages. Each subcircuit is a node in the graph for receiving event messages from its fanin subcircuits. Each voltage point is also a node for propagating its state. The mapping of circuits to event graph is illustrated in Figure 3. Evidence of the power of distributed data structures comes from our programming experience. The event-graph took a couple weeks to design, debug, and optimize. Given this, and a familiarity with

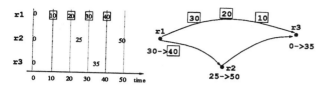

Figure 4: Deadlock due to storage constraints.

SWEC, the parallel conservative simulator took only a few days to implement.

Since we target combinational circuits, the data dependence graph of the circuit is acyclic, and deadlock due to the lack of global information cannot occur. However, because the event graph places an upper bound on the number of outstanding messages, the subcircuits sending new events may have to wait for its fanout subcircuits to release buffer space. This creates cycles in the overall dependence graph. Therefore, deadlock due to storage constraints can still occur. The problem is illustrated in Figure 4. The left-hand picture shows a sequential simulation with numbers denoting simulation steps and boxed numbers denoting steps that produced an event. In the parallel version on the right, the edge capacity is three events. The parallel simulation is deadlocked because **r2** has not produced an event to tell **r3** that it can proceed beyond time zero. Therefore, **r1** blocks because event 40 cannot be propagated due to the lack of space and **r2** blocks waiting for **r1**. If buffer space were unbounded, **r1** and **r2** would continue running, eventually producing an event from **r2** that would allow **r3** to proceed.

We resolve deadlocks using a combination of null messages and deadlock detection and recovery. A null message is an event that carries only the simulation time. A subcircuit sends a null message if it proceeds for several time steps and blocks without producing any event message. Our heuristics produce enough null messages to eliminate deadlocks in practice.

5 Performance Results

In this section, we report on the performance of the parallel timing simulator on various platforms, and provide statistics to show the effectiveness of our approach.

We have ported the runtime system to the CM5, SP1/SP2 and Paragon. The CM5 port is built on the Thinking Machines active message layer, called **CMAML**, the SP port uses the IBM message passing

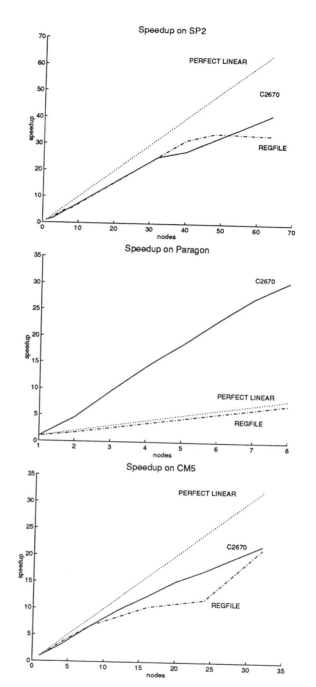

Figure 5: Speedups of the conservative parallel SWEC. The speedups are with respect to the 1 processor parallel implementation.

	subcircuits	mosfets	time steps	events	SP2	Paragon	CM5
C2670	2033	5364	2510047	760166	402 (243)	2159 (3046.4)	3916 (84 0.5)
REGFILE	325	4832	177987	957412	112.4 (153.7)	324.7 (460.3)	622 .7 (518.48)

Table 1: Statistics from the benchmark circuits. The execution times of the sequential SWEC on a single processor of machine is shown, with the running time of the one-processor parallel code in parentheses. All times are in seconds.

library MPL, and the paragon port is based on the NX communication library. We use the active message developed at Berkeley for both the SP and the NX ports [Lun94]. The SP and Paragon have higher bandwidth networks and faster processors than CM5. In general, the gap between the computation and communication performance is more pronounced for the SP and Paragon than for the CM5.

Table 1 describes the benchmark circuits. Due to differences in float-point characteristics, the number of time steps performed and consequently the number of events are slightly different among the three machines. REGFILE is a 32-bit register file, and C2670 is an unknown circuit from the ISCAS benchmark suite. Both circuits are combinational. C2670 has quite uniform computation granularities for the subcircuits, which usually contain only a few transistors. The computation of REGFILE is dominated by 32 very large subcircuits which represent the 32 bit slices of the register file. Notice that parallel SWEC on 1 node sometimes outperforms the sequential SWEC. Although parallelization causes communication and threading overhead, the distributed, data-driven approach of parallel SWEC alleviates the bottleneck of a centralized priority queue in SWEC, whose accessing cost grows with the number of subcircuits.

The speedups of our parallel implementation is show in Figure 5.[1] The curves are nearly linear for C2670, indicating good scalability for larger circuits. Both C2670 and REGFILE show speedup of over 5 on the 8-node Paragon, over 20 on the 32-node CM5, and over 41 and 33 on the 64-node SP2. The superlinear speedup of C2670 on the Paragon will be explained later.

Two program parameters can be set to fit the machine characteristics for better performance: degree of message aggregation and edge capacity. Figure 6 shows the impact of these two parameters on per-

formance. The running times in the graphs are normalized with respect to the minimum time in each curve to show the percentage of increase in time.

Message aggregation is shown to be essential for performance. For C2670, the running time can increase by more than a factor of 3 when aggregation is not performed. The effect of aggregation is not as significant for REGFILE, which has very large subcircuits so that the communication time occupies a smaller portion of the execution time. In most cases, the running time increases when the degree of aggregation exceeds a certain threshold. This is because aggregation delays the progress of the simulation, so it becomes counterproductive unless there is sufficient parallelism. The threshold is about 2K bytes for C2670.

The effect of edge capacity is dependent on the input as well as the machine configuration. A large capacity improves performance by increasing concurrency, and consequently reduces the number of null messages that have to be sent to avoid deadlock. The increase in concurrency also provides more threads for hiding latency, as well as more opportunities for message aggregation. The results show that C2670 is less sensitive to edge capacity than REGFILE, because it has more parallelism (more subcircuits). The increase in running time for REGFILE can be as much as 60% on the SP1 and Paragon when the edge capacity is too small.

A large edge capacity, however, may degrade the performance of the memory hierarchy because it requires more memory. This is demonstrated by the running time of C2670 on Paragon, which increases by a factor of 3 when the edge capacity is too large. The simulation of C2670 may require more than 300M bytes of memory, and our local Paragon configuration provides only about 10M bytes of physical memory per node. This also explains the superlinear speedup within the parallel implementation for C2670 in Figure 5, since for a smaller number of processors the machine may spend a significant amount of time paging.

---

[1] We were not able to obtain some of the SP2 results. For processor numbers 1 through 8, an SP1 was used with performances scaled to reflect the speed difference.

6   Related Work

Parallel asynchronous simulation is a well studied problem. Chandy and Misra developed the conservative method[KC81]. Jefferson introduced the optimistic method, also known as the time-warp algorithm[Jef85]. In our prior work[WY93], we compared the potential of the both methods for parallelizing SWEC[LKMS91], and developed a CM5 specific implementation using the optimistic method.

Recent research has produced a variety of runtime support such as the Chare kernel [SK91], Nexus[FKOT91], and the compiler-controlled threaded abstract machine (TAM) [CSS+91]. Nexus provides mechanisms similar to remote thread invocation, but is more heavyweight than our runtime system, because it supports arbitrary thread suspension and heterogeneous computing. The threads in our runtime system are similar in spirit to the TAM threads, but do not require compiler support for static thread allocation, and are intended for coarser grained, dynamically generated threads.

The idea of aggregating small messages to reduce communication overhead can be found in the Fortran-D compiler[HKT91], which uses message vectorization in parallel loops to reduce overhead, and in PARTI[BSS91], which uses runtime preprocessing to pre-allocate storage for aggregated array accesses in bulk synchronous programs. It is a common technique for regular, array-based computations and for data structures that are irregular in time space but not time.

There have been a number of attempts at developing application-specific distributed data structures such as irregular grids[BSS91, KB94] and oct-trees[WS93]. As with aggregation, these data structures are targeted toward loosely synchronous, rather than asynchronous applications.

7   Conclusions

We have presented a runtime communication layer for a general class of irregular problems, and a distributed even graph data structure for asynchronous simulation. The runtime layer provides and portability layer for the Multipol library. Performance portability is obtained using split-phase operations, multithreading, and message aggregation. We demonstrated good performance on two nontrivial circuits. We also quantified some of the performance trade-offs that make portability difficult.

The combination of runtime support and distributed data structures were shown to be effective for writing portable parallel programs for asynchronous simulation. Future work includes the development of an optimistic parallel implementation using the infrastructure described in this paper, application of these data structures to other problems, and the automatic tuning of program and data structure parameters for improving performance.

References

[BC90]   Guy E. Blelloch and Siddhartha Chatterjee. VCODE: A data-parallel intermediate language. In *Frontiers '90*, pages 471–480, College Park, MD, October 1990.

[BCF+93]   Zeki Bozkus, Alok Choudhary, Geoffrey Fox, Tomasz Haupt, and Sanjay Ranka. A compilation approach for Fortran 90D/HPF compilers on distributed memory MIMD computers. In *Languages and Compilers for Parallel Computing*, pages h1–h23, Portland, OR, August 1993.

[Ble93]   Guy E. Blelloch. NESL: A nested data-parallel language (version 2.6). Technical Report CMU-CS-93-129, CMU School of Computer Science, Pittsburgh, PA, April 1993. Updated version of CMU-CS-92-103, January 1992.

[BSS91]   H. Berryman, J. Saltz, and J. Scroggs. Execution time support for adaptive scientific algorithms on distributed memory multiprocessors. *Concurrenty: Practice and Experience*, pages 159–178, June 1991.

[CDG+93]   David E. Culler, Andrea Dusseau, Seth Copen Goldstein, Arvind Krishnamurthy, Steven Lumetta, Thorsten von Eicken, and Katherine Yelick. Parallel programming in Split-C. In *Supercomputing '93*, Portland, Oregon, November 1993.

[CDI+95]   Soumen Chakrabarti, Etienne Deprit, Eun-Jin Im, Jeff Jones, Arvind Krishnamurthy, Chih-Po Wen, and Katherine Yelick. Multipol: A distributed data structure library. Technical Report To appear, University of California at Berkeley, Computer Science Division, 1995.

[CSS+91] D. Culler, A. Sah, K. Schauser, T. von Eicken, and J. Wawrzynek. Fine-grain Parallelism with Minimal Hardware Support: A Compiler-Controlled Threaded Abstract Machine. In *Proc. of 4th Int. Conf. on Architectural Support for Programming Languages and Operating Systems*, Santa-Clara, CA, April 1991.

[FKOT91] Ian Foster, Carl Kesselman, Robert Olson, and Steve Tuccke. Nexus: An interoperability toolkit for parallel and distributed computer systems. Technical Report ANL/MCS-TM-189, Argonne National Laboratory, 1991.

[Fox91] G. C. Fox. The architecture of problems and portable parallel software systems. Technical Report SCCS-134, Syracuse Center for Computational Science, 1991.

[Hig93] High Performance Fortran Forum. High Performance Fortran language specification version 1.0. Draft, May 1993.

[HKT91] Seema Hiranandani, Ken Kennedy, and Chau-Wen Tseng. Compiler optimizations for Fortran D on MIMD distributed-memory machines. In *Supercomputing '91*, New Mexico, November 1991.

[Jef85] D.R. Jefferson. Virtual time. *ACM Transactions on Programming Languages and Systems*, 7(3), July 1985.

[KB94] Scott Kohn and Scott Baden. A robut parallel programming model for dynamic non-uniform scientific computations. In *Proceedings of the Scalable High Performance Computing Conference*, Knoxville, TN, May 1994.

[KC81] J. Misra K.M. Chandy. Asynchronous distributed simulation via a sequence of parallel computations. *Communications of the ACM*, 24(11), April 1981.

[LKMS91] S. Lin, E. Kuh, and M. Marek-Sadowska. SWEC: A stepwise equivalent conductance simulator for cmos vlsi circuits. In *Proc. of European Design Automation conference*, February 1991.

[Lun94] Steve Luna. Implementing an efficient portable global memory layer on distributed memory multiprocessors. Master's thesis, Computer Science Division, University of California at Berkeley, 1994.

[SBW91] Joel Saltz, Harry Berryman, and Janet Wu. Multiprocessors and run-time compilation. *Concurrenty: Practice and Experience*, December 1991.

[SK91] Wei Shu and L.V. Kalé. Chare kernel – a runtime support system for parallel computations. *Journal of Parallel and Distributed Computing*, 11:198–211, 1991.

[vECGS92] Thorsten von Eicken, David E. Culler, Seth Copen Goldstein, and Klaus Erik Schauser. Active messages: a mechanism for integrated communication and computation. In *International Symposium on Computer Architecture*, 1992.

[WS93] M.S. Warren and J.K. Salmon. A parallel hashed oct-tree n-body algorithm. In *Supercomputing '93*, pages 12–21, Portland, Oregon, November 1993.

[WY93] Chih-Po Wen and Katherine Yelick. Parallel timing simulation on a distributed memory multiprocessor. In *International Conference on CAD*, Santa Clara, CA, November 1993. An earlier version appeared as UCB Technical Report CSD-93-723.

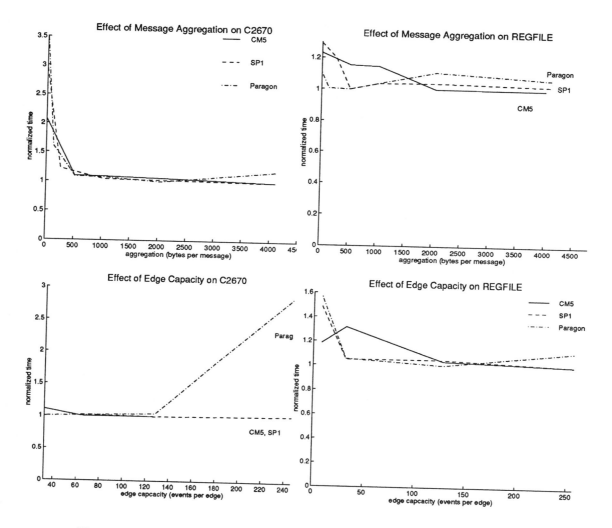

Figure 6: Impact of message aggregation and edge capacity on performance.

Task Allocation for Safety and Reliability in Distributed Systems*

Santhanam Srinivasan and Niraj K. Jha
Department of Electrical Engineering
Princeton University, Princeton, NJ 08544
{srinivas,jha}@ee.princeton.edu

Abstract

Distributed systems are increasingly being employed in a number of critical applications like aircraft control and banking systems. The inherent complexity of distributed systems can increase the potential for system failures. Some work has been done in the past on allocating tasks to distributed systems with the aim of maximizing the **reliability** *(the probability that no component fails while processing), which, however, does not give any guarantees about the system performance when a failure does occur. Such systems are* **unsafe**. *In this work we propose heuristic techniques which, using the concept of assertion checking, introduce* **safety** *into the system and simultaneously attempt to maximize the reliability. We present a number of simulation results to show the efficacy of our scheme as compared to previously known methods.*

1 Introduction

Distributed systems consist of loosely coupled processors which communicate with one another only by message passing and which do not share any common memory. The increased commercialization of such systems means that ensuring system reliability is of critical importance. The traditional hardware redundancy schemes for introducing system-level fault tolerance are very expensive. Hence, we have to turn to software techniques for hardware fault tolerance. Our work is motivated by this need to improve the reliability of the distributed system with no extra hardware cost.

In order to exploit the parallelism inherent in the job to be executed on the distributed system, it is partitioned into a set of intercommunicating *tasks*. These tasks and the inter-task communication then have to be assigned or *allocated* to the different processors and inter-processor links, respectively, in the system. Work in [1] uses task redundancy as a means of increasing system reliability. In [2], the authors

present a scheme called *task based fault tolerance*, to introduce *fault security/safety* into the system. This scheme uses the concept of *assertion checking* which is more general and has much lower overhead as compared to task replication. *Assertion tasks*, which check some property of the output of the original tasks in the system, are used to detect faults in the system.

In [2], however, no explicit reliability model is assumed and minimization of schedule length is used as the objective. Reliability has been considered as the objective function for allocation in [3] and an explicit measure of the reliability of an allocation has been introduced. *Reliability* is defined to be the probability that the system will not fail during the time that it is processing the job [3]. This, however, does not give any guarantees about the behavior of the system when a failure does occur. Such failures, if not caught immediately, could be catastrophic in safety-critical applications.

Our main motivation in this paper is to introduce *safety* and *reliability* into a heterogeneous distributed system through software techniques. Hardware techniques for safety and the tradeoff between reliability and safety have been addressed in the literature [4, 5]. Some techniques for reliability and safety analysis have been described in [6]. In introducing safety, we would like to encompass both transient and permanent faults. When a transient fault is detected one can retry the job or just the affected tasks till the fault goes away. If a permanent fault is detected, one can, after locating the fault, either reconfigure the system around the fault or replace the faulty component and then rerun the job. We need to note that a system can have very high safety but very low reliability. Clearly, this is not satisfactory. In this paper, we discuss techniques to introduce safety in the system such that the reliability is maximized.

The first phase of our two-pass scheme allocates the tasks of the original task graph with reliability as the objective. Our allocation procedure uses the concept of *clustering based on static levels*. In [3] the task graphs considered do not have any precedence constraints. We show that for task graphs with

---

*Acknowledgments: This work was supported by the Office of Naval Research under Contract no. N00014-91-J-1199.

precedence constraints, our method performs better in terms of reliability and is comparable in terms of CPU time. In the second phase, the fault tolerance tasks are allocated such that the decrease in reliability is minimized. We introduce a new task-level allocation procedure for the extra tasks that produces results of the same quality as those in [3] but runs an order of magnitude faster.

2 Definitions

The *job* to be processed by the distributed system is assumed to be decomposed into indivisible *tasks*, which are represented as a directed graph called the *task graph*. The process of *task allocation* takes as input the task graph and the *target architecture* specified by the processor nodes, their interconnection, and a number of parameters defining the relationship between the tasks in the task graph and the processors. It then attempts to find a mapping of the tasks to the processors using some optimization criterion. In this work we address the problem of task allocation in order to introduce safety in the system and at the same time maximize the reliability.

2.1 Task Graph

The task graph is a directed acyclic graph $G = \{T, E\}$, where $T = \{t_1, t_2, \ldots, t_n\}$ represents a set of n tasks and E represents the set of weighted and directed edges among the tasks. The communication from task t_i to task t_j is represented by a directed edge (t_i, t_j) between the two tasks. The weight on this edge, c_{ij}, represents the volume of data being transmitted from task t_i to task t_j. Each node in the task graph is labeled with a vector called *Exec_cost*:

$$Exec\_cost(t_i) \quad = \quad [e_{i1}, \ldots, e_{im}]$$

The jth element of this vector, e_{ij}, represents the execution time of t_i on processor p_j. If task t_i cannot be executed on processor p_j, the corresponding execution time e_{ij} is ∞. A *source (sink) task* in a task graph is defined as a task with no fanins (fanouts).

Constraints on the allocation are represented by two matrices. The *exclusion matrix Ex* [7] has tasks as rows and columns. Entry $Ex(i, j)$ is a '1' if task t_i cannot be on the same processor as task t_j, else it is a '0'. The *preference matrix Pr* [7] has tasks as rows and processors as columns. Entry $Pr(i, j)$ in the matrix is a '0' if task t_i cannot be allocated to processor p_j and '1' otherwise. A point to note is that apart from arising as natural system constraints, the concepts of task exclusion and preference also give a handle to the user to provide hints to the system in order to fine tune the performance with respect to second order effects that might not be captured by our model of the system. Each node in the task graph is also assigned a label called *comp_cost*. This is the average of the execution costs of the task on those processors on which it can be executed.

2.2 System Model

The distributed system is assumed to consist of a set $P = \{p_1, p_2, \ldots, p_m\}$ of heterogeneous processors connected by an arbitrary interconnection network. The processors only have local memory and do not share any global memory. The *processor graph* is a convenient abstraction of the processors together with the interconnection network. It has processors as nodes and there is a weighted edge between two nodes if the corresponding processors can communicate with each other. The weight w_{ij} on the edge between processors p_i and p_j represents the delay involved in sending or receiving a message of unit length from one processor to another. In order to have an approximate estimate of this delay, irrespective of the two processors under consideration, we use the average of the weights on all the edges in the processor graph. This is called the *average unit delay*.

Interprocessor communication (IPC) has conventionally been viewed as a major factor inhibiting parallelism. Most schemes try to reduce IPC with the aim of decreasing schedule length. As will be evident from the discussion in Section 2.3, IPC is also very detrimental to the reliability of the system due to the unreliable nature of the communication links.

Based on the parameters of the underlying architecture, our algorithm tries to minimize the IPC. We assume that once a task has completed, the processor computing the task stores the output data in its local memory. If the data is needed by a task being computed on the same processor, it reads it from the local memory. The overhead incurred by this is negligible, so for all practical purposes we will consider it to be zero. Using this fact, our algorithm tries to group together heavily communicating tasks and allocate them to the same processor. In case two communicating tasks are allocated to different processors, IPC overhead is incurred in communicating data between these two processors. In such cases, our algorithm tries to allocate the heaviest inter-task communication onto the most reliable links. We must note, however, that in order to improve the reliability the other constraint is that the tasks with higher computation times have to be allocated to more reliable processors. Concentrating on just one of the two factors may actually adversely affect reliability. Hence our algorithm factors in both constraints in attempting to determine the most reliable allocation.

The *allocation matrix X* is an $n \times m$ binary matrix representing the mapping of the n tasks to the m processors. Element x_{ij} is '1' if task t_i has been assigned to processor p_j and is '0' otherwise.

Example 1 Example task and processor graphs are shown in Figure 1(a). The task graph has 5 tasks $a, \ldots, e$ and the processor graph has 3 processors p_0, p_1, p_2. The set of source and sink tasks are $\{a, c\}$ and $\{e\}$ respectively. Each node in the task graph is labeled with its comp_cost. The average unit delay is 1.0. □

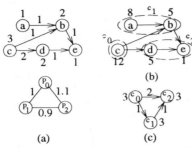

Figure 1: (a) example task/processor graphs (b) static levels/clustering (c) cluster graph

2.3 Reliability Model

We use the same model for reliability as in [3]. In this model the processor and link failures are considered time-dependent and treated in a uniform fashion. An explicit cost function is derived for the reliability of an allocation. We will reproduce the important notations and definitions here for completeness.

Components are either *operational* or *failed*. Components failing during idle periods, when no jobs are running, are replaced and such failures are not considered critical. The future life of all components are assumed to follow a Poisson process with a constant failure rate. Failures of components are assumed to be statistically independent.

Reliability Analysis
A closed form expression has been derived in [3] for the reliability of a given allocation. This expression can be used to drive our algorithms in searching for an allocation that maximizes the reliability. Consider an n-node task graph and m-node processor graph. We denote the failure rate of processor p_i by λ_i and the failure rate of the link between p_i and p_j by μ_{ij}. Given the allocation matrix X, we can write the reliability cost, $relcost(X)$, as [3]:

$$relcost(X) = \sum_{j=1}^{m}\sum_{i=1}^{n}\lambda_j x_{ij} e_{ij}$$
$$+ \sum_{k=1}^{m-1}\sum_{b>k}\sum_{i=1}^{n}\sum_{j=1}^{n} 2\mu_{kb} x_{ik} x_{jb} c_{ij} w_{kb} (1)$$

where the reliability is given by $e^{-relcost(X)}$. The '2' in the above equation is due to the fact that each item of data being transmitted along a link incurs a *send* and a *receive* cost.

In order to increase the reliability, we have to reduce *relcost* as much as possible. The first term in this equation tells us that allocating tasks with greater execution times to more reliable processors might be a good heuristic to increase the reliability. From the second term we see that IPC needs to be reduced as

much as possible. Of course, we cannot avoid IPC altogether. However, allocating larger volumes of data to more reliable links is a good heuristic to increase the overall reliability. We shall shortly quantify these ideas and show how to effectively handle both constraints simultaneously.

We shall be using expressions derived from Equation (1) to drive our allocation algorithm for safety and reliability.

2.4 Assertions

We assume that a fault in a processor will result in an error in at least one task computation performed on it and that the failure of a link will corrupt at least one data item being sent along the link. In order to introduce safety in the system, each task in the task graph is either checked by an assertion task, whenever possible, or else is subject to duplication and comparison [2]. For assertion checking, the original task is modified by encoding its data elements using a system-level code. The encoded output data elements are given to the assertion task which checks their encoding. If the encoding is not satisfied, it indicates the existence of an error and correspondingly, a faulty processor or link. An assertion can be of any type, e.g. checksums in a matrix-matrix multiplication [8], sum-of-squares in FFT [9], type and range check for certain data elements etc. Usually, assertions are much less expensive than duplications.

2.5 Exclusion & Preference Compatibility

Each cluster in the system has a label which gives the *execution cost* of the cluster. This is the sum of the comp_costs of all the tasks in the cluster. The clusters and the dependencies among them can be represented by a *cluster graph*. This graph is obtained from the task graph by creating weighted edges between communicating clusters. The weights on the edges indicate the total volume of data being transferred from tasks in one cluster to tasks in the other. Each cluster and task has an *exclusion vector* and a *preference vector* attached to it. Exclusion vector $excl(t)$ of task t is the row corresponding to t in the exclusion matrix. This vector stores the information regarding which tasks cannot be allocated to the same processor as t. The exclusion vector, $excl(c)$, of a cluster c is the logical OR of the exclusion vectors of all the tasks in the cluster. The preference vector $pref(t)$ of task t is the row corresponding to task t in the preference matrix. The preference vector $pref(c)$ of cluster c is obtained by taking the logical AND of the preference vectors of the component tasks. The exclusion and preference vectors of a cluster represent the combined effect of the exclusion and preference vectors of the component tasks and are used to evaluate whether the cluster can be allocated to a particular processor. The binary nature of the exclusion and preference vectors and the use of logical operators for manipulating them enables a very efficient implementation (for example using a packed array of bits, otherwise known as sets, cf. Pascal). Note that a preference vector of all zeros is not allowed as this precludes the task or cluster from being allocated to any processor.

Definition 1 *Task t_1 is said to be exclusion compatible with cluster c (task t_2) if the exclusion vector of c (t_2) does not have a '1' entry in the column corresponding to t_1.*

Definition 2 *Task t_1 is said to be preference compatible with cluster c (task t_2) if, when t_1 is added to c (t_2), the resulting preference vector does not have all '0' entries.*

Definition 3 *Task t_1 is said to be fully compatible with cluster c (task t_2) if it is both exclusion compatible and preference compatible with c (t_2).*

Definitions 1, 2 and 3 can be applied to determine the compatibility of two clusters too. If all the tasks in cluster c_1 are fully compatible with cluster c_2 or *vice versa*, then clusters c_1 and c_2 are said to be fully compatible.

2.6 Problem Definition

We are given a job in the form of a task graph, the target architecture in the form of a processor graph, the exclusion and preference matrices and the fault tolerance information. Our objective is to find a *safe mapping* of the tasks to the processors of the system such that the overall reliability of the allocation is maximized. Since the allocation problem is known to be NP-complete, our aim is to devise good heuristics for the problem. We will explore several heuristics to find the approach yielding the best results.

3 Allocation for Reliability

We next show how allocation can be done to meet our desired objective.

3.1 Clustering

The first step of our algorithm, called *clustering*, uses techniques from linear clustering [10] and groups tasks into *clusters* to form the *cluster graph*. First, each task t_i is assigned a *static level*, $SL(t_i)$, which is the sum of comp_costs and communication costs along the longest path from t_i to a sink task in the system [2]. The unclustered tasks on the longest directed path in the task graph starting from the unclustered task with the highest static level, such that all these tasks are mutually compatible and such that the sum of their execution costs does not exceed a *threshold* are grouped into a cluster. We chose the sum of the average execution cost of all tasks divided by the number of processors as the threshold, to appropriately *size* the clusters. This clustering algorithm is essentially the same as the three-pass procedure proposed in [2] with the three passes consolidated into one.

Figures 1(b) and 1(c) show the result of assigning static levels to the tasks in the task graph of Figure 1(a), performing clustering using our clustering algorithm and the resulting cluster graph. The sum of the average execution costs of all the tasks in this example is 9. Averaging this over 3 processors, we get a threshold of 3.

3.2 Allocation of Original Tasks

After the clustering stage the task graph has been partitioned into clusters. The next step is to allocate the clusters to the processors in the system. Our objective for allocation is to minimize the *relcost*. In this section we present a new cluster-based allocation strategy to maximize reliability. Task-level allocation strategies for maximizing reliability have been proposed in [3]. We first give our allocation method and then the best method from [3]. In Section 5, we compare the performance of these two methods.

3.2.1 Cluster Allocation

In order to represent the combined effect of the exclusion vectors of all the tasks allocated to a processor, each processor in the system also has an exclusion vector associated with it, which is the logical OR of the exclusion vectors of all the tasks allocated to this processor. We initialize the exclusion vector of all processors to all '0's. Before a fresh cluster is allocated to a processor, we have to verify whether the processor and cluster are exclusion and preference compatible, *i.e.* to verify that (a) no task in the cluster excludes any task already allocated to the processor and (b) all tasks in the cluster are preference compatible with the processor. A processor is not exclusion compatible with a cluster if the exclusion vector of the processor has a '1' entry for some task which belongs to the cluster or *vice versa*. Similarly, a processor is not preference compatible with a cluster if the cluster's preference vector has a '0' in the column corresponding to that processor.

First, all the clusters containing tasks with specific preference are allocated. For this, the preference vectors of the clusters are checked. If any cluster has a '1' in some single position j then that cluster has to be allocated to processor p_j. It might, however, happen that although preference is forcing a cluster to go to some processor, exclusion is preventing it. This means that some task in the cluster is required to be mapped to a particular processor with which some other task in it is exclusion incompatible. If this happens, the cluster is broken up into smaller clusters such that the set of tasks contributing the '1' to the preference vector are all exclusion compatible with that processor. Now the clusters can be mapped to the appropriate processors. After any cluster is mapped to a processor, the exclusion and preference vectors of the cluster and the exclusion vector of the processor are updated.

Next, the remaining clusters have to be allocated. The clusters are sorted in the decreasing order of their *total cost*, which is the sum of their execution and communication costs, and considered for allocation in that order.

The cluster allocation algorithm is a function of a user specified parameter, which we shall denote by k_a. We refer to this as the *order* of the algorithm. The allocation algorithm keeps track of the *relcost* for the allocated clusters using Equation (1). After any cluster is allocated, this is updated as follows. Suppose cluster c_u is allocated to processor p_v then the increase in *relcost*, $\Delta relcost(c_u, p_v)$, can be calculated in the following manner. The increase in *relcost*, $\Delta relcost(t_i, p_v)$, when a task t_i is allocated to p_v is

$$\Delta relcost(t_i, p_v) = e_{iv}\lambda_v + \sum_{t_j} 2\mu_{vw}c_{ij}w_{vw} \quad (2)$$

where t_j is a fanout of t_i such that t_j has been previously allocated to a processor p_w distinct from p_v. Now we can write

$$\Delta relcost(c_u, p_v) = \sum_{t_i \in c_u} \Delta relcost(t_i, p_v) \quad (3)$$

At any time the algorithm considers a group of k_a clusters and tries to find the best allocation for them. This is done by considering all possibilities for allocating the k_a clusters and for each choice, marking the clusters as temporarily allocated and then computing the increase in *relcost*. Then, that allocation which results in the least increase in *relcost* (i.e. least decrease in reliability) is chosen. Then the algorithm moves on to the next set of k_a clusters. As soon as a cluster c is allocated to a processor p, all tasks in c, processor p and all the (relevant) clusters in the system are updated as follows. The execution cost of all the tasks in c is set to their execution cost on p and the total cost of c is updated. The weights on the edges of the cluster graph are modified such that edges between clusters allocated to the same processor have a weight of 0 and the edge between clusters c_u and c_v allocated to distinct processors p_i and p_j has a weight equal to the product of w_{ij} and the volume of communication between c_u and c_v. The execution costs of all tasks in c are added to the *bin cost* of p (*i.e.* sum of the execution and communication costs of all the tasks allocated to p). When k_a is set to the number of clusters, the algorithm explores all combinations of clusters and processors, and when k_a is set to '1', it is a greedy algorithm with a lookahead of just '1'. Thus, k_a controls the fraction of the search space that the algorithm explores.

3.2.2   Task Level Allocation Scheme

In [3], several task-level allocation schemes were proposed for maximizing reliability. We shall be using the method that performed the best (in our experiments), for comparison with our cluster-level allocation scheme.

We have to note that the notion of task exclusion and preference is not used in [3]. Also, the task graphs considered do not have any precedence constraints. For our experiments, we have adapted the spirit of this algorithm to accommodate these differences. The adapted allocation scheme works as follows. First the tasks with specific preferences are allocated. The remaining tasks are sorted in the decreasing order of their *communication cost* (as compared to the total cluster costs used in our algorithm) and are considered for allocation in that order. For each possible processor that the first task can be allocated to, the remaining tasks are allocated in a single-step greedy process, that is, to that processor that leads to the minimum increase in *relcost*. Equation (2) can be used to compute the incremental increase in *relcost*. Of each of these possible allocations, the one that yields the lowest *relcost* (highest reliability) is chosen.

In order to be fair in the comparison of our k_a-step algorithm with this one, we next need to introduce the *order* concept in this method too. This can be done quite naturally by taking all combinations of processors for the first k_a tasks and then allocating the rest in a single-step greedy process and then picking the best solution.

4   Safety

After the cluster allocation stage, the reliability of the allocation has been maximized. However, no guarantees can be given about system behavior when a failure occurs. As remarked earlier, the allocation is unsafe. The next step is to add safety to the allocation. We apply the principle of task-based fault tolerance [2] to achieve this.

Each task has five types of fault tolerance information associated with it [2].

1. Is the task assertible, *i.e.* does an assertion particular to the task exist?

2. Exec_cost vector of the assertion task, if the task is assertible.

3. Is the assertion *costly*, *i.e.* does it require as inputs all the inputs of the task that it is checking?

4. Volume of data being communicated from the task to the assertion task.

5. Exec_cost of the comparison task in the case when duplication is done.

Once the allocation of the original tasks is done, the task graph is taken through two fault tolerance stages. The first stage is called *ft_task creation*, in which the nodes required for fault tolerance are added to the task graph. The next stage is *ft_task allocation* in which these new nodes are allocated to the processors under the fault tolerance constraints.

Ft_task Creation: This step is described in detail in [2]. We shall briefly describe the concepts here for completeness. In this step, an assertion node is added to the task graph for each task that is assertible. The assertion node has to be allocated to a processor different from the original task. Hence the assertion is made preference incompatible with the processor to which the original task has been allocated. If the task is costly, all the fanins of the original task are duplicated as fanins of the assertion. Next, corresponding to each non-assertible task, a duplicate and a comparison task are created. The duplicate is made exclusion incompatible with the original in order to force it to be allocated to a different processor from the original. The comparison node has as its fanins the outputs from the original and the duplicate. The comparison task cannot be on the same processor as the original but may be mapped to the same processor as the duplicate (since the duplicate's output does not fan out to any other task). So, the comparison is made preference incompatible with the processor executing the original.

Ft_task Allocation for Safety to Maximize Reliability: After all the extra tasks needed to introduce safety have been created, the next step is to allocate them so that the overall increase in *relcost* is minimized in order to maximize the final reliability. We shall present a task-level heuristic which, for this purpose, performs as well, in terms of the quality of the results, as the task-level heuristic from [3] described in Section 3.2.2, but runs an order of magnitude faster.

Our task-level heuristic is similar to the "order-k_a" cluster allocation method and has an "order" parameter which we denote by k_b. The extra tasks added by the Ft_task_creation stage are sorted in the decreasing order of their total costs. The parameter k_b is a user-specified entity (the value of k_b used in this step may be distinct from the value of k_a used in the cluster allocation stage). The tasks are now considered in batches of at most k_b at a time. For these k_b tasks, from all the possible choices of processors to allocate, the allocation that leads to the least increase in *relcost* is chosen. Then the next set of k_b tasks is considered, and so on.

The increase in *relcost*, $\Delta relcost(t_i, p_v)$, when task t_i is allocated to processor p_v, can be computed using Equation (2).

5   Experimental Results

In this section, we describe the performance of our algorithms on a large number of examples including a real-life digital signal processing (DSP) example.

5.1   Performance on Synthetic Workloads

Due to a lack of benchmarks for validating our algorithm, we have tested it on a variety of randomly generated task and processor graphs. Random graphs have been used by several researchers in the past (see for example [3, 11]). We have tried to choose random workloads with characteristics similar to available real-life examples.

5.1.1   Generation of Random Examples
We chose several different types of task and processor graphs with structures which occur commonly in a variety of real-life algorithms. The chosen types of task graphs were (a) binary trees (BTREE) (b) lattices (LATTICE) (c) randomly interconnected chains (CHAIN) (d) server-client formations (SERVERCLIENT)

Each example had 100 tasks in the task graph and 10 processors in the processor graph. The processor graph connectivity (the probability of a given edge in the graph) was chosen to be 0.9. 50% of the tasks were chosen to be assertible (at random) in each task graph with 10% of these assertions being costly. The execution cost of the tasks was chosen uniformly between [5, 195] units. The communication weights for the tasks were taken uniformly between [1, 10] units. The link failure rates were chosen uniformly from the range 7.5×10^{-6} to 12.5×10^{-6} per hour and the processor failure rates between 0.95×10^{-6} to 1.05×10^{-6} per hour. Each data point is an averaged value over 250 random graphs.

5.1.2   Performance on Random Examples
As has been explained in previous sections, there are two stages in the allocation algorithm. The first stage allocates the original tasks with the objective of maximizing the reliability. Our algorithm is cluster-based and has an "order" parameter, k_a. We compare the performance of this algorithm with the best task-level allocation algorithm from [3], as presented in Section 3.2.2. In order to have a fair comparison, we have introduced the "order" parameter in this algorithm as well, as a means of forcing the algorithm to search through more or less of the search space. Our first set of experiments compare the performance of these two algorithms with respect to the reliability achieved and the amount of CPU time[1] consumed in searching for the solution. In the tables, for brevity, we shall refer to the task-level algorithm for allocating the original tasks as T0 and our cluster-based algorithm for this purpose as T1.

The second stage allocates the tasks needed to introduce safety. We compare the performance of the modified version, with "order" k_b, of the task-level allocation algorithm from [3] (denoted by F0) with the performance of our "order" k_b task-level allocation algorithm (denoted by F1). Note that increasing k_b increases the complexity of both methods but in different ways.

In each case, we present the results for the four different types of workloads and for different values of k_a and k_b. Our objective is to determine that combination of T0, T1, F0 and F1 and the choice of values for k_a and k_b which yields the best results in terms of reliability and is most cost-effective in terms of CPU

---

[1]All CPU times in this paper were measured on a SPARCstation-1 with a 16MB RAM.

time.

Comparison of cluster-based and task-based allocation schemes: In Table 1, the results of the simulation runs comparing T0 and T1 are presented. Simulations were conducted for k_a set to 1 and 2 for both methods. In the table, T1 with k_a set to 2 is chosen as the base case. The "% extra" columns show the percent unreliability greater than the base case that is obtained by the particular method, for the particular type of workload. Also presented is the time taken in seconds by each method to complete the allocation of the original tasks.

As can be seen, T1 with k_a set to 1 performs consistently better than T0, for both k_a set to 1 and to 2. Also, T1 with k_a equal to 1 takes about the same amount of CPU time as T0. T0 shows a slight improvement with respect to itself on increasing k_a from 1 to 2. T1 shows hardly any improvement on increasing k_a and takes a lot more time to complete. (The results for both T0 and T1 saturate for values of k_a greater than 2, which is not shown in these tables.)

From this table we can conclude that T1 with k_a equal to 1 is the most cost-effective in terms of minimizing the unreliability and consumes about the same amount of CPU time as T0, $k_a = 1$, in searching for a solution.

Comparison of F0 and F1 for allocating extra tasks: Tables 2 and 3 compare the performance of our algorithm for allocating the extra tasks and the algorithm presented in [3], adapted to fit our assumptions and scenario, in order to make a fair comparison.

Having decided that increasing k_a beyond 1 does not help much in decreasing the unreliability, we have set k_a to 1 for this step, for both T0 and T1. Choosing T1 with k_a equal to 1 as the base case, Table 2 shows the % increase in unreliability obtained by the different combinations of T0 and T1 for the allocation of the original tasks and F0 and F1 for allocating the extra tasks. Our objective, of course, is to determine that method which leads to the least increase in unreliability.

As can be seen from the table, unlike k_a, increasing k_b from 1 to 2 does decrease the unreliability for both F0 and F1. (Increasing it further does not help; this is not shown in the table.) Also, although F0 does a little better than F1 for k_b equal to 1, both produce solutions of the same quality when k_b is increased to 2.

Table 3 shows the number of CPU seconds consumed by both the methods for both values of k_b. The CPU time taken by T0 and T1 is reproduced from Table 1 for comparison. As can be seen, when k_b is 1, F1 runs in less than $\frac{1}{5}$th the time taken by F0. It is an order of magnitude faster when k_b is increased to 2.

From these results, we can conclude that the combination of T1 with k_a equal to 1 for allocating the original tasks and F1 with k_b equal to 2 for introducing safety form the best combination, in achiev-

ing the most reliable allocation and in requiring the fewest number of CPU seconds in doing so. Also, using this combination of algorithms, we see that the average increase in unreliability due the addition of safety is around 48.4% for the different types of workloads. The reliabilities of the system before and after the addition of safety were 0.99901 and 0.99846 for LATTICE, 0.99906 and 0.99863 for BTREE, 0.99905 and 0.99859 for CHAIN, and 0.99902 and 0.99859 for SERVERCLIENT. Clearly, this small decrease in reliability is not too high a price to pay for the addition of safety. These experimental results validate the use of our technique to add safety.

5.2 Performance on a DSP Example

To show the performance of our scheme for reliable and safe allocation, we ran our algorithm on a sizeable DSP example from [12], which gives the details of the task graph for the example (there are 119 tasks in this task graph). About 33% of the tasks were assertible [2] and none of the assertible tasks were costly.

We generated a number of heterogeneous processor graphs with 7 fully connected processors, with processor failure rates chosen randomly between 3.2×10^{-3} and 4.0×10^{-3} per hour and link failure rates between 1.8×10^{-2} and 5.4×10^{-2} per hour. We performed our experiments with T1 ($k_a = 1$) for allocating the original tasks and F1 ($k_b = 2$) for allocating the tasks required to add safety. On an average, the reliability before safety was added was 0.999988 and the reliability after the addition of safety was 0.999969. This is a very marginal decrease in reliability which is not much of a price to pay for the addition of safety. The cluster allocation stage of the algorithm took 7.9 seconds of CPU time and the addition of safety took an extra 2.6 seconds.

6 Conclusions

In the past, methods have been presented for task allocation to maximize reliability. However, such schemes do not give any guarantees about the system behavior when a failure occurs. This is a scenario for potentially undetected faults and catastrophe. In this work, we have addressed the problem of introducing safety into the system during task allocation. We have presented a new cluster-based allocation technique to maximize reliability and we have shown that it performs better than previously known techniques for this purpose. Then, using the concept of task-based fault tolerance, we have devised a technique to introduce safety. We have presented a new task-level allocation scheme for allocating the extra tasks needed for this purpose. This method produces solutions of quality equal to the best known reliability-driven task-level allocation heuristic and, for our purposes, runs an order of magnitude faster. Our experimental results show that the decrease in reliability due to the introduction of safety is very small which proves the efficacy of our scheme.

Table 1: Comparison between T0 and T1

| | $k_a = 1$ | | | | $k_a = 2$ | | | |
| | T0 | | T1 | | T0 | | T1 | |
	% extra	CPU time (seconds)	% extra	CPU time (seconds)	% extra	CPU time (seconds)	% extra	CPU time (seconds)
LATTICE	22	6.1	1	6.8	20	10.9	0	16.2
BTREE	7	5.7	0	6.1	6	8.8	0	23.7
CHAIN	8	3.5	0	3.7	5	5.2	0	6.0
SERVER-CLIENT	4	4.9	0	5.1	4	6.7	0	11.3

Table 2: Comparison of % overhead in unreliability after addition of safety

| | $k_b = 1$ $(k_a = 1)$ | | | | $k_b = 2$ $(k_a = 1)$ | | | |
	T0 + F0	T0 + F1	T1 + F0	T1 + F1	T0 + F0	T0 + F1	T1 + F0	T1 + F1
LATTICE	82	89	60	61	75	81	56	56
BTREE	57	61	52	52	54	58	56	56
CHAIN	57	60	51	51	52	58	48	48
SERVER-CLIENT	53	58	49	49	51	56	44	44

Table 3: CPU seconds needed to add safety

| | Original Tasks | | Addition of Safety | | | |
| | $k_a = 1$ | | $k_b = 1$ | | $k_b = 2$ | |
	T0	T1	F0	F1	F0	F1
LATTICE	6.1	6.8	5.7	1.0	46.2	4.0
BTREE	5.7	6.1	5.8	0.9	45.2	3.9
CHAIN	3.5	3.7	3.6	0.6	27.7	2.5
SERVER-CLIENT	4.9	5.1	5.1	1.0	49.3	4.1

References

[1] S. Tridandapani and A.K. Somani, "Efficient utilization of spare capacity for fault detection and location in multiprocessor systems," in Proc. Int. Symp. Fault-Tolerant Comput., Montreal, pp. 440-447, July 1992.

[2] S. Yajnik, S. Srinivasan and N.K. Jha, "TBFT: A task-based fault tolerance scheme for distributed systems," in Proc. Int. Conf. Parallel & Distr. Comput. Systems, Las Vegas, Nevada, Oct. 1994.

[3] S.M. Shatz, J.P. Wang and M. Goto, "Task allocation for maximizing reliability of distributed computer systems," IEEE Trans. Comput., vol. 41, no. 9, pp. 1156-1168, Sept. 1992.

[4] N.H. Vaidya and D.K. Pradhan, "Fault-tolerant design strategies for reliability and safety," IEEE Trans. Comput., vol. 42, no. 10, pp. 1195-1206, Oct. 1993.

[5] M. Mullazani, "Reliability and safety," in Proc. 4th IFAC Workshop Safety Comput. Contr. Systems, pp. 141-146, 1985.

[6] B.W. Johnson and J.H. Aylor, "Reliability & safety analysis of a fault-tolerant controller," IEEE Trans. Reliab., vol. 35, no. 10, pp. 355-362, Oct. 1986.

[7] P.Y.R. Ma, E.Y.S. Lee and M. Tsuchiya, "A task allocation model for distributed computing systems," IEEE Trans. Comput., vol. C-31, no. 1, pp. 41-47, Jan. 1982.

[8] K.H. Huang and J.A. Abraham, "Algorithm-based fault tolerance for matrix operations," IEEE Trans. Comput., vol. C-33, no. 6, pp. 518-528, June 1984.

[9] A.L.N. Reddy and P. Banerjee, "Algorithm-based fault detection for signal processing applications," IEEE Trans. Comput., vol. 39, no. 10, pp. 1304-1308, Oct. 1990.

[10] G.C. Sih and E.A. Lee, "Declustering: a new multiprocessor scheduling technique," IEEE Trans. Parallel & Distr. Syst., vol. 4, no. 6, pp. 625-637, June 1993.

[11] V.M. Lo, "Heuristic algorithms for task assignments in distributed systems," IEEE Trans. Comput., vol. 37, no. 11, pp. 1384-1397, Nov. 1988.

[12] C.M. Woodside and G.G. Monforton, "Fast allocation of processes in distributed and parallel systems," IEEE Trans. Parallel & Distr. Syst., vol. 4, no. 2, pp. 164-174, Feb. 1993.

Time Sharing Massively Parallel Machines*

Brent Gorda - brent@nersc.gov
National Energy Research Supercomputer Center
Livermore, CA 94550 U.S.A

Rich Wolski - rich@cs.ucsd.edu
UCSD Computer Science and Engineering Department

INTRODUCTION

The development of performance efficient applications on today's parallel machines continues to be hampered by a lack of good scheduling methodologies. Interactive development and production computing are often best served by different scheduling policies. The most widely-used solution is to space-share the machine, giving each user exclusive access to a subset of the resources. Unfortunately since resources are idle much of the time during development, space-sharing can cause the overall machine to be under-utilized. Alternatively, production use of a machine in a space-sharing environment tends to monopolize system resources for long periods of time. The result is poor response time for interactive use.

As part of the Massively Parallel Computing Initiative (MPCI) at the Lawrence Livermore National Laboratory [1] (LLNL), we have developed a simple, effective and portable time sharing mechanism which addresses these issues by scheduling gangs of processes as a unit on tightly coupled parallel machines. By time-sharing the resources, our system interleaves production and interactive jobs. Immediate priority is given to interactive use, maintaining good response time. Production jobs are scheduled during idle periods, making use of the otherwise unused resources. In this paper we discuss our experience with gang scheduling over the 3 year life-time of the project. In the next section we motivate the project and discuss some of its details. We then describe the general scheduling problem and related efforts. The next section describes our implementation. We conclude this paper with some observations and possible future directions.

BACKGROUND

In 1989 the Lawrence Livermore National Laboratory (LLNL), began a study of parallel computing and its applicability to the scientific computational requirements of the Department of Energy. The purpose of the Massively Parallel Computing Initiative (MPCI), was to investigate parallel distributed memory computing as a combination of parallel machine architecture, programming models, coding strategies, and algorithms. A 126 processor BBN TC2000 built by Bolt Beranek and Neuman Advanced Computers Inc. (BBN ACI) was installed at LLNL and staffed with computer and computational science personnel. Collaborations with other DOE Laboratories, industrial R&D groups and universities were encouraged by making the project's resources accessible. Reports detailing the experiences and research conducted by MPCI participants can be found in [1, 2, 3].

Space Sharing

As delivered, the TC2000 supports space-sharing in the form of clusters. An arbitrary sub-set of available processors may be joined together to operate on an application when the application is initiated. Parallel applications execute by forming sub-clusters out of the available resources in the free cluster (a pool of available processors maintained by the system).

Experience both at LLNL and elsewhere soon showed that space-sharing alone was inadequate in a production environment because it does not accommodate both production and interactive users. Those developing applications want quick response for short periods of time, flexibility in choosing the number of processors, and a completely reproducible execution environment. Frequently, race conditions are dependent on the number of processors or the processor identities (due to their location within the system's topology) making it important to control the execution environment. Production users, in contrast, want long periods dedicated of time. Often several hours or days while using all available resources.

We developed the MPCI Gang Scheduler (GS) in response to these conflicting needs. In consideration of the vendor, who was weary of external modifications to its operating system, we developed the GS to run as a privileged daemon outside of the OS (in the public cluster). To manage resources on a parallel machine, the GS enforces a protocol for resource access that all applications must observe. We made the GS transparent to the end-user, however, by adding interface code to the start-up libraries used by the various parallel languages available on the system.

---

*Work performed under the auspices of the U. S. Department of Energy by the Lawrence Livermore National Laboratory under contract No. W-7405-ENG-48.

The GS maintains a list of processes associated with currently executing applications. It implements time sharing by permitting groups of processes belonging to the same application to access the CPUs in turn. To avoid starvation, the scheduling policy is round robin. The GS combines space- and time-sharing by "packing" jobs onto the processors. Applications are assigned processors according to the current load and potential for available cpu time.

Related Research

Gang scheduling processes on a parallel machine can be done in a variety of ways. Other variants of gang scheduling include family scheduling [6], and co-scheduling [4]. In family scheduling, parallel threads from a single application form a family. When the application is scheduled, processors choose from a queue of threads belonging to a single family. Hence the family is scheduled as a unit. Family scheduling is a generalization of the techniques we have implemented. The chief difference is that our decision not to modify the operating system kernel prevents us from implementing the process migration of family scheduling. Under co-scheduling, tasks are scheduled as a unit on only those processors available at the time of their initiation. Since system load can affect the number of processors available at any given time, the parallelism available to an application varies from run to run. The goal of co-scheduling is to keep the machine fully utilized at the possible expense of parallelism.

One of the advantages of gang scheduling is that application execution times are more consistent. The GS assigns and removes jobs from processors as a an indivisible unit. Thus, whenever the application is running it has all of its processors, and when it goes idle (as it may do frequently during development or debugging), its processors become available to other applications. Another advantage is that applications can spin-wait during their timeslice yielding lower message passing overhead. Since the number of processors used by the application is not dependent on system load, the applications can make better use of static load-balancing techniques[5]. Our implementation of gang scheduling is non-pre-emptive further enhancing reproducible execution. However, one drawback of our implementation is that all resources are consumed indivisibly. The result is that if two jobs share any single processor, then they must run at mutually exclusive times. As shown in Figure 1, job "a" cannot run with either job "b" or "c" due to conflicts on processors 3 and 4. Jobs "b" and "c", can be gang scheduled to run at the same time.

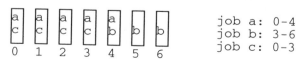

job a: 0-4
job b: 3-6
job c: 0-3

Figure 1. Gang Scheduling: procs given/taken as a group

To mitigate this, we have outfitted the GS with a packing algorithm that attempts to keep all processors utilized. All processors are shared via a global timeslice. Two independent jobs using disjoint processor sets may be scheduled during the same global timeslice and hence, may share the machine. Alternatively, if two processor sets have a non-zero intersection, then they must be scheduled during different global timeslices. The GS will, whenever possible, try to assign independent jobs to disjoint processors so they may share the machine simultaneously.

On the TC2000 the GS is implemented as a daemon running as root in the public cluster (we partitioned the machine into a small public cluster, and a large free cluster). It, therefore does not negatively impact parallel applications which draw resources exclusively from the free cluster. The machine setup thus appears to the user as in Figure 2. Users familiar with the architecture see the GS cluster in place of the free cluster.

Public Cluster   Gang Scheduled Cluster

Figure 2. The TC2000 setup at Livermore

IMPLEMENTATION

The GS supports different execution classes, each governed by a tunable set of scheduling parameters. Priorities can be assigned to the different classes as desired. For MPCI, we specified four such classes: interactive, production, standby, and benchmark. The GS attempts to service interactive jobs first assuming that they will execute for a short time and then voluntarily relinquish their processors. During idle times, production jobs are used to fill the gaps. Interactive jobs that execute past a configurable threshold time are automatically "demoted" into the production class to prevent monopolization of the machine. Production jobs receive longer timeslices to help amortize the overhead of context switching, but are only scheduled alongside, or in the absence of runnable interactive jobs. As we intend to develop a fair share charging algorithm as part of our future efforts, we created a standby class, wherein programs receive any free cycles that the machine has to offer. We also specified a benchmark class to allow single-user access for timing or scalability tests. An application may request a block of benchmark time in which no other applications will be allowed to execute. Requests for 60 seconds (the default) or less are serviced immediately in our current configuration. Programs requesting more than 60 seconds are required to wait until midnight to run. This way, programmers can submit their benchmark runs at any time during the day, and receive their results the next morning.

Machine Support

The TC2000 system supports several features that aided the implementation of the GS. Each processor runs a copy of the nX operating system, and thus acts as a stand-alone system. On each processor, Unix time-sharing takes place if there are multiple processes. Process identification (pid) numbers are allocated in a global space, ensuring uniqueness. The only modification we required was a way for the GS daemon process to make user processes either runnable or blocked. To facilitate this, the SIGUSR1 and SIGUSR2 signals are used (BBN made this modification to nX at our request). The scheduler can stop a process by sending it a SIGUSR1. SIGUSR2 is used to make the process once again runnable. In general there is only one process ready to run on each processor so that process receives all cycles while it is executing. As a result, context switching is simple: sends a group of running processes SIGUSR1 to end a time-slice. Then send the new group of processes the SUIGUSR2 start signal.

To keep track of the placement of processes, the GS enforces a strict protocol. When an application starts up or wants to increase its processor count it asks for a set of processors, either telling the GS their IDs or letting the GS optimize the selection. To populate processors, the application uses the nX system call `fork_and_bind`.

During the applications existence, the number of processors may vary, as may the number of processes. As long as a single process remains the GS keeps the job in the runnable state. To track of processes the GS reads the kernel process table during idle times, verifying the existence and ownership of processes in the GS cluster. Stray jobs, and jobs without an owner are removed.

Gangster

We provide users with an interactive tool, the gangster tool for observing the system and controlling some aspects of their jobs. An example display is shown in Figure 3 (truncated for space reasons). Gangster communicates with the GS daemon to query the state of the machine, and to make requests on the users' behalf. The tool is so popular that users think of the scheduling system as simply *gangster*.

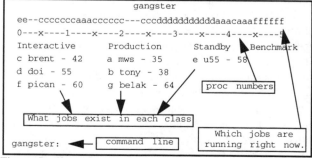

Figure 3. Gangster display

In Figure 3, the line after the title describes jobs that are currently running and their processors. Each letter represents a processor, and corresponds to a job below. Users interact gangster on the command line. By specifying the job's id, a user may query wall-clock execution time, accrued CPU time, and processor information for a job. They may also stop or change the run class of their jobs.

Operating Parameters and Robustness

At the outset of the project, we were uncertain as to how to optimize the operating parameters of the system to provide maximum throughput while maintaining a reasonable level of underactivity. Therefore, we provided an mechanism to allow dynamic changes to the GS daemon without interrupting service. For example an operator can specify how long to wait before demoting a job from interactive to production, and the number of times that a user can bump the job back up. Other parameters include a limit on the number of jobs any single user is allowed to submit, timeslices for classes and for benchmarking.

Since all parallel jobs must interface with the GS, robustness is paramount. We included a mechanism for the GS to be upgraded while running without a noticeable degradation of service. Every transaction between the GS daemon and client applications uses a well-known socket. If that socket is busy or unattached the interface code loaded with the application retries. Meanwhile, the operator interface allows the GS to be shut down and restarted. When the GS receives a shutdown command it ensures that there are no outstanding transactions, writes a checkpoint file, and exits. A new version can then be started using this file. The application retrying then gets service and internal consistency-checking corrects any inconsistent data (such as jobs exiting in the interim) in a few seconds. From the users' perspective, the GS software never goes down, but does acquires more features over time.

RESULTS

The GS has been exceptionally stable and performs well. While we did not consider scalability issues in our initial implementation, we do not believe there are any significant performance bottlenecks. In particular, the socket interface used between a client and the scheduler has not proven to be a noticeable performance problem. Since installation, the scheduler has had no significant instances of failure. During its use in the MPCI project (mid 91 - late 92), the GS has handled over 35,000 job requests of various sizes. We have recently examined the logfiles in an attempt to understand the job mix the machine during its lifetime. The graph in Figure 4 depicts the size distribution for a total of 35,848 job requests.

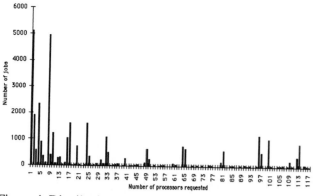

Figure 4. Distribution of job sizes.

As expected, there are peaks centering around popular processor counts such as powers of 2. The reasons for other commonly chosen processor counts are not so obvious. Processor counts of 25, 49 and 100 are due to the use of square grid within applications. Processor counts of 26, 50, and 65 are due to a square grid with an added *master*. On the high end, 112 and 113 are the typical GS cluster size. (The largest cluster ever is 117 processors).

It is important to notice that many of the jobs were run on smaller processor counts. More than 70% of the programs in this data used less than 32. This is a surprising result as the environment rewards highly parallel applications (the charging algorithm charges time on a per job basis, not per CPU). A case could be made that many of the applications were executed while in development, so they were run on less CPUs, however production use (scheduled primarily during non-business hours) should see higher processor counts. This was not the case.

Finally, we are interested in the difference between wall-clock and CPU time as a measure of the impact of time-sharing. Figure 5 shows this data for job sizes. The average wall clock time, (solid line) is the amount of time from start to job completion. Average cpu time (dotted) depicts the amount of machine time dedicated to the job. The difference in these times shows the amount of time the job was swapped out.

CONCLUSION

During the MPCI project, time-sharing of the BBN TC2000 was controlled exclusively by a locally developed Gang Scheduler. The goals project were to provide a robust scheduling environment that enabled time and space sharing of the machine. This software ran a real-world work load for several years without a single significant failure. We described the features and implementation of the system. We also discussed the user interface and some options. Finally we codified some of the data gathered and highlighted some interesting trends.

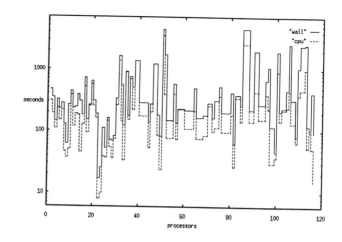

Figure 5. cpu and wall time vs. processors.

FUTURE

The High Performance Parallel Processing Project (H4P), is a DOE project involving the LLNL, Los Alamos National Laboratory, Cray Research Incorporated, Thinking Machines Corporation and various industrial partners. The project is a collection of over 20 efforts with industrial partners aimed at proving the applicability of parallel processing to industry. We will be developing the Gang Scheduler as one of these.

REFERENCES

[1] Eugene D. Brooks, Karen H. Warren, ed. *The 1991 MPCI Yearly Report: The Attack of the Killer Micros*, LLNL,1991 - UCRL-ID-107022

[2] Eugene D. Brooks et. al., editors.*The 1992 MPCI Yearly Report: Harnessing the Killer Micros* - LLNL, 1992 - UCRL-ID-107022-92

[3] Dale E. Nielsen, Jr.*General Purpose Parallel Supercomputing*, LLNL UCRL-ID-108228.

[4] J.K. Ousterhout, *Scheduling Techniques for Concurrent Systems*, Proceedings of the Distributed Systems Conference, IEEE Computer Society, 1982 pp. 22-30.

[5] Eugene D. Brooks, et al. *The Parallel C Preprocessor*, Scientific Programming, Vol. 1 / Number 1, Fall 1992

[6] R.M. Bryant et. al. *Operating system support for parallel programming on RP3*, IBM J. Research and Development vol.35 No 5/6 Sept./Nov 1991.

COST-EFFECTIVE DATA-PARALLEL LOAD BALANCING

James P. Ahrens
Dept. of Computer Science & Engineering
University of Washington
Seattle, Washington 98195
ahrens@cs.washington.edu

Charles D. Hansen
Advanced Computing Laboratory
Los Alamos National Laboratory
Los Alamos, New Mexico 87545
hansen@acl.lanl.gov

Abstract – Load balancing algorithms improve a program's performance on unbalanced datasets, but can degrade performance on balanced datasets, because unnecessary load redistributions occur. This paper presents a cost-effective data-parallel load balancing algorithm which performs load redistributions only when the possible savings outweigh the redistribution costs. Experiments with a data-parallel polygon renderer show a performance improvement of up to a factor of 33 on unbalanced datasets and a maximum performance loss of only 27 percent on balanced datasets when using this algorithm.

1. INTRODUCTION

Load balancing algorithms provide the basis for efficient parallel solutions to many important computational problems including the n-body problem, polygon and volume rendering, and optimization problems. The completion time of these parallel solutions depends on the completion time of the processor with the maximum computational workload. Load balancing algorithms attempt to distribute the computational workload evenly among all processors. This reduces the maximum workload on any processor and thus reduces the completion time of the parallel solutions.

A significant problem when using a load balancing algorithm is the possibility that along with improving performance on some datasets it will degrade performance on others. Wikstrom et al. [1] use a computation model and experimental results to present evidence that using a load balancing algorithm does not always improve a program's performance. The authors show the execution time of a load-balanced version of a program can substantially exceed the execution time of the original version of the same program. This is because the costs of load redistributions can exceed the savings achieved by the redistributions.

Other researchers have studied the problem of deciding when to balance with different workloads and problem types. Nicol and Townsend [2] describe a partitioning strategy which uses performance measurements to decide how to partition an irregular grid among processors. Nicol and Reynolds [3] describe a data-parallel load balancing algorithm which is targeted for applications with uncertain behavior. The algorithm uses a probabilistic model of the cost of delay and the benefits of balancing to decide when

to run a single balancing operation.

In this paper, a load balancing algorithm is described which uses a prediction of the costs and calculation of the possible savings to decide when redistribution is cost-effective. A major advantage of utilizing a cost-effective load balancing algorithm is that the execution time of a load-balanced version of a program is never significantly worse than the execution time of the original version of the same program. This result depends only on never underestimating the costs of load balancing.

In Section 2, the type of workloads and programs the load balancing algorithm works with is presented along with a high-level description of the load balancing algorithm. Section 2 also describes when and how load balancing occurs. Section 3 presents a performance study of a data-parallel rendering program to which the load balancing algorithm has been applied. Section 4 concludes.

2. THE LOAD BALANCING ALGORITHM

The load balancing algorithm can be applied to data-parallel programs which compute the solutions of a collection of independent tasks. The tasks are independent in time; they do not have to execute in any specific order and there are no data dependencies between the tasks. A set of tasks can share the same read-only problem data. An example of how a set of tasks share the same problem data occurs in the polygon rendering application. Multiple tasks are used to process a row of pixels. Each task computes the solution for one pixel in the row. The pixel's location is computed by adding an offset to the row's initial pixel location. The initial pixel location is stored as part of the problem data.

For the following discussion, we assume a virtual processing facility provides the abstraction of having one virtual processor assigned to each data element of a parallel array. In a prototypical program, each processor is assigned problem data and its associated tasks. A processor's workload is the number of tasks associated with its assigned problem data. Values in the range $1..number\_of\_tasks$ are used as indices to refer to these tasks. A task's index is used to calculate the specific problem data on which the task computes. In the pixel processing example described above, the task index is used as an offset; task i processes the ith pixel location. The prob-

```
<initial instructions>
forloop index = 1, MAX(workload) {
    WHERE (index ≤ workload) {
        Solution Phase(index)}}
<further instructions>
```

Figure 1: A prototypical program to be augmented with the load balancing algorithm

lem data and workload are stored in parallel arrays named *problem_data* and *workload*.

In order to process the tasks, the program increments a global task index counter, *index*, which starts at 1 and ends at the maximum workload of all the processors. During each iteration of the global index, each processor checks if they have a task with that index, and if they do, they compute a solution for the task. The instructions used to compute the solution of a multiple tasks in parallel are called the **solution phase**. A pseudo-code description of a prototypical program is shown in Figure 1. Note that the WHERE statement activates processors for which the test is true and idles processors elsewhere.

As the computation proceeds, more and more processors complete the processing of their tasks and remain idle for the rest of the loop iterations. These processors are then termed *idle* processors. Processors which have tasks to complete are termed *active* processors. Processors are idled because all processors must process tasks with the same task index at the same time.

To improve the program's performance, the program is modified so that tasks with different indices can be processed at the same time. The load balancing algorithm then distributes tasks from heavily loaded active processors to idle processors and tries to balance the workload among all processors.

How a program is augmented with the load balancing algorithm is now described. The load balancing algorithm consists of three distinct phases: the **information gathering phase**, the **decision phase** which decides when load balancing should occur and the **redistribution phase** which distributes tasks from active processors to idle processors. The basic iteration structure of the program is preserved. At the beginning of each iteration, the information gathering phase is executed. Then the decision phase is run, utilizing the gathered information to decide if balancing should occur during this iteration. If the decision is to balance, the redistribution phase is run, moving problem data from active to idle processors and assigning these idle processors new task indices to process. When tasks are distributed, the task indices originally assigned to an active processor can be assigned to multiple idle processors. Thus, different processors can work on different task indices during the same iteration. In the pixel processing example, this could mean, for example, that the first 3 pixels of one processor's row are processed along with the first 5

```
<initial instructions>
loop {
    Information Gather Phase
    IF (Decision Phase returns TRUE) THEN {
        Redistribution Phase }
    WHERE (workload > 0) {
        Solution Phase(parallel_index)
        workload = workload - 1 }}
until (All elements of workload = 0)
<further instructions>
```

Figure 2: A prototypical load-balanced program

pixels of another processor's row. A parallel array, named *parallel_index*, is used to keep track of the current task index computed by each processor. A pseudo-code description of a prototypical load-balanced program is shown in Figure 2.

Note that the execution time of both the original and load-balanced prototypical programs presented in Figures 1 and 2 is proportional to the maximum number of tasks assigned to any processor. The major difference between the programs, is the load-balanced program can reduce the maximum number of tasks on any processor by redistributing the workload.

2.1 When to load balance

The information gathering phase creates a trial workload, named *new_workload*, which is used by the decision phase to decide if redistributing load will be cost-effective. This new workload contains a more balanced distribution of tasks and therefore has a smaller maximum number of tasks on any processor than the original workload. From this new workload a measure of the possible savings is calculated as the maximum number of tasks on any processor in the original workload minus the maximum number of tasks on any processor in the new workload. This is shown in the equation below:

$$savings \text{ in iterations} =$$
$$MAX(workload) - MAX(new\_workload)$$

The savings are measured in terms of the number of future iterations of the loop which will execute the solution phase. The maximum number of tasks in a workload dictates the number of iterations that must be executed to process these tasks. If the new distribution is used then these "saved" iterations will not have to be executed.

The costs of the load balancing algorithm are also measured. Since the savings are measured in terms of the number of iterations, the costs are converted to this unit as well. The costs of the load balancing algorithm are incurred during the execution of the information gathering phase and redistribution phase. In order to quantify the costs of these phases, during each iteration their execution times are measured. The execution time of the solution phase during each iteration is also measured. $time_{info}$, $time_{redis}$ and $time_{soln}$ are the execution times of one execution of the information gathering, redistribution and solution phases. An estimate of the load balancing cost in terms of number of iterations can then be calculated by multiplying the sum of the execution time of the information gathering and redistribution phases by the inverse of the execution time of the solution phase, as shown in the equation below:

$$costs \text{ in iterations} = (time_{info} + time_{redis}) \times \frac{1 \text{ iteration}}{time_{soln}}$$

In order to provide a guarantee that the load balancing algorithm will always make cost-effective load balancing decisions, this cost measure must not be underestimated. Initial runs of the load-balanced program on various datasets are used to compute an overestimated cost measurement. The longest information gathering time of any iteration and redistribution time of any iteration are then divided by the shortest solution time of any iteration for each dataset. The largest of the resulting cost measures provides an estimate of an upper bound on the load balancing cost in terms of iterations. To this initial estimate a constant is added to assure the cost measure will always be an overestimate. This overestimate is then used in all future runs of the load balanced program.

Utilizing the overestimated costs and the calculated savings a cost-effective load balancing decision is then made by the decision phase. If the savings are greater than the costs then the redistribution phase is executed. Since each load balancing decision results in a cost-effective iteration, the sum of these decisions results in a cost-effective program execution.

2.2 How to load balance

An efficient redistribution algorithm is essential for good performance. Biagioni and Prins [4] and Nicol [5] describe efficient data-parallel redistribution algorithms which use scan communication routines to organize data movement. Our algorithm also uses scans to organize data movement. In our redistribution algorithm, workload is distributed by copying problem data from heavily loaded active processors to idle processors. The workload, in the form of task indices, is then divided up and assigned to the active and idle processors with copies of the problem data. Each active processor's workload is assigned some number of idle processors. This assignment is computed by assigning the idle processors in proportion to the workload on each active processor. Thus, heavy workloads are assigned more idle processors than light workloads and load is balanced evenly. A more detailed description of the redistribution algorithm is presented in [6].

3. A PERFORMANCE STUDY

The load balancing algorithm has been added to a data-parallel polygon renderer [7]. A series of experiments were executed using the original and a load balanced version of the renderer. The steps taken by the renderer include: *scan conversion*, which maps the polygons onto the rows (*scan lines*) of the resulting image and *z-buffering*, which maps the scan lines into pixels of the image. In the original version of the renderer, the scan conversion and z-buffering steps were implemented using data-parallel loops of the form shown in Figure 1. In the load balanced version, the steps were implemented using load balanced data-parallel loops of the form shown in Figure 2. For the scan conversion loop, the problem data is polygons and each task consists of creating and processing a scan line from a polygon. The number of scan lines in a polygon is dependent upon its image-space height. In the z-buffering loop, the problem data is scan lines and each task consists of creating and processing pixels from a scan line. The number of pixels in a scan line is dependent upon its length. For all the experiments, the renderer generates output images which are 512×512 pixels in size. The experiments were executed on the Advanced Computing Laboratory's 1024 processor CM-5 at Los Alamos National Laboratory.

Table 1: The Maximum and Average Workloads for the Scan Conversion Loop for the Balanced and Unbalanced Datasets

View	Balanced		Unbalanced	
	Max	Avg	Max	Avg
(0,0)	8	5	231	6
(45,45)	10	5	169	6
M(0,0)	13	5	512	8
M(45,45)	21	1	468	4

In the first experiment, performance data for the original and load balanced renderers is presented. The polygon datasets used in this experiment are a balanced and unbalanced version of the same scientific output, a hydrodynamics simulation of an oil well perforator. The relative balance of a polygon dataset is dependent upon the type of algorithm used to generate the dataset, the viewing and magnification transformations applied to the dataset, and the number of its polygons which have been clipped from view. Table 1 presents the maximum and average workloads of these datasets for the scan conversion loop. The first column of the table lists the viewing and magnification transformations that were applied to the datasets. The "M" before the viewing angle means the dataset has been magnified. Notice in the balanced datasets the difference between the maximum and average workload is small, whereas in the unbalanced datasets the difference is large.

On the unbalanced datasets, the decision phase and the redistribution phase work together to effectively to improve the renderer's performance. Table 2 shows the results of rendering the unbalanced polygons with (**LB**) and without (**OR**) the assistance of the load balancing algorithm on 32, 64, 128, 256 and 512 processors of the CM-5. Notice the poor performance of the original renderer on the unbalanced datasets and the improvement obtained when

Table 2: Rendering of Unbalanced Datasets in Seconds

View	32		64		128		256		512	
	OR	LB	OR	LB	OR	LB	OR	LB	OR	LB
(0,0)	48.91	6.21	28.98	3.57	18.18	2.12	12.76	1.35	9.99	0.92
(45,45)	38.03	5.33	22.37	3.00	14.27	1.78	10.45	1.15	8.24	0.81
M (0,0)	109.64	10.62	65.50	5.90	41.74	3.40	30.12	2.16	23.92	1.46
M (45,45)	99.13	4.42	57.91	2.42	37.19	1.46	27.33	0.94	22.40	0.67

Table 3: Rendering of Balanced Datasets in Seconds

View	32		64		128		256		512	
	OR	LB	OR	LB	OR	LB	OR	LB	OR	LB
(0,0)	4.71	5.16	2.67	2.95	1.57	1.72	0.88	1.00	0.54	0.64
(45,45)	5.49	6.12	2.98	3.33	1.69	1.89	0.99	1.13	0.62	0.73
M (0,0)	10.61	12.47	5.80	6.77	3.18	3.73	1.82	2.21	1.11	1.40
M (45,45)	12.27	13.43	6.69	7.34	3.74	4.12	2.24	2.50	1.51	1.65

using the load balancing algorithm. The performance of the load-balanced renderer provides a factor of 8 to 33 improvement over the performance of the original renderer on the unbalanced datasets.

Table 3 shows the results of rendering the balanced polygons with and without the assistance of the load balancing algorithm on 32, 64, 128, 256 and 512 processors. Notice that when the load balanced renderer is applied to the balanced datasets its performance is approximately the same as the original renderer. It is difficult for a load balancing algorithm to provide good performance on a balanced dataset since any redistribution steps will simply waste time. The worst case empirical performance loss is only 27 percent on balanced datasets when using the load balancing algorithm.

The original renderer's performance on the balanced datasets provides an estimate of the target performance we would like to achieve with the addition of a load balancing algorithm. The performance of the load-balanced renderer on the unbalanced datasets is within 70 percent of the performance of the original renderer on the balanced datasets.

In a second experiment, three other polygon datasets were tested. Two datasets were generated from different outputs of a fluid-dynamics simulation and the other from the output of a particle interaction simulation. Two of the the datasets are balanced and one is unbalanced. In summary, performance improvements ranged from a factor of 4 to 33 and the worst case empirical performance loss is only 25 percent.

4. CONCLUSIONS

A significant problem when using a load balancing algorithm is the possibility that along with improving performance on some datasets it will degrade performance on others. In this paper, a data-parallel load balancing algorithm was described which will not substantially degrade a program's performance on any dataset. This property re-

sults from utilizing an empirical measurement of the cost of load balancing along with a calculation of the possible savings to restrict load balancing to only when it is cost-effective.

Acknowledgments This research was performed at the Advanced Computing Laboratory of Los Alamos National Laboratory, Los Alamos, NM 87545.

REFERENCES

[1] M. C. Wikstrom, G. M. Prabhu, and J. L. Gustafson. Myths of load balancing. In *Parallel Computing '91*, pages 531–549, 1991.

[2] D. M. Nicol and J. C. Townsend. Accurate modeling of parallel scientific computation. In *Proceedings of the 1989 SIGMETRICS Conference*, pages 165–170, May 1989.

[3] D. M. Nicol and P. F. Reynolds Jr. Optimal dynamic remapping of data parallel computations. *IEEE Transactions on Computers*, 39(2):206–219, February 1990.

[4] E. S. Biagioni and J. F. Prins. Scan directed load balancing for highly parallel mesh-connected parallel computers. In *Unstructured Scientific Computation on Scalable Multiprocessors*, pages 371–95, October 1990.

[5] D. M. Nicol. Communication efficient global load balancing. In *Proceedings of the Scalable High Performance Computing Conference*, pages 292–299, April 1992.

[6] J. P. Ahrens and C. D. Hansen. Cost-effective data-parallel load balancing. Technical Report TR-95-04-02, University of Washington, 1995.

[7] F. A. Ortega, C. D. Hansen, and J. P. Ahrens. Fast data parallel polygon rendering. In *Proceedings of Supercomputing '93*, pages 709–718, November 1993.

TABLE OF CONTENTS- FULL PROCEEDINGS

(R): Regular Papers
(C): Concise Papers